Advanced Strategies for Catalyst Design

Advanced Strategies for Catalyst Design

Editor

Laura Orian

MDPI • Basel • Beijing • Wuhan • Barcelona • Belgrade • Manchester • Tokyo • Cluj • Tianjin

Editor
Laura Orian
University of Padova
Italy

Editorial Office
MDPI
St. Alban-Anlage 66
4052 Basel, Switzerland

This is a reprint of articles from the Special Issue published online in the open access journal *Catalysts* (ISSN 2073-4344) (available at: https://www.mdpi.com/journal/catalysts/special_issues/catalyst_design).

For citation purposes, cite each article independently as indicated on the article page online and as indicated below:

LastName, A.A.; LastName, B.B.; LastName, C.C. Article Title. *Journal Name* **Year**, *Volume Number*, Page Range.

ISBN 978-3-0365-0556-5 (Hbk)
ISBN 978-3-0365-0557-2 (PDF)

Cover image courtesy of Laura Orian.

© 2021 by the authors. Articles in this book are Open Access and distributed under the Creative Commons Attribution (CC BY) license, which allows users to download, copy and build upon published articles, as long as the author and publisher are properly credited, which ensures maximum dissemination and a wider impact of our publications.

The book as a whole is distributed by MDPI under the terms and conditions of the Creative Commons license CC BY-NC-ND.

Contents

About the Editor vii

Laura Orian
Editorial: Special Issue on "Advanced Strategies for Catalyst Design"
Reprinted from: *Catalysts* **2021**, *11*, 38, doi:10.3390/catal11010038 1

Lei Ma, Shiyan Yuan, Hanfei Zhu, Taotao Jiang, Xiangming Zhu, Chunshan Lu and Xiaonian Li
Pd_4S/SiO_2: A Sulfur-Tolerant Palladium Catalyst for Catalytic Complete Oxidation of Methane
Reprinted from: *Catalysts* **2019**, *9*, 410, doi:10.3390/catal9050410 5

Fengqin Wang, Tiankui Huang, Shurong Rao, Qian Chen, Cheng Huang, Zhiwen Tan, Xiyue Ding and Xiaochuan Zou
Synthesis of GO-SalenMn and Asymmetric Catalytic Olefin Epoxidation
Reprinted from: *Catalysts* **2019**, *9*, 824, doi:10.3390/catal9100824 23

Yoshinao Kobayashi and Yusuke Sunada
A Four Coordinated Iron(II)-Digermyl Complex as an Effective Precursor for the Catalytic Dehydrogenation of Ammonia Borane
Reprinted from: *Catalysts* **2020**, *10*, 29, doi:10.3390/catal10010029 37

Naomi N. González Hernández, José Luis Contreras, Marcos Pinto, Beatriz Zeifert, Jorge L. Flores Moreno, Gustavo A. Fuentes, María E. Hernández-Terán, Tamara Vázquez, José Salmones and José M. Jurado
Improved NOx Reduction Using C_3H_8 and H_2 with Ag/Al_2O_3 Catalysts Promoted with Pt and WOx
Reprinted from: *Catalysts* **2020**, *10*, 1212, doi:10.3390/catal10101212 49

Xiaopeng Zhang, Xiangkai Han, Chengfeng Li, Xinxin Song, Hongda Zhu, Junjiang Bao, Ning Zhang and Gaohong He
Promoting Effect of the Core-Shell Structure of $MnO_2@TiO_2$ Nanorods on SO_2 Resistance in Hg^0 Removal Process
Reprinted from: *Catalysts* **2020**, *10*, 72, doi:10.3390/catal10010072 73

Yanxia Liu, Lin Zhao, Yagang Zhang, Letao Zhang and Xingjie Zan
Progress and Challenges of Mercury-Free Catalysis for Acetylene Hydrochlorination
Reprinted from: *Catalysts* **2020**, *10*, 1218, doi:10.3390/catal10101218 87

Yanxia Liu, Yagang Zhang, Lulu Wang, Xingjie Zan and Letao Zhang
The Role of Iodine Catalyst in the Synthesis of 22-Carbon Tricarboxylic Acid and Its Ester: A Case Study
Reprinted from: *Catalysts* **2019**, *9*, 972, doi:10.3390/catal9120972 117

Claudia Carlucci, Michael Andresini, Leonardo Degennaro and Renzo Luisi
Benchmarking Acidic and Basic Catalysis for a Robust Production of Biofuel from Waste Cooking Oil
Reprinted from: *Catalysts* **2019**, *9*, 1050, doi:10.3390/catal9121050 131

Giovanni Battista Alteri, Matteo Bonomo, Franco Decker and Danilo Dini
Contact Glow Discharge Electrolysis: Effect of Electrolyte Conductivity on Discharge Voltage
Reprinted from: *Catalysts* **2020**, *10*, 1104, doi:10.3390/catal10101104 143

Zhishan Su, Changwei Hu, Nasir Shahzad and Chan Kyung Kim
Asymmetric Cyanation of Activated Olefins with Ethyl Cyanoformate Catalyzed by Ti(IV)-Catalyst: A Theoretical Study
Reprinted from: *Catalysts* **2020**, *10*, 1079, doi:10.3390/catal10091079 159

Shah Masood Ahmad, Marco Dalla Tiezza and Laura Orian
In Silico Acetylene [2+2+2] Cycloadditions Catalyzed by Rh/Cr Indenyl Fragments
Reprinted from: *Catalysts* **2019**, *9*, 679, doi:10.3390/catal9080679 173

About the Editor

Laura Orian is currently Associate Professor of Physical Chemistry at the University of Padova, where she graduated in 1997 and received her Ph.D. in 2001. Her main research interest is in the theoretical rationalization of the chemical reactivity, particularly of metal-based catalysts and biological and bioinspired systems. Her studies are focused on the comprehension of the reaction mechanisms of different systems, ranging from small model molecules to enzymes. The major goal of her research is the rational computer-assisted design of functional molecules aimed at reactivity prediction in the advance of the experiment. Her computational investigation is also applied in different scientific areas, including redox biology, medicine, and toxicology, and is systematically based on a rigorous approach to chemical structure and reactivity shared with numerous national and international collaborators. L.O. has participated as speaker at more than 40 national and international meetings, she has given more than 10 invited/keynote lectures at conferences and at foreign universities, she is currently member of four editorial boards and has co-authored more than 80 peer-reviewed papers and two book chapters. In 2018, L.O. has obtained the National Scientific Habilitation as Full Professor in the disciplinary area 03-A2/Models and methodologies for chemical sciences. L.O. is also active in scientific divulgation in the schools for pupils and in STEM activities for secondary school students and shares numerous chemistry projects with their teachers.

Editorial

Editorial: Special Issue on "Advanced Strategies for Catalyst Design"

Laura Orian

Dipartimento di Scienze, Chimiche Università degli Studi di Padova Via Marzolo 1, 35131 Padova, Italy; laura.orian@unipd.it

Received: 24 December 2020; Accepted: 28 December 2020; Published: 31 December 2020

The word catalyst comes from the Greek κατα'λυσις, which means dissolution and was introduced in 1836 by the Swedish Berzelius. All chemists know that the availability of an efficient catalyst is extremely valuable to enhance the reaction rate and optimize the regio- and stereoselectivity. The discovery of novel catalysts is based on different approaches, like simulation of different catalyzed reactions, trial and error procedures, screening of existing libraries of catalysts, and last but not least, good chemical intuition and knowledge. More often, a combination of these ways is employed, and serendipity plays an important role too. Anyway, the highly ambitious goal is the control of chemical reactions using accurately designed catalysts (i.e., molecules or materials that not only are efficient but also may have additional important requirements, like low cost and low environmental impact).

To quantify the importance of rationally tailored catalysts, a search in the Scopus database was performed, which returned 43,897 entries associated with the string "Catalyst AND Design" in the period from 1928 to 2021; among these, 20,838 entries belong to the period 2016–2021, clearly suggesting that the interest and the effort in catalyst design have been rapidly increasing in the last 5 years (Figure 1a). When strategy is included in the search by typing the complete string "Catalyst AND Design AND Strategy," 4087 entries are found in the period 2016–2021 (Figure 1b). Importantly, in both graphs in Figure 1, the number of works is increasing year after year, indicating the interest of the scientific community in the rational and guided search for novel catalytic systems. In this scenario, computational methods represent a valid support for different reasons. First, in the last 2 decades, accurate quantum chemistry protocols have provided mechanistic details of elementary and complex reactions, thus providing quantitative energetic and kinetic insight. Taking advantage of the impressive silicon technology development, chemical reactions involving complex systems can nowadays be investigated using (super)computers. Finally, machine-assisted screening of large datasets of chemical compounds is considered a common good practice to explore in silico the potential activity or extract those features that seem relevant to design a functional molecule. All these observations let us foresee that, in the very near future, chemists will be able to design efficient catalysts and then prepare them in the lab, minimizing synthetic effort and costs.

In this Special Issue, 11 contributions dealing with different problems of catalysis are gathered, and the main topics are summarized here.

Lei Ma et al. [1] have reported on a strategy to protect palladium catalysts by sulfur species and have applied it to the catalytic oxidation of methane.

The problem of asymmetric olefin epoxidation has been investigated and thoroughly discussed by Zou et al. [2], who employed salenMn immobilized on graphene oxide as catalyst.

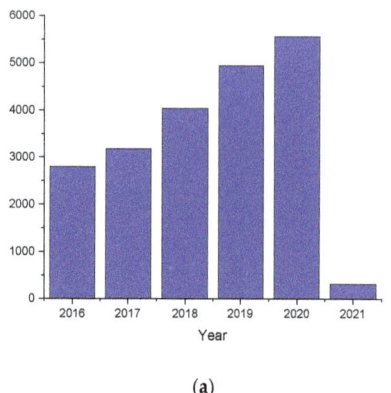 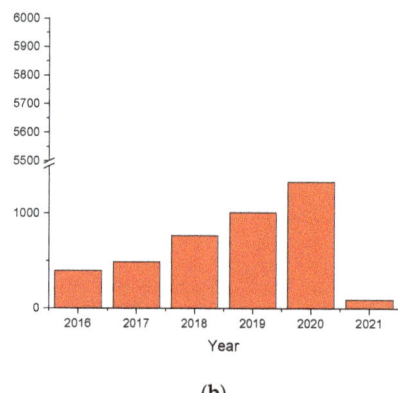

Figure 1. Results of a search in the Scopus database limited to the period 2016–2021: (**a**) search string "Catalyst AND Design"; (**b**) Search string "Catalyst AND Design AND Strategy." Last access to the database on 22 December 2020.

Kobayashi and Sunada have described the synthesis of a four coordinated Fe(II) digermyl complex, inspired by the silicon analog, which is used as catalyst in the dehydrogenation of ammonia borane [3].

Contreras et al. have reported on improvements in the reduction of NO_x using C_3H_8 and H_2 by adding Pt to the Ag/Al_2O_3-WO_x catalyst [4].

The work by He et al. [5] deals with the removal of elementary mercury in the presence of SO_2 using Mn/Ti nanorods, showing how the coating with TiO_2 protects Mn by the unwanted deactivation by SO_2.

The review article by Zhang et al. [6] focuses on acetylene hydrochlorination catalyzed by activated carbon-supported $HgCl_2$ and the challenge of finding a nontoxic catalyst. Particularly, noble and non-noble metal and nonmetal catalysts are considered alternative candidates, and advantages and issues are critically discussed.

The paper by Zan et al. have reported on their research of substitutes for fossil fuel-based petroleum products [7]. They have presented the synthesis of 22-carbon tricarboxylic acid and its ester via the Diels-Alder reaction starting from PUFAs and their esters and fumaric acid and fumarate, respectively. Iodine has been used as catalyst.

Carlucci et al. have been working on strategies to produce biofuels using as source waste cooking oil, and they succeeded in optimizing the reaction conditions (acid catalysis) to reach high yields, up to 99% [8].

Dini and et al. have carried out a study on the disposal of chemical waste from wastewaters, proposing the technique of contact glow discharge electrolysis (CGDE) with a promising low-cost implication [9].

Finally, Kim et al. and Orian et al. have contributed with two theoretical studies [10,11]. In the former, the asymmetric cyanation of olefins with ethyl cyanoformate catalyzed by Ti(IV) has been explored, while in the latter, the acetylene [2+2+2] cycloaddition to benzene catalyzed by Rh/Cr indenyl fragments has been investigated. The accurate description of the reaction mechanisms combined with the activation strain model nicely demonstrates that times are mature to perform a rigorous and quantitative catalyst design in silico.

I wish to express my gratitude to all the authors who have contributed to this Special Issue, demonstrating that a rational design of a catalyst can be pursued in very different approaches and fields. Special thanks also to the editorial office for the efficient support.

Funding: This research was funded by the Università degli Studi di Padova, thanks to the P-DiSC (BIRD2018-UNIPD) project MAD^{3}S; P.I.: L.O.

Conflicts of Interest: The author declares no conflict of interest.

References

1. Ma, L.; Yuan, S.; Jiang, T.; Zhu, X.; Lu, C.; Li, X. Pd$_4$S/SiO$_2$: A Sulfur-Tolerant Palladium Catalyst for Catalytic Complete Oxidation of Methane. *Catalysts* **2019**, *10*, 410. [CrossRef]
2. Wang, F.; Huang, T.; Rao, S.; Chen, Q.; Huang, C.; Tan, Z.; Ding, X.; Zou, X. Synthesis of GO-SalenMn and Asymmetryc Catalytic Olefin Epoxidation. *Catalysts* **2019**, *9*, 824. [CrossRef]
3. Kobayashi, Y.; Sunada, Y. A Four Coordinated Iron(II)-Digermyl Complex as an Effective Precursor for the Catalytic Dehydrogenation of Ammonia Borane. *Catalysts* **2020**, *10*, 29. [CrossRef]
4. González Hernández, N.N.; Contreras, J.L.; Pinto, M.; Zeifert, B.; Flores Moreno, J.L.; Fuentes, G.A.; Hernández-Terán, M.E.; Vázquez, T.; Salmones, J.; Jurado, J.M. Improved NOx Reduction Using C$_3$H$_8$ and H$_2$ with Ag/Al$_2$O$_3$ Catalysts Promoted with Pt and WOx. *Catalysts* **2020**, *10*, 1212. [CrossRef]
5. Zghang, X.; Han, X.; Li, C.; Song, X.; Zhu, H.; Bao, J.; Zhang, N.; He, G. Promoting Effect of the Core-Shell Structure of MnO$_2$@TiO$_2$ Nanorods on SO$_2$ Resistance in Hg^0 Removal Process. *Catalysts* **2020**, *10*, 72. [CrossRef]
6. Liu, Y.; Zhao, L.; Zhang, Y.; Zhang, L.; Zan, X. Progress and Challenges of Mercury-Free Catalysis for Acetylene Hydrochlorination. *Catalysts* **2020**, *10*, 1218. [CrossRef]
7. Liu, Y.; Zhang, Y.; Wang, L.; Zan, X.; Zhang, L. The Role of Iodine Catalyst in the Synthesis of 22-carbon Tricarboxilic Acid and its Ester: A Case Study. *Catalysts* **2019**, *9*, 972. [CrossRef]
8. Carlucci, C.; Andresini, M.; Degennaro, L.; Luisi, R. Benchmarking Acidic and Basic Catalysis for a Robust Production of Biofuel from Waste Cooking Oil. *Catalysts* **2019**, *9*, 1050. [CrossRef]
9. Alteri, G.B.; Bonomo, M.; Decker, F.; Dini, D. Contact Glow Discharge Electrolysis: Effect of Electrolyte Conductivity on Discharge Voltage. *Catalysts* **2020**, *10*, 1104. [CrossRef]
10. Su, Z.; Hu, C.; Shahzad, N.; Kim, C.K. Asymmetric Cyanation of Activated Olefins with Ethyl Cyanoformate catalyzed by Ti(IV)-Catalyst: A Theoretical Study. *Catalysts* **2020**, *10*, 1079. [CrossRef]
11. Ahmad, S.M.; Dalla Tiezza, M.; Orian, L. In Silico Acetylene [2+2+2] Cycloadditions Catalyzed by Rh/Cr Indenyl Fragments. *Catalysts* **2019**, *9*, 679. [CrossRef]

© 2020 by the author. Licensee MDPI, Basel, Switzerland. This article is an open access article distributed under the terms and conditions of the Creative Commons Attribution (CC BY) license (http://creativecommons.org/licenses/by/4.0/).

Article

Pd_4S/SiO_2: A Sulfur-Tolerant Palladium Catalyst for Catalytic Complete Oxidation of Methane

Lei Ma [1,*], Shiyan Yuan [1], Hanfei Zhu [1], Taotao Jiang [1], Xiangming Zhu [2,*], Chunshan Lu [1] and Xiaonian Li [1,*]

[1] Industrial Catalysis Institute, Laboratory Breeding Base of Green Chemistry-Synthesis Technology, Zhejiang University of Technology, Hangzhou 310014, China; yuanshiyan318420@163.com (S.Y.); hanfeizhu0701@foxmail.com (H.Z.); jiangtaotao2016@outlook.com (T.J.); lcszjcn@zjut.edu.cn (C.L.)
[2] Centre for Synthesis and Chemical Biology, UCD School of Chemistry, University College Dublin, Belfield, Dublin 4, Ireland
* Correspondence: malei@zjut.edu.cn (L.M.); xiangming.zhu@ucd.ie (X.Z.); xnli@zjut.edu.cn (X.L.); Tel.: +86-571-88320920 (L.M.)

Received: 4 April 2019; Accepted: 29 April 2019; Published: 30 April 2019

Abstract: Sulfur species (e.g., H_2S or SO_2) are the natural enemies of most metal catalysts, especially palladium catalysts. The previously reported methods of improving sulfur-tolerance were to effectively defer the deactivation of palladium catalysts, but could not prevent PdO and carrier interaction between sulfur species. In this report, novel sulfur-tolerant SiO_2 supported Pd_4S catalysts (5 wt. % Pd loading) were prepared by H_2S–H_2 aqueous bubble method and applied to catalytic complete oxidation of methane. The catalysts were characterization by X-ray diffraction, Transmission electron microscopy, X-ray photoelectron Spectroscopy, temperature-programmed oxidation, and temperature-programmed desorption techniques under identical conditions. The structural characterization revealed that Pd_4S and metallic Pd^0 were found on the surface of freshly prepared catalysts. However, Pd_4S remained stable while most of metallic Pd^0 was converted to PdO during the oxidation reaction. When coexisting with PdO, Pd_4S not only protected PdO from sulfur poisoning, but also determined the catalytic activity. Moreover, the content of Pd_4S could be adjusted by changing H_2S concentration of H_2S–H_2 mixture. When H_2S concentration was 7 %, the Pd_4S/SiO_2 catalyst was effective in converting 96% of methane at the 400 °C and also exhibited long-term stability in the presence of 200 ppm H_2S. A Pd_4S/SiO_2 catalyst that possesses excellent sulfur-tolerance, oxidation stability, and catalytic activity has been developed for catalytic complete oxidation of methane.

Keywords: sulfur-tolerance; Pd_4S; catalytic oxidation of methane; sulfur poisoning

---

1. Introduction

Methane as the main component of natural gas is playing an increasingly important role in the global energy structure [1,2]. Effective catalytic complete oxidation of CH_4 can improve combustion efficiency and reduce air pollutants, such as CO, NO_x, and unburned hydrocarbon [3–8]. It has therefore found great applications in modern industry, such as catalytic exhaust converters aimed to reduce methane emission and catalytic gas turbine combustors designed to combust fuel under mild conditions [9,10]. Supported PdO catalysts have shown excellent catalytic property in methane oxidation, and currently are under extensive study [11–18]. However, once sulfur species (e.g., H_2S or SO_2) are present in the reaction atmosphere, the poisoning of PdO catalyst which would lead to inactive PdO-SO_x is irreversible and the activity of the catalyst cannot be recovered at relatively low temperature [19–25].

Hence, many efforts have been devoted in the past decade to enhance sulfur resistance of supported PdO catalysts. Currently there are two primary approaches to improve the performance of palladium catalysts against sulfur poisoning: the first is to use alkaline carriers (e.g., γ-Al_2O_3) and enhance their adsorption ability toward acidic sulfur species to protect the active phase of PdO [23,26,27], and the second is to introduce an extra active ingredient to form a catalyst system with dual activity, such as Pd–Pt complex, to avoid sulfur poisoning [24,28]. However, these methods can only defer the deactivation of palladium catalysts as neither the alkaline carrier nor the extra active ingredient can prevent the interaction between PdO and sulfur species. In other words, sulfur poisons would eventually encroach PdO and make it inactive. As such, these catalyst systems did not really possess the ability of sulfur-tolerance.

The research on sulfur-tolerant palladium catalyst system in methane catalytic oxidation reaction is rare and lack of substantial progress in the literature [20,26–28]. Therefore, development of a reliable sulfur-tolerant catalyst system seems to be of great importance. In view of the great potential of catalytic oxidation of methane, we decided to pursue a new catalytic system that could be of high tolerance of sulfur poisons. We speculated Pd_xS_y in combination with acidic carrier would be an ideal system for methane oxidation based on the following two points. First, it would be the best scenario for sulfur-tolerant catalysts that neither the support nor the palladium active phase itself could interact with sulfur species. Pd_xS_y and acidic carrier such as SiO_2 would likely meet the requirement. Second, seeing that sulfur and oxygen are located in the adjacent position of the same main group in the periodic table, i.e., the chalcogen group, the designed Pd_xS_y might possess similar catalytic activity to PdO.

In the literature, the majority of Pd_xS_y (Pd_4S, Pd_3S, $Pd_{2.8}S$, $Pd_{16}S_7$, PdS or PdS_2, etc.) were prepared by gas sulfuration with H_2S–H_2 [25,29–35]. Herein, we would like to present a new procedure for the preparation of Pd_xS_y/SiO_2 catalysts via aqueous bubble sulfuration of Pd/SiO_2 with H_2S–H_2 with varied H_2S concentration. Compared with the gas sulfuration, the aqueous bubble method was performed in much milder conditions. Besides the sulfur resistance and catalytic activity, the stability of Pd_xS_y/SiO_2 catalyst was the other focus of our investigation.

2. Results and Discussion

2.1. Fresh Catalysts

2.1.1. Catalyst Characterization

Freshly prepared catalysts were first tested by X-ray diffraction (XRD) as showed in Figure 1. The freshly prepared catalyst is named (fV), where V represents H_2S concentration. No other palladium species were found except for metallic Pd^0. The sulfuration of Pd to form Pd_4S under H_2S–H_2 atmosphere was via the reaction: $4Pd + H_2S \leftrightarrow Pd_4S + H_2$ [36–40]. This reversible reaction meant that metallic Pd^0 could not be fully sulfurized to Pd_4S. Therefore, it was reasonable to detect metallic Pd^0 on the surface of freshly sulfurized catalyst. However, it was worth noting that the intensity of the diffraction peak of metallic Pd^0 changed regularly with the increase of H_2S concentration. The average particle size of metallic Pd^0 calculated by Scherrer formula was further shown in Table 1. With the increase of H_2S concentration, the change of average particle size of metallic Pd^0 could be divided into three stages. They were gradually reduced (0 → 5%), stable (5% → 7%), and increased (7% → 10%). These results indicated the Pd/SiO_2 precursor had been changed by sulfuration.

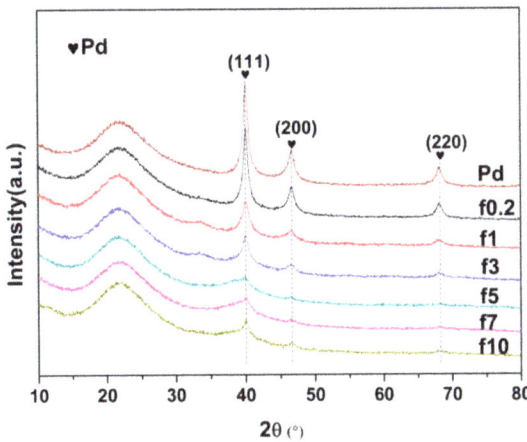

Figure 1. X-ray diffraction (XRD) patterns of freshly prepared catalysts sulfurized with different H_2S concentration.

Table 1. Average particles size of freshly prepared catalysts sulfurized with different H_2S concentration.

Catalyst	FWHM	Size (nm)
Pd	0.761	12.2
f0.2	0.803	11.5
f1	1.328	6.9
f3	1.297	6.7
f5	3.329	2.6
f7	3.262	2.6
f10	2.163	4.0

FWHM: full-width at half-maximum

Since XRD characterization could not directly provide evidence of the existence of Pd_xS_y, the freshly prepared catalysts were further characterized by high resolution transmission electron microscopy (HR-TEM) and illustrated in Figure 2. The lattice spacing distance of Pd species particles on freshly prepared catalyst are 0.225 nm and 0.245 nm, corresponding to the Pd(111) and Pd_4S(102), respectively [JCPDS No. 73-1387, JCPDS No. 46-1043]. In addition to the metallic Pd(111), the Pd_4S(102) was found on the surface of all the freshly prepared catalysts. The literature on the preparation of Pd_xS_y under H_2S-H_2 atmosphere showed that the Pd_4S with the highest proportion of Pd/S was the first Pd_xS_y species [38,41]. Therefore, it was reasonable to firstly form the Pd_4S under the condition of 60 °C aqueous bubble sulfuration.

The subsequent characterization by X-ray photoelectron spectroscopy (XPS) could provide further evidence for the existence of Pd_xS_y, as shown in Figure 3. After curve fitting analysis, the S2p$_{3/2}$ of freshly prepared catalyst could be deconvoluted into a large peak at 168.5 eV and a small peak at 165.1 eV, which corresponded to S^0 [42] and Pd_4S [42–44], respectively. With the increase of H_2S concentration, the XPS peak intensity of sulfur species increased gradually. At the same time, the Pd3d$_{5/2}$ could be deconvoluted into two peaks at ~335.6 eV and ~337.5 eV which corresponded to metallic Pd^0 [45,46] and Pd_4S [47,48], respectively. This indicated that Pd_4S and metallic Pd^0 were the primary palladium species on the freshly prepared catalysts. These results are in good agreement with the characterization results of XRD and HR-TEM. If the Pd_4S had similar properties to PdO, the amount of Pd_4S would be the key factor influencing the performance of the catalyst. The XPS data (Figure 3) were further analyzed to provide Table 2 that listed the composition ratio of palladium species of freshly prepared catalysts. It could be noted that the sulfuration process was the transition of metallic Pd^0 to Pd_4S. With the increase of H_2S concentration, the change of Pd_4S ratio could also be divided into three stages. They were gradually increased (0 → 5%), stable (5% → 7%), and reduced (7% → 10%). Variation of metallic Pd^0 content was opposite to that of Pd_4S. This variation was entirely consistent with the average particle size of metallic Pd^0. It was obvious that the variation of the average particle size of metallic Pd^0 was caused by the change of metallic Pd^0 content on the catalyst surface.

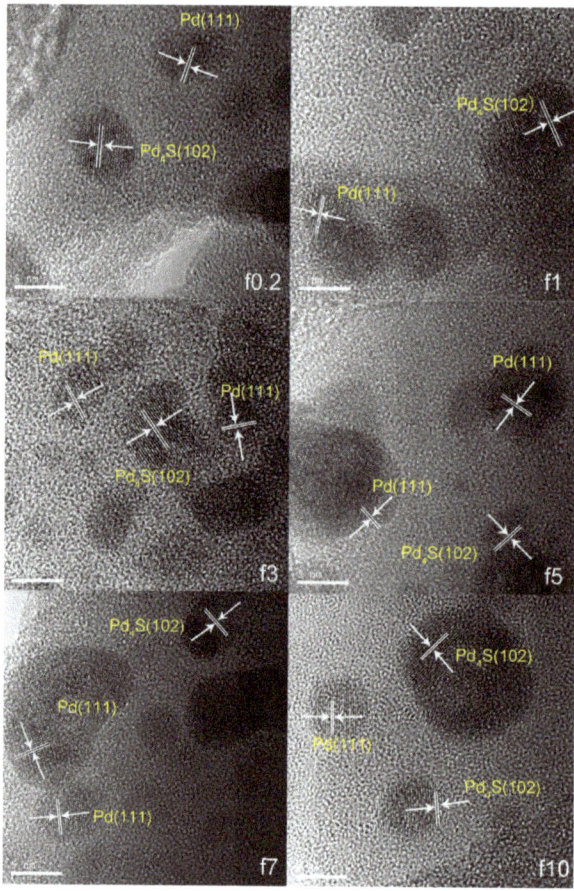

Figure 2. High resolution transmission electron microscopy (HR-TEM) images of freshly prepared catalysts with different H_2S concentration.

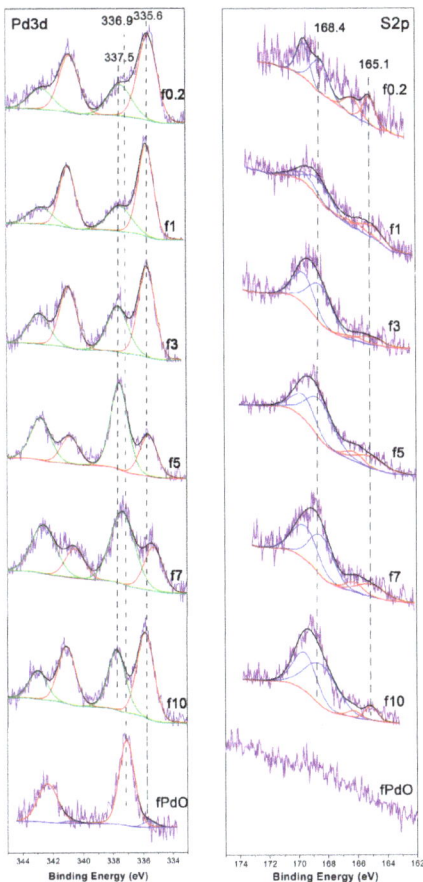

Figure 3. X-ray photoelectron spectroscopy (XPS) Pd3d and S2p of freshly prepared catalysts with different H_2S concentration.

Table 2. Palladium species content of freshly prepared catalysts with different H_2S concentration [a].

Catalyst	Composition Ratio of Palladium Species (%)	
	Pd^0	Pd_4S
f0.2	68.9	31.1
f1	68.6	31.4
f3	61.6	38.4
f5	33.4	66.6
f7	32.1	67.9
f10	57.9	42.1

[a]: Calculated by Pd fitted peak area by XPS.

2.1.2. Activity Studies

In order to test whether the Pd_4S possessed the desired sulfur-tolerance and catalytic activity, its catalytic property for methane oxidation in the presence of 200 ppm H_2S was then examined by measuring methane conversion against the reaction time at different reaction temperatures. As shown in Figure 4, when the temperature was 500 °C, the presence of H_2S had no impact on all the catalysts. However, once the reaction temperature dropped to 400 °C, the compared PdO/SiO_2 catalyst was deactivated very rapidly. We estimated that at this temperature sulfur species could turn the active PdO into inactive PdO–SO$_x$ species, which could then hardly decompose at low temperature [23,24,49]. In contrast, under the same conditions, all the catalysts prepared by our own procedure did not show any deactivation at 400 °C. This clearly stated that the combination of Pd$_4$S and SiO$_2$ was effective in resisting sulfur poisoning. At the same time, the activity of the catalyst with 7% H$_2$S concentration was even equal to that of PdO catalyst. This further stated that the combination of Pd$_4$S and SiO$_2$ had high enough catalytic activity. Furthermore, it was found that the methane conversion of all catalysts decreased to less than 10% at 370 °C. This dramatic decrease in catalytic activity was not due to sulfur poisoning, but the high reaction temperature required for methane activation [50–52]. Therefore, it was best to set the reaction temperature at 400 °C. This temperature could not only reflect the sulfur-tolerance, but also compare the catalytic activity.

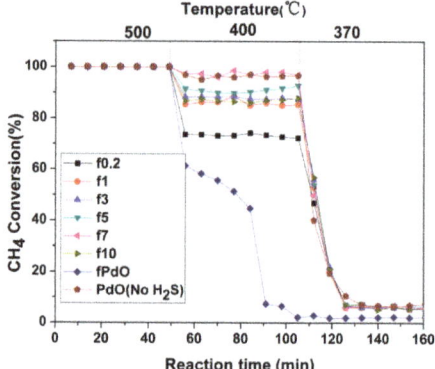

Figure 4. Catalytic performance of freshly prepared catalysts with different H$_2$S concentration. Gas composition: CH$_4$ (v% = 2%), O$_2$ (v% = 8%), H$_2$S (v% = 0.02%), and N$_2$ (v% = 89.8%); Flow rate = 200 mL/min; and GHSV (gas hourly space velocity) = 60000 mL/(g·h).

Meanwhile, it should be pointed out here that although all the sulfurized catalysts exhibited the similar sulfur-tolerance, the catalytic activity at 400 °C was not same. As the H$_2$S concentration increased from 0.2% to 7%, the catalytic activity increased gradually. However, when the H$_2$S concentration increased to 10%, the catalytic activity decreased. This variation was very similar to the variation of Pd$_4$S content on the surface of freshly prepared catalyst. However, we could not simply assume that catalytic activity depended solely on Pd$_4$S. The reason was that the stability of Pd$_4$S and metallic Pd^0 on the surface of freshly prepared catalyst was still undiscovered under high reaction temperature and presence of oxygen. Under such reaction condition, metallic Pd^0 was liable to be oxidized to PdO. Therefore, palladium species on the surface of freshly prepared catalyst might not be the true palladium species involved in the reaction. Next, we would characterize the used catalyst to determine the actual palladium species on the catalyst surface.

2.2. Used Catalyst

2.2.1. Catalyst Characterization

The XRD patterns of used catalysts were given in Figure 5. The used catalyst is named (uV), where V represents H$_2$S concentration. The characteristic diffraction peak attributed to sulfur species was found on the compared PdO/SiO$_2$ catalyst. However, no diffraction peaks attributed to sulfur species were found in all sulfurized catalysts after the reaction. The reason should be that the sulfurized catalyst had the ability to resist sulfur poisoning. Not surprisingly, almost all the diffraction peaks observed in the sulfurized catalyst were attributed to PdO after the reaction. A weak diffraction peak of metallic Pd^0 could be observed only when H$_2$S concentration was 0.2%. This implied that most of the metallic Pd⁰ on the freshly prepared catalyst had been oxidized to PdO under reaction conditions. However, the above results did not show whether the most important Pd_4S remains stable during the reaction and this needed to be identified by other characterization.

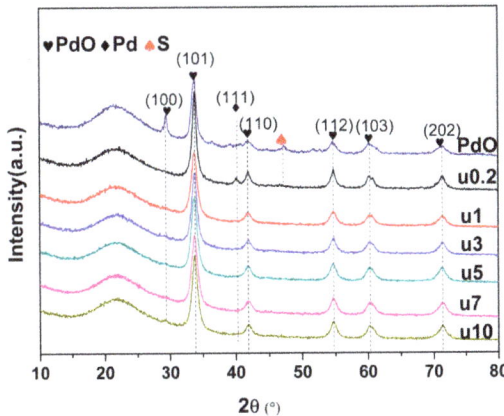

Figure 5. XRD patterns of used catalysts with different H_2S concentration.

The presence of PdO confirmed our previous speculation that the palladium species on the surface of freshly prepared catalysts were not entirely stable during the vigorous oxidation reaction. However, XRD characterization could only reveal that PdO species were formed during the reaction of freshly prepared catalysts. However, whether the essential Pd_4S species could stably exist in large quantities would be the main factor affecting the sulfur-tolerance of the catalyst. Figure 6 shows HR-TEM images of the used catalysts. It could be found that the palladium species on the surface of the used catalyst were more complicated than the freshly prepared catalyst. Figure 6 clearly indicates that in addition to Pd⁰(111), PdO(101) and PdO(110) also appeared in large numbers on the surface of all the used catalysts [JCPDS No.06-0515]. At the same time, the most interesting thing was that Pd_4S(102) species could be found on all catalyst surfaces. This indicated that the Pd_4S species could remain stable during the high temperature and high oxygen reaction of methane oxidation.

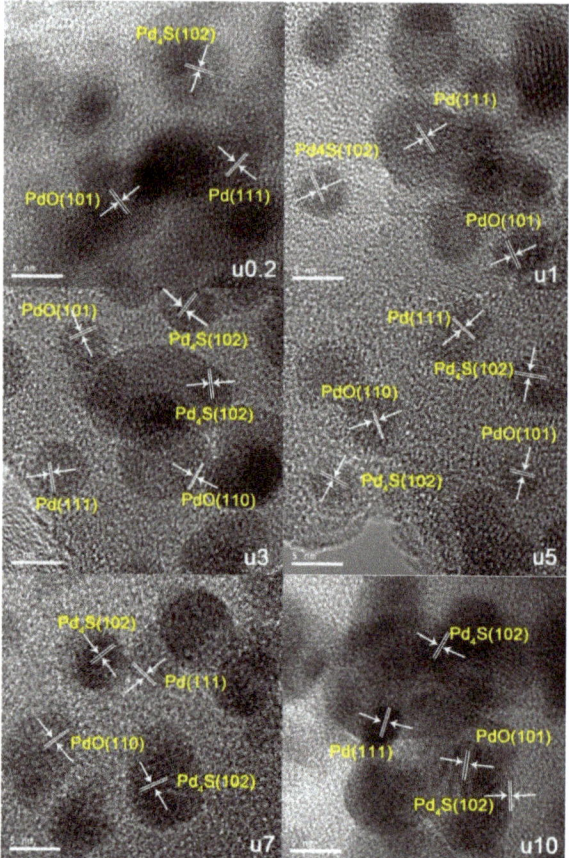

Figure 6. HR-TEM images of the used catalysts with different H_2S concentration.

Owing to the lack of data on the relative content of PdO on the surface of used catalyst, when Pd_4S and PdO coexist, we could not distinguish whether Pd_4S and PdO affect the catalytic activity alone or in combination. Therefore, used catalysts were further analyzed by XPS characterization, as shown in Figure 7. After curve fitting analysis, the $Pd3d_{5/2}$ could be deconvoluted into three peaks at ~337.5 eV, ~336.9 eV, and ~335.6 eV, which were assigned to Pd_4S [47,48], PdO [48], and metallic Pd^0 [45,46], respectively. The above characterization results revealed the fact that under the reaction conditions, the actual palladium species on the used catalyst surface should be Pd_4S, PdO, and metal Pd^0. At the same time, the S2p peak attributed to S^0 was not observed (see Figure 3), only the very weak S2p peak attributed to Pd_4S could be observed. In contrast, the Pd3d peak at ~337.3 eV and S2p peak at 168.1 eV of used PdO/SiO_2 could be attributed to $PdO-SO_x$. This comparison results indicated that Pd_4S could protect other palladium species from sulfur poisoning.

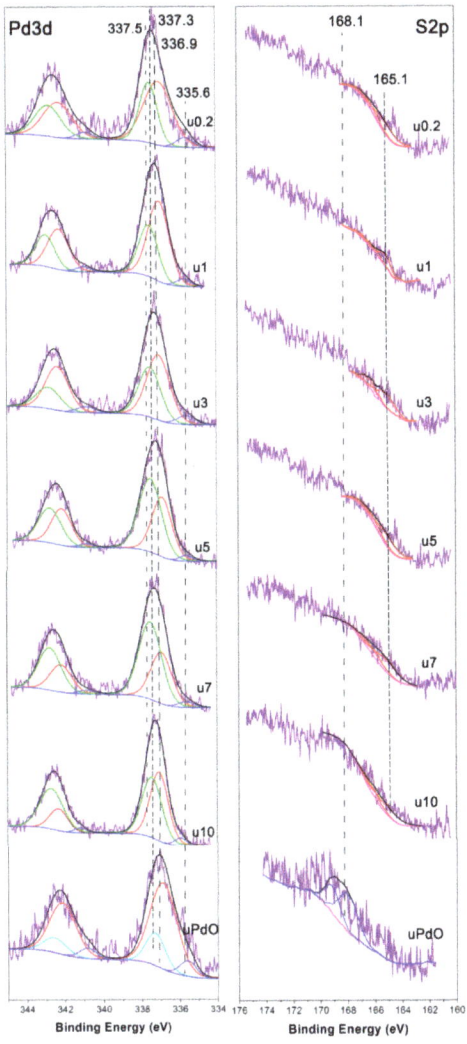

Figure 7. XPS Pd3d and S2p of the used catalysts with different H₂S concentration.

Table 3 listed the composition ratio of palladium species of used catalysts. Compared with the relative content of palladium species on the surface of fresh catalyst (Table 1), it could be noted that a large number of PdO species appeared on the surface of the used catalyst, while the Pd^0 content of the metal decreased significantly. This was clearly the oxidation of metal Pd^0 to PdO during the reaction. At the same time, the relative content of Pd_4S was basically stable during the reaction. This indicated that the PdO formed during the reaction was mainly derived from the metal Pd^0, but not derived from Pd_4S. Temperature-programmed Oxidation (TPO) of fresh Pd/SiO_2, f7 and u7 catalysts with the same mass was investigated by thermogravimetric analysis. The oxidation stability of the catalyst was examined by recording the mass change in the process of air oxidation. The results are presented in Figure 8. Three samples had a similar dehydration process before 100 °C. However, with the increase of temperature, the three samples showed completely different changes. Freshly prepared Pd/SiO_2 had the greatest mass increase, which should be related to the oxidation of metallic Pd^0 to PdO. The mass increase of f7 with the same mass was significantly smaller than that of freshly prepared Pd/SiO$_2$, while u7 with the same mass had no significant mass increase. f7 was composed of metallic Pd^0 and Pd$_4$S, while u7 was mainly composed of PdO and Pd$_4$S. This stated that the mass increase of f7 was mainly related to the oxidation of metallic Pd^0, while Pd$_4$S maintained sufficient oxidation stability even under temperature higher than 500 °C. This meant that Pd$_4$S could not be oxidized into palladium oxide nor into palladium sulfate. Therefore, it could be possible to conclude that in the low-temperature catalytic oxidation of methane, Pd$_4$S had sufficient stability to resist the oxidation of oxygen.

Table 3. Palladium species content of used catalysts with different H$_2$S concentration [a].

Catalyst	Composition Ratio of Palladium Species (%)		
	Pd	PdO	Pd$_4$S
u0.2	5.3	59.5	35.2
u1	4.0	60.4	35.6
u3	5.9	54.2	39.9
u5	2.7	40.4	56.9
u7	3.1	36.6	60.3
u10	3.6	48.9	47.5

[a]: Calculated by Pd fitted peak area by XPS.

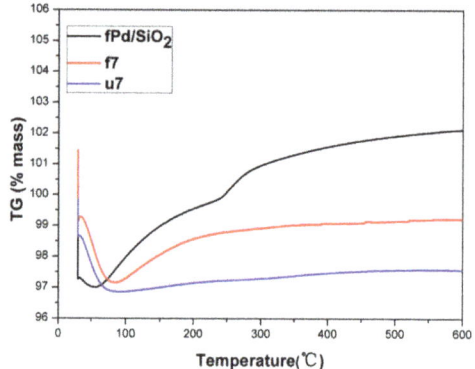

Figure 8. Temperature-programmed Oxidation (TPO) profiles of the (u7, f7) catalyst and freshly prepared Pd/SiO$_2$.

2.2.2. Activity Studies

Figure 9 covered the evaluation of the used catalysts and the evaluation method was completely consistent with the freshly prepared catalysts. To our delight, the sulfur-tolerance and catalytic activity of each used catalyst was almost unchanged with the fresh catalyst at 400 °C, which indicated that the composition of the catalyst had remained stable after the initial evaluation. The previous characterization results had confirmed that Pd$_4$S, PdO, and metal Pd^0 were the three palladium species actually involved in the reaction. In the case that PdO and metal Pd^0 did not have sulfur-tolerance, it was the existence of Pd$_4$S that gave the palladium active component the ability to resist the poisoning of sulfur species. Moreover, to our surprise, Pd$_4$S was not merely resistant to sulfur itself, its presence could even protect PdO and metal Pd^0 from the poisoning of sulfur species.

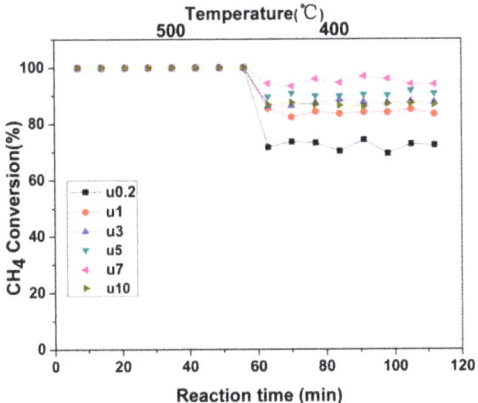

Figure 9. Catalytic performance of used catalysts with different H_2S concentration. Gas composition: CH_4 (v% = 2%), O_2 (v% = 8%), H_2S (v% = 0.02%), and N_2 (v% = 8 9.8%); Flow rate: 200 mL/min; and GHSV = 60000 mL/(g·h).

The sulfur-tolerance of the catalyst depended on Pd_4S, but it remained to be confirmed if catalytic activity was related to the type of palladium species. To this end, we correlated the relative amounts of Pd_4S and PdO species on the used catalyst surface with the catalytic activity of the used catalyst. The secondary reaction performance of the catalyst was selected at 400 °C and the results are shown in Figure 10. It could be noted that the variation of catalytic activity at 400 °C was closely related to the change of Pd_4S content with H_2S concentration increase. As the H_2S concentration increased from 0.2% to 7%, the relative content of Pd_4S and the catalytic activity of the used catalyst gradually increased. Compared with content change of (Pd_4S + PdO) and PdO, the content of (Pd_4S + PdO) did not change with the concentration of H_2S, but the change of PdO content was opposite to the catalytic activity of the used catalyst. The above results showed clearly that not only the sulfur-tolerance, but also the catalytic activity relied completely on Pd_4S. Because the coexistence of PdO did not play any decisive role, it was completely reasonable to use Pd_4S/SiO_2 as the catalyst prepared by the aqueous bubble sulfuration method. The consequences of 72-hour stability test of the u7 catalyst with and without H_2S are shown Figure 11. It could be found that the catalyst could maintain long-term stability regardless of the presence of H_2S in the reaction atmosphere.

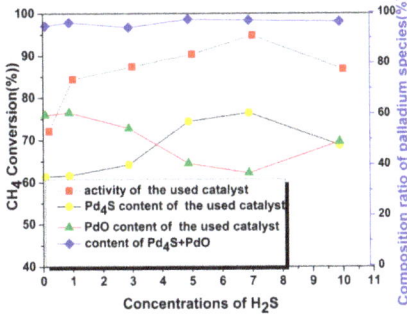

Figure 10. Relationship profile between composition ratio of Pd_4S and PdO on used catalysts and second evaluation of methane conversion. The composition ratio of palladium species was calculated through Pd fitted peak area by XPS.

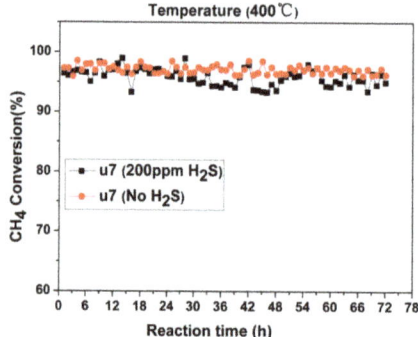

Figure 11. Long-term stability test of the u7 catalyst with and without H_2S. Gas composition: CH_4 (v% = 2%), O_2 (v% = 8%), H_2S (v% = 0.02%), and N_2 (v% = 89.8%); Flow rate: 200 mL/min; and GHSV = 60000 mL/(g·h).

2.3. Mechanism of Sulfur-Tolerance

Next, we turned to investigate the cause of sulfur-tolerance of the Pd_4S/SiO_2 using the technique of hydrogen sulfide temperature-programmed desorption (H_2S-TPD). The u7 catalyst and the freshly prepared PdO/SiO_2 catalyst were chosen for comparison and the results are presented in Figure 12. SO_2 and H_2S were the primary desorption species detected in the study. The PdO/SiO_2 catalyst released SO_2 in relatively large quantity, which suggested that PdO could readily assimilate H_2S to form PdO–SO_x. As a result, PdO/SiO_2 catalyst was poisoned and deactivated by H_2S. In contrast, desorption products were barely detected from the u7 catalyst. Seeing that Pd_4S was the only discrepancy between the two catalysts, i.e., the presence of Pd_4S in u7 and its absence in PdO/SiO_2, we were convinced that Pd_4S played a vital role in the sulfur-tolerance of u7 catalyst, as depicted in Scheme 1. Aside from not adsorbing H_2S, Pd_4S also prevented PdO from adsorbing H_2S onto its surface. Therefore, it was the Pd_4S that blocked the adsorption of H_2S and endowed the catalyst with sulfur-tolerance.

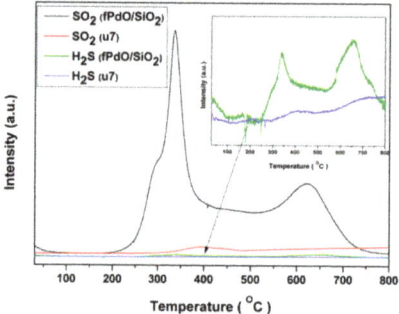

Figure 12. Hydrogen sulfide temperature-programmed desorption (H_2S-TPD) profiles of the u7 catalyst and freshly prepared PdO/SiO_2 catalyst.

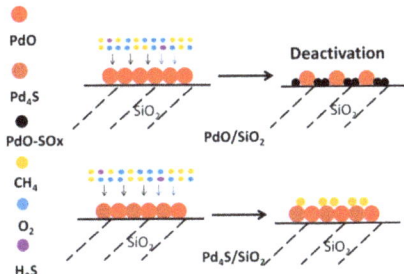

Scheme 1. Schematic diagram of methane catalytic oxidation on the surface of Pd$_4$S/SiO$_2$ and PdO/SiO$_2$ in presence of H$_2$S.

3. Materials and Methods

3.1. Catalyst Preparation

The PdO/SiO$_2$ catalyst was prepared by isovolumetric impregnation. Typically, weigh a proper quantity of 40~60 mesh SiO$_2$ (M = 10 g Vg = 0.94 cm^3/g, R$_d$ = 10.6 nm, S = 353.0028 m^2/g, Qingdao Baisha Catalyst Factory, Qingdao, China) was dispersed in palladium acetate (Aladdin Industrial Corporation, Shanghai, China) aqueous solution (V$_{Pd}$ = 0.05 g/mL) overnight. Then the sample was dried in an oven at 110 °C for 4 h and calcined at 500 °C for 4 h in the air. The theoretical loading of Pd was 5 wt. %.

The PdO/SiO$_2$ was first reduced to Pd/SiO$_2$ by H$_2$ at 500 °C for 1h. Pd$_x$S$_y$/SiO$_2$ catalyst was prepared by using above Pd/SiO$_2$ as precursor. 2.0 g Pd/SiO$_2$ and 100 ml deionized water were stirred at 350 rpm in a three-necked flask. Adjust the flow rate of H$_2$S and H$_2$ by flow controller (D07-19B, Beijing Sevenstart Electronics Co. Ltd, Beijing, China) to configure different concentrations of H$_2$S/H$_2$ mixture. Then 30 mL/min of H$_2$S–H$_2$ (V$_{H2S}$% = 0.2, 1, 3, 5, 7, 10) was fed into the suspension at 60 °C for 1 h. Afterwards, the sample was filtered and rinsed with distilled water till neutral. Finally, the sample was dried at 110 °C for 4 h. The freshly prepared catalyst is named (fV), and the used catalyst is named (uV), where V represents H$_2$S concentration.

3.2. Catalyst Characterization

X-Ray Diffraction (XRD) was carried out on a Thermo ARL X'TRA diffractometer (PNAlytical Co. Holland) using Cu K-a radiation (45 kV, 40 mA). Average particle size was determined from XRD measurements using the Scherrer formula:

$$d = \frac{K\lambda}{\beta \cos\theta},$$

where d is the average particle size(nm), K is the Scherrer constant and the diffraction angle is denoted θ, λ = 0.154 nm stands for the wavelength of Cu K-a radiation, β denotes the full-width at half-maximum (FWHM) of diffraction peak.

High Resolution Transmission Electron Microscopy (HR-TEM) images were taken by a Philips-FEI Tecnai G2 F30 S-Twin transmission electron microscope operated at 300 kV (Philips-FEI Co. Holland).

X-Ray Photoelectron Spectroscopy (XPS) was measured on a Thermo ESCALAB 250 Axis Ultra (KRATOS, Kanagawa, Japan) using a monochromated Al Kα X-ray source (hv = 1485.6 eV) with a fixed analyzer pass energy of 80 eV. The binding energy values were referenced to the Si 2p as internal standard (Si 2p = 103.4 eV) and the maximum deviation value was 2.6 eV in the sample. After subtraction of the Shirley-type background, the core-level spectra were decomposed into their component with mixed Gaussian-Lorentzan lines by a non-linear least squares curve fitting procedure, using the public software package XPSPEAK4.1. The corresponding atomic ratio in different chemical environment was obtained from the fitted XPS spectra of Pd 3d and S 2p. The XPS profiles were fitted by the software named "XPS peak". Then the fitting peak area was used to calculate the proportion of different species.

Temperature-programmed Oxidation (TPO) was performed on a NETZSCH STA 409 PC/PG instrument (NETZSCH Co. Selbc, Germany) The oxidation stability of the catalysts was investigated by recording the mass change in the process of air oxidation. TPO was performed by heating 0.01 g sample from room temperature to 600 °C in air atmosphere with a heating rate of 5 °C/min and a flow rate of 40 mL/min.

Temperature-programmed Desorption (TPD) of hydrogen sulfide experiments were performed in a self-made tubular quartz reactor (5mm i.d.). The sample (0.2 g) was first swept at 110 °C for 1 h using pure He and cooled to room temperature in the same atmosphere. Then the sample was in situ treated with H_2S (0.2% in N_2) at a flow rate of 30 mL/min for 0.5 h and swept 1 h to remove physisorbed and/or weakly bound species. TPD was performed by heating the sample from room temperature to 800 °C with constant increase of 5 °C/min in pure He. The TPD spectra were recorded by a quadrupole mass spectrometer (QMS 200 Omnistar, Pfeiffer Co. Germany).

3.3. Evaluation of Catalysts

The catalytic activity of the freshly prepared and used catalysts was tested with 0.2 g sample in a self-made continuous fixed bed reactor (8 mm i.d.) at atmospheric pressure. Gases consisting of CH_4 (v% = 2 %), O_2 (v% = 8 %), H_2S (v% = 0.02 %), and N_2 were fed through the flow controller (D07-19B, Beijing Sevenstart Electronics Co. Ltd., Beijing, China) at 200 mL/min and used in all the experiments. Finally, the temperature control controller (AI808PK1L2, XiaMen YuDian Automation Technology Co., Ltd., XiaMen, China) is used to control different reaction temperatures. The effluent gases were sampled and simultaneously analyzed online by gas chromatographs (GC-9790, Zhejiang Fuli Analytical Instruments Corp., Hangzhou, China). Exhaust gases were analyzed using a PoraPak Q column and a thermal conductivity detector (TCD). Methane conversions were calculated for outlet CO_2 concentrations.

4. Conclusions

Through the characterization of freshly prepared catalyst and used catalyst, we confirmed that Pd_4S, PdO, and metallic Pd^0 were the actual palladium species involved in the oxidation of methane. Sulfur-tolerance and catalytic activity was completely dependent on Pd_4S, and independent on other palladium species. Moreover, Pd_4S had sufficient oxidation stability under reaction conditions. Therefore, even in the presence of PdO, there are enough reasons to represent the catalyst with Pd_4S/SiO_2. In summary, we have developed a powerful sulfur-tolerant catalyst for methane oxidation by incorporating Pd_4S into SiO_2, and to the best of our knowledge, this is the first report on Pd catalysts with such property. In view of the excellent sulfur-tolerant property, the excellent oxidation stability and the high catalytic activity, the Pd_4S/SiO_2 catalyst will definitely find valuable and versatile use in catalytic chemistry.

At the same time, as the Pd_xS_y species with the highest proportion of Pd/S, we cannot determine whether the outstanding performance of Pd_4S in methane catalytic oxidation is only a special case. Furthermore, sulfur-tolerant mechanism of Pd_4S needs further theoretical research.

Author Contributions: Conceptualization, L.M.; Data curation, S.Y. and H.Z.; Formal analysis, L.M., S.Y., and H.Z.; Funding acquisition, L.M. and C.L.; Investigation, T.J.; Methodology, X.Z.; Project administration, L.M. and X.L.; Writing—original draft, S.Y.; Writing—review & editing, L.M. and X.Z.

Funding: This work is financially supported by National Natural Science Foundation of China (Grant No.21473159, 21476208).

Conflicts of Interest: The authors declare no conflict of interest.

References

1. Trinchero, A.; Hellman, A.; Grönbeck, H. Methane oxidation over Pd and Pt studied by DFT and kinetic modeling. *Surf. Sci.* **2013**, *616*, 206–213. [CrossRef]
2. Mahara, Y.; Ohyama, J.; Tojo, T.; Murata, K.; Ishikawa, H.; Satsuma, K. Enhanced activity for methane combustion over a Pd/Co/Al$_2$O$_3$ catalyst prepared by a galvanic deposition method. *Catal. Sci. Technol.* **2016**, *6*, 4773–4776. [CrossRef]
3. Chen, J.; Arandiyan, H.; Gao, X.; Li, J. Recent Advances in Catalysts for Methane Combustion. *Catal. Surv. Asia* **2015**, *19*, 140–171. [CrossRef]
4. Colussi, S.; Gayen, A.; Camellone, M.F.; Boaro, M.; Llorca, J.; Fabris, S.; Trovarelli, A. Nanofaceted Pd-O Sites in Pd-Ce Surface Superstructures: Enhanced Activity in Catalytic Combustion of Methane. *Angew. Chem.* **2009**, *121*, 8633–8636. [CrossRef]
5. Ji, Y.; Guo, Y.B. Nanostructured perovskite oxides as promising substitutes of noble metals catalysts for catalytic combustion of methane. *Chin. Chem. Lett.* **2018**, *29*, 252–260.
6. Bossche, M.V.D.; Gronbeck, H. Methane Oxidation over PdO(101) Revealed by First-Principles Kinetic Modeling. *J. Am. Chem. Soc.* **2015**, *137*, 12035–12044. [CrossRef] [PubMed]
7. Monai, M.; Montini, T.; Gorte, R.J.; Fornasiero, P. Catalytic Oxidation of Methane: Pd and Beyond. *Eur. J. Inorg. Chem.* **2018**, *25*, 2884–2893. [CrossRef]
8. Jiang, L.; Zheng, Y.; Chen, X.; Xiao, Y.; Cai, G.; Zheng, Y.; Zhang, Y.; Huang, F. Catalytic Activity and Stability over Nanorod-Like Ordered Mesoporous Phosphorus-Doped Alumina Supported Palladium Catalysts for Methane Combustion. *ACS Catal.* **2018**, *8*, 11016–11028.
9. Fouladvand, S.; Skoglundh, M.; Carlsson, P.A. A transient in situ infrared spectroscopy study on methane oxidation over supported Pt catalysts. *Catal. Sci. Technol.* **2014**, *4*, 3463–3473. [CrossRef]
10. Florén, C.R.; Bossche, M.V.D.; Creaser, D.; Grobeck, H.; Carlsson, P.A.; Korpi, H.; Skoglundh, M. Modelling complete methane oxidation over palladium oxide in a porous catalyst using first-principles surface kinetics. *Catal. Sci. Technol.* **2018**, *8*, 508–520. [CrossRef]
11. Banerjee, A.C.; Golub, K.W.; Abdul, Md. H.; Billor, M.Z. Comparative study of the characteristics and activities of Pd/γ-Al$_2$O$_3$ catalysts prepared by Vortex and Incipient Wetness Methods. *Catalysts* **2019**, *9*, 336. [CrossRef]
12. Banerjee, A.C.; McGuire, M.M.; Lawnick, O.; Bozack, M.J. Low-temperature activity and PdO/PdOx transition in methane combustion by a PdO-PdOx/γ-Al$_2$O$_3$ catalyst. *Catalysts* **2018**, *8*, 266. [CrossRef]
13. Ciuparu, D.; Lyubovsky, M.R.; Altman, E.; Pfefferle, L.D.; Datye, A. Catalytic combustion of methane over palladium based catalysts. *Catal. Rev. Sci. Eng.* **2002**, *44*, 593–649. [CrossRef]
14. Persson, K.; Pfefferle, L.D.; Schwartz, W.; Ersson, A.; Jaras, S.G. Stability of palladium-based catalysts during catalytic combustion of methane: The influence of water. *Appl. Catal. B* **2007**, *74*, 242–250. [CrossRef]
15. Liu, Y.; Wang, S.; Gao, D.; Sun, T.; Zhang, C.; Wang, S.D. Influence of metal oxides on the performance of Pd/Al$_2$O$_3$ catalysts for methane combustion under lean-fuel conditions. *Fuel Process. Technol.* **2013**, *111*, 55–61. [CrossRef]
16. Baylet, A.; Royer, S.; Marecot, P.; Tatibouet, J.M.; Duprez, D. High catalytic activity and stability of Pd doped hexaaluminate catalysts for the CH$_4$ catalytic combustion. *Appl. Catal. B* **2008**, *77*, 237–247. [CrossRef]
17. Ma, J.; Lou, Y.; Cai, Y.; Zhao, Z.; Wang, L.; Zhan, W.; Guo, Y.; Guo, Y. The relationship between the chemical state of Pd species and the catalytic activity for methane combustion on Pd/CeO$_2$. *Catal. Sci. Technol.* **2018**, *8*, 2567–2577. [CrossRef]
18. Mihai, O.; Smedler, G.; Nylén, U.; Olofsson, M.; Olsson, L. The effect of water on methane oxidation over Pd/Al$_2$O$_3$ under lean, stoichiometric and rich conditions. *Catal. Sci. Technol.* **2017**, *7*, 3084–3096. [CrossRef]
19. Zi, X.; Liu, L.; Xue, B.; Dai, H.; He, H. The durability of alumina supported Pd catalysts for the combustion of methane in the presence of SO$_2$. *Catal. Today* **2011**, *175*, 223–230. [CrossRef]
20. Wilburn, M.S.; Epling, W.S. Sulfur deactivation and regeneration of mono- and bimetallic Pd-Pt methane oxidation catalysts. *Appl. Catal. B* **2017**, *206*, 589–598. [CrossRef]
21. Monai, M.; Montini, T.; Melchionna, M.; Ducchon, T.; Kus, P.; Chen, C.; Tsud, N.; Nasi, L.; Prince, K.C.; Veltruska, K.; et al. The effect of sulfur dioxide on the activity of hierarchical Pd-based catalysts in methane combustion. *Appl. Catal. B* **2017**, *202*, 72–83. [CrossRef]

22. Yin, F.; Ji, S.; Wu, P.; Zhao, F.; Liu, H.; Li, C. Preparation of Pd-Based Metal Monolithic Catalysts and a Study of Their Performance in the Catalytic Combustion of Methane. *ChemSusChem* **2008**, *1*, 311–319. [CrossRef]
23. Hoyos, L.J.; Praliaud, H.; Primet, M. Catalytic combustion of methane over palldium supported on alumina and silica in presence of hydrogen-sulfie. *Appl. Catal. A* **1993**, *98*, 125–138. [CrossRef]
24. Venezia, A.M.; Carlo, G.D.; Liotta, L.F.; Pantaleo, G.; Kantcheva, M. Effect of Ti(IV) loading on CH_4 oxidation activity and SO_2 tolerance of Pd catalysts supported on silica SBA-15 and HMS. *Appl. Catal. B* **2011**, *106*, 529–539. [CrossRef]
25. Ortloff, F.; Bohnau, J.; Kramar, U.; Graf, F.; Kolb, T. Studies on the influence of H_2S and SO_2 on the activity of a PdO/Al_2O_3 catalyst for removal of oxygen by total oxidation of (bio-)methane at very low O_2:CH_4 ratios. *Appl. Catal. B* **2016**, *182*, 550–561. [CrossRef]
26. Meeyoo, V.; Trimm, D.L.; Cant, N.W. The effect of sulfur containing pollutants on the oxidation activity of precious metals used in vehicle exhaust catalysts. *Appl. Catal. B* **1998**, *16*, L101–L104. [CrossRef]
27. Ordóñez, S.; Hurtado, P.; Díez, F.V. Methane catalytic combustion over Pd/Al_2O_3 in presence of sulfur dioxide: development of a regeneration procedure. *Catal. Lett.* **2005**, *100*, 27–34. [CrossRef]
28. Corro, G.; Cano, C.; Fierro, J.L.G. A study of Pt–Pd/γ-Al_2O_3 catalysts for methane oxidation resistant to deactivation by sulfur poisoning. *J. Mol. Catal. A: Chem* **2010**, *315*, 35–42. [CrossRef]
29. Simon, L.J.; Ommen, J.G.V.; Jentys, A.; Lercher, J.A. Sulfur-Tolerant Pt-Supported Zeolite Catalysts for Benzene Hydrogenation: I. Influence of the Support. *J. Catal.* **2001**, *201*, 60–69. [CrossRef]
30. Ferrer, I.J.; Diazchao, P.; Pascual, A.; Sánchez, C. An investigation on palladium sulphide (PdS) thin films as a photovoltaic material. *Thin Solid Films* **2007**, *515*, 5783–5786. [CrossRef]
31. Xu, W.; Ni, J.; Zhang, Q.F.; Feng, F.; Xiang, Y.Z.; Li, X.N. Tailoring supported palladium sulfide catalysts through H_2-assisted sulfidation with H_2S. *J. Mater. Chem. A* **2013**, *1*, 12811–12817. [CrossRef]
32. Zhang, Q.F.; Xu, W.; Li, X.N.; Jiang, D.H.; Xiang, Y.Z.; Wang, J.G.; Cen, J.; Romano, S.; Ni, J. Catalytic hydrogenation of sulfur-containing nitrobenzene over Pd/C catalysts: In situ sulfidation of Pd/C for the preparation of $PdxSy$ catalysts. *Appl. Catal. A* **2015**, *497*, 17–21. [CrossRef]
33. Mccue, A.J.; Guerrero-Ruiz, A.; Rodríguez-Ramos, I.; Anderson, J.A. Palladium sulphide – A highly selective catalyst for the gas phase hydrogenation of alkynes to alkenes. *J. Catal.* **2016**, *340*, 10–16. [CrossRef]
34. Menegazzo, F.; Canton, P.; Pinna, F.; Pernicone, N. Bimetallic Pd–Au catalysts for benzaldehyde hydrogenation: Effects of preparation and of sulfur poisoning. *Catal. Commun.* **2008**, *9*, 2353–2356. [CrossRef]
35. Mccue, A.J.; Anderson, J.A. Sulfur as a catalyst promoter or selectivity modifier in heterogeneous catalysis. *Cheminform* **2014**, *4*, 272–294. [CrossRef]
36. O'Brien, C.P.; Gellman, A.J.; Morreale, B.D.; Miller, J.B. The hydrogen permeability of Pd_4S. *J. Membr. Sci.* **2011**, *371*, 263–267. [CrossRef]
37. O'Brien, C.P.; Howard, B.H.; Miller, J.B.; Morreale, B.D.; Gellman, A.J. Inhibition of hydrogen transport through Pd and $Pd_{47}Cu_{53}$ membranes by H_2S at 350 degrees. *J. Membr. Sci.* **2010**, *349*, 380–384. [CrossRef]
38. Morreale, B.D.; Howard, B.H.; Iyoha, O.; Enick, R.M.; Ling, C.; Sholl, D.S. Experimental and Computational Prediction of the Hydrogen Transport Properties of Pd_4S. *Ind. eng. chem. res* **2007**, *46*, 6313–6319. [CrossRef]
39. Zubkov, A.; Fujino, T.; Sato, N.; Yamada, K. Enthalpies of formation of the palladium sulphides. *J. Chem. Thermodyn.* **1998**, *30*, 571–581. [CrossRef]
40. Iyoha, O.; Enick, R.; Killmeyer, R.; Morreale, B. The influence of hydrogen sulfide-to-hydrogen partial pressure ratio on the sulfidization of Pd and 70 mol% Pd–Cu membranes. *J. Membr. Sci.* **2007**, *305*, 77–92. [CrossRef]
41. Hensen, E.J.M.; Brans, H.J.A.; Lardinois, G.M.H.J.; deBeer, V.H.J.; vanVeen, J.A.R.; van Santen, R.A. Periodic Trends in Hydrotreating Catalysis: Thiophene Hydrodesulfurization over Carbon-Supported 4d Transition Metal Sulfides. *J. Catal.* **2000**, *192*, 98–107. [CrossRef]
42. Gerson, A.R.; Bredow, T. Interpretation of sulfur 2p XPS spectra in sulfide minerals by means of ab initio calculations. *Surf. Interface Anal.* **2000**, *29*, 145–150. [CrossRef]
43. Bhatt, R.; Bhattacharya, S.; Basu, R.; Singh, A.; Deshpande, U.; Surger, C.; Basu, S.; Aswal, D.K.; Gupta, S.K. Growth of Pd_4S, PdS and PdS_2 films by controlled sulfurization of sputtered Pd on native oxide of Si. *Thin Solid Films* **2013**, *539*, 41–46. [CrossRef]
44. Senftle, T.P.; Van Duin, A.C.T.; Janik, M.J. The Role of Site Stability in Methane Activation on $PdxCe1-xO\delta$ Surfaces. *ACS Catal.* **2015**, *5*, 6187–6199. [CrossRef]

45. Beketov, G.; Heinrichs, B.; Pirard, J.P.; Chenakin, S.; Kruse, N. XPS structural characterization of Pd/SiO$_2$ catalysts prepared by cogelation. *Appl. Surf. Sci.* **2013**, *287*, 293–298. [CrossRef]
46. Romanchenko, A.S.; Mikhlin, Y.L. An XPS study of products formed on pyrite and pyrrhotine by reacting with palladium (II) chloride solutions. *J. Struct. Chem.* **2015**, *56*, 531–537. [CrossRef]
47. Corro, G.; Vázquez-Cuchillo, O.; Banuelos, F.; Fierro, J.L.G.; Azomoza, M. An XPS evidence of the effect of the electronic state of Pd on CH$_4$ oxidation on Pd/gamma-Al$_2$O$_3$ catalysts. *Catal. Commun.* **2007**, *8*, 1977–1980. [CrossRef]
48. Chenakin, S.P.; Melaet, G.; Szukiewicz, R.; Kruse, N. XPS study of the surface chemical state of a Pd/(SiO$_2$+TiO$_2$) catalyst after methane oxidation and SO$_2$ treatment. *J. Catal.* **2014**, *312*, 1–11. [CrossRef]
49. Weng, X.; Ren, H.; Chen, M.; Wan, H. Effect of Surface Oxygen on the Activation of Methane on Palladium and Platinum Surfaces. *ACS Catal.* **2014**, *4*, 2598–2604. [CrossRef]
50. Cargnello, M.; Jaén, J.J.D.; Garrido, J.C.H.; Bakhmutsky, K.; Montini, T.; Gámez, J.J.C.; Gorte, R.J.; Fornasiero, P. Exceptional Activity for Methane Combustion over Modular Pd@CeO2 Subunits on Functionalized Al$_2$O$_3$. *Science* **2012**, *337*, 713–718. [CrossRef] [PubMed]
51. Widjaja, H.; Sekizawa, K.; Eguchi, K.; Arai, H. Oxidation of methane over Pd/mixed oxides for catalytic combustion. *Catal. Today* **1999**, *47*, 95–101. [CrossRef]
52. Sekizawa, K.; Widjaja, H.; Maeda, S.; Ozawa, Y.; Eguchi, K. Low temperature oxidation of methane over Pd catalyst supported on metal oxides. *Catal. Today* **2000**, *59*, 69–74. [CrossRef]

© 2019 by the authors. Licensee MDPI, Basel, Switzerland. This article is an open access article distributed under the terms and conditions of the Creative Commons Attribution (CC BY) license (http://creativecommons.org/licenses/by/4.0/).

Article

Synthesis of GO-SalenMn and Asymmetric Catalytic Olefin Epoxidation

Fengqin Wang [1,†], Tiankui Huang [1,†], Shurong Rao [1,†], Qian Chen [1], Cheng Huang [1], Zhiwen Tan [2], Xiyue Ding [3] and Xiaochuan Zou [1,*]

[1] Department of Biological and Chemical Engineering, Chongqing University of Education, Nan'an 400067, China; wangfengqin0611@163.com (F.W.); xiaokui46485@163.com (T.H.); rao1334898571@163.com (S.R.); qian13340380042@163.com (Q.C.); HccVictory@163.com (C.H.)
[2] College of Chemistry, Chongqing Normal University, Chongqing 401331, China; tzw15310181725@163.com
[3] College of Food and Biological Engineering, Jimei University, Xiamen 361021, China; xiyue772751020@163.com
* Correspondence: zouxc@cque.edu.cn; Tel.: +86-23-8630-7018; Fax: +86-23-6163-8000
† F.Q. Wang, T.K. Huang and S.R. Rao. contributed equally to this works.

Received: 9 September 2019; Accepted: 26 September 2019; Published: 30 September 2019

Abstract: Graphene oxide (GO) was used as a catalyst carrier, and after the hydroxyl group in GO was modified by 3-aminopropyltrimethoxysilane (MPTMS), axial coordination and immobilization with homogeneous chiral salenMnCl catalyst were carried out. The immobilized catalysts were characterized in detail by FT–IR, TG–DSC, XPS, EDS, SEM, X-ray, and AAS, and the successful preparation of GO-salenMn was confirmed. Subsequently, the catalytic performance of GO-salenMn for asymmetric epoxidation of α-methyl-styrene, styrene, and indene was examined, and it was observed that GO-salenMn could efficiently catalyze the epoxidation of olefins under an m-CPBA/NMO oxidation system. In addition, α-methyl-styrene was used as a substrate to investigate the recycling performance of GO-salenMn. After repeated use for three times, the catalytic activity and enantioselectivity did not significantly change, and the conversion was still greater than 99%. As the number of cycles increased, the enantioselectivity and chemoselectivity gradually decreased, but even after 10 cycles, the enantiomeric excess was 52%, which was higher than that of the homogeneous counterpart under the same conditions. However, compared to fresh catalysts, the yield decreased from 96.9 to 55.6%.

Keywords: graphene oxide; 3-aminopropyltrimethoxysilane; heterogeneous catalyst; asymmetric epoxidation

---

1. Introduction

Chiral salenMnCl catalyst (Jacobsen's catalyst) has been proved to be one of the most effective catalysts for asymmetric catalytic epoxidation of alkenes [1,2]. The chiral epoxides obtained are widely used in the synthesis of fine chemicals including pesticides, flavorings, and pharmaceuticals, such as the key intermediates for anti-hypertensive drugs and for the side chain of the anti-cancer drug paclitaxel. In most cases, homogeneous catalytic reactions are highly efficient. However, because it is difficult to separate and recycle the expensive chiral catalysts after the reaction, it is impossible to realize continuous flow reactors and large-scale production, which increases operating costs and wastes limited resources. Heterogeneity of homogeneous catalysts is an important strategy to enable the reuse of expensive chiral catalysts and large-scale synthesis [3]. Common catalytic carrier materials include organic polymer, inorganic solidseries, and organic polystyrene/inorganic hydrogen phosphate (Zr, Al, Zn, Ca) [4–15]. Unfortunately, immobilized catalysts still have some deficiencies such as low catalytic efficiency and poor reusability. Therefore, developing a class of highly-efficient heterogeneous catalysts is still the focus of current studies.

In recent years, GO has been widely researched as a promising catalyst carrier [16], mainly because of its unique two-dimensional planar structure (which is beneficial to improve the dispersion of catalytic active species on the surface), higher specific surface area (which is beneficial to increase its forces with active species and, to a certain extent, prevent the migration and leaching of active species in the reaction process), and the nanometer size effect (which can improve the catalytic reaction rate). In addition, the surface of GO contains a large amount of oxygen-containing active groups, such as carboxyl, hydroxyl, and epoxy groups, which are easy to chemically modify [17] and can react with noble metal catalysts and homogeneous ligands to obtain various supported catalysts with different coordination modes. Hence, GO is widely used in organic catalysis [18,19], photocatalysis [20,21], and electrocatalysis [22]. Noble metal catalysts, such as Pd^{2+}, Pd^0 [18], and Au [19], can coordinate with GO to obtain composite catalysts that can be used for a cross-coupling reaction and an addition reaction between phenylacetylene and hydrogen, achieving efficient catalysis of the catalyst (up to 99%) with excellent catalyst stability. At the same time, GO successfully immobilizes Schiff-base ligands [23] and L-proline [24] homologous ligands through covalent grafting and clever use of hydrogen bonds and evaluates the catalytic activity of the Suzuki reaction in the water phase and the asymmetric catalytic aldol reaction, respectively. GO/Schiff-base catalysts can efficiently catalyze the Suzuki reaction of aryl halide in the water phase and aryl boric acid under mild conditions, and it was observed that the activity of the catalysts remained nearly unchanged after five repeated uses. GO/L-proline catalysts possess high catalytic activity and good reuse performance, and the asymmetric selectivity of catalytic reactions increases compared with homogeneous catalysts due to the unique layered structure of GO. It is considered that an obvious spatial confinement effect [25] appears in the nanosphere, which enhances the chiral inducibility of heterogeneous asymmetric reactions.

To continue to study the potential performance of GO as a catalyst carrier and further develop a simple heterogeneous epoxidation catalysts with good catalytic efficiency and high stability, we modified the hydroxyl groups in GO that were functionalized by MPTMS. Then, axial coordination fixation was carried out with the homogeneous chiral salenMnCl catalyst, to investigate in detail the catalytic performance of the supported catalyst for asymmetric epoxidation of α-methyl-styrene, styrene, and indene compared with the homogeneous catalyst. In addition, the catalytic oxidation of α-methyl-styrene was used as a template reaction to investigate the cyclic performance of the supported catalyst.

2. Results and Discussion

2.1. FT–IR Analysis

FT-IR spectroscopy results of GO, salenMnCl, and catalyst GO-salenMn are shown in Figure 1a–c. Figure 1a shows the FT-IR spectrum of GO. A very wide and strong absorption peak appeared between 3100 and 3500 cm^{-1}, which denoted the stretching vibration peak of hydroxyl groups (COO–H/O–H), indicating that GO contained hydroxyl and carboxyl groups. The absorption peak at approximately 1724 cm^{-1} indicated the stretching vibration peak of the carboxyl group (C=O). In addition, the absorption peaks at 1218 and 1050 cm^{-1} denoted the stretching vibration peaks of hydroxyl groups (C–OH) and the C–O bond of epoxy groups (C–O–C), respectively. The results showed that GO was successfully prepared by the Hummer method. Figure 1b shows the FT-IR spectrum of the homogeneous salenMnCl catalyst, showing that 2963–2864 cm^{-1} are the stretching vibration absorption peaks of methyl and methylene groups $\nu_{(C-H)}$, respectively. Characteristic absorption peaks at 1610 cm^{-1} and 522 cm^{-1} were indicative of C=N and Mn–N bonds, respectively. In Figure 1c, 2936–2853 cm^{-1} were the stretching vibration absorption peaks of methyl and methylene groups $\nu_{(C-H)}$, respectively. The absorption peak at approximately 1620 cm^{-1} in GO-salenMn denoted the stretching vibration of $\nu_{(C=N)}$, which indicated that chiral heterogeneous catalysts possessed structures similar to those of homogeneous chiral catalysts [26]. These results preliminarily confirmed the heterogeneity of homogeneous salenMnCl catalyst.

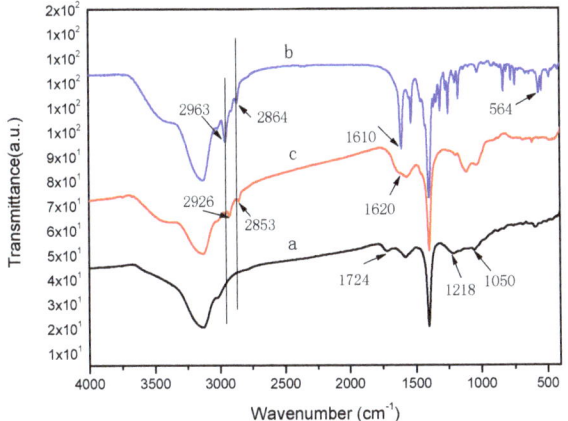

Figure 1. IR spectra of (**a**) GO, (**b**) salenMnCl, and (**c**) GO-salenMn.

2.2. TG–DSC Analysis

TG–DSC analysis of catalyst GO-salenMn is shown in Figure 2. Weightlessness was divided into three stages. The first stage was the heat absorption process at 120 °C room temperature. In this stage, the weight loss of water was approximately 15%, which was mainly due to the removal of water attached to the surface of the catalyst. The second stage was the strong exothermic process at 120 °C–500 °C. The weight loss of 32% during this stage was mainly due to the decomposition of oxygen-containing groups on graphene oxide and the decomposition of salenMn complexes. After 500 °C, weight loss slowed as the temperature increased. Finally, until 700 °C, the remaining mass of carbon skeleton and manganese oxide was approximately 38%.

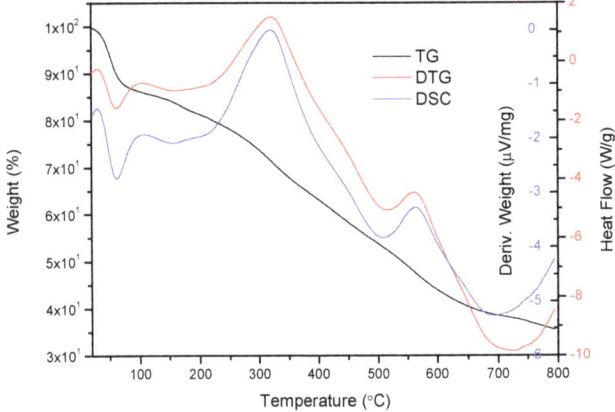

Figure 2. TG–DSC spectra of GO-salenMn.

2.3. XPS Analysis

XPS is an important method for analyzing metal valence on the surface of materials. The binding energy of the electrons in the center of the metal is affected by the differences in the electron environment, such as the oxidation state or the spin state. Figure 3 shows the full spectra of XPS scans of salenMn and

GO-salenMn. Figure 3a clearly shows the characteristic peaks of C1s, O1s, N1s, Mn 2p$^{1/2}$, and Mn 2p$^{3/2}$ in homogeneous salenMn catalyst. In Figure 3b, in addition to the characteristic peaks in homogeneous salenMnCl, new-additional characteristic peaks, Si 2p and Si 2s, proved that GO successfully achieved the support of homogeneous salenMn after MPTMS modification. In addition, as seen in Figure 4, the electron binding energy of Mn2p$^{3/2}$ in catalyst GO-salenMn was 641.4 eV, which was slightly higher than that of Mn2p$^{3/2}$ in the homogeneous salenMnCl complex. Thus, the electron cloud density of manganese atoms decreased after support, which was similar with the reported value [8,27], possibly because the micro environment changed after metallic Mn was immobilized with 3-aminopropyl trimethoxy silane and modified GO. A similar observation has also been made for a chiral salenMn^{III} catalyst, which was axially grafted on an amine (–NH$_2$) group modified organic polymer/inorganic zirconium hydrogen phosphate through N–Mn bonding [28].

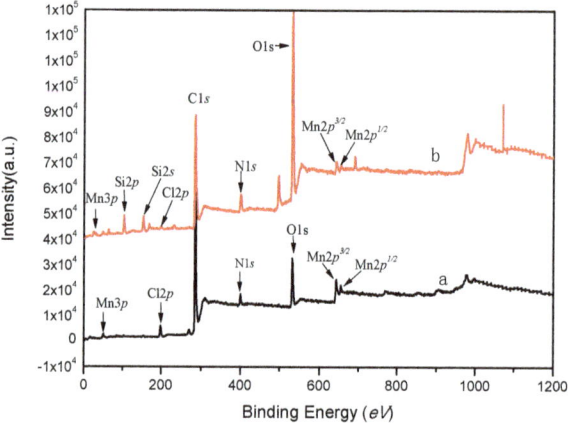

Figure 3. XPS spectra of (**a**) salenMn and (**b**) GO-salenMn.

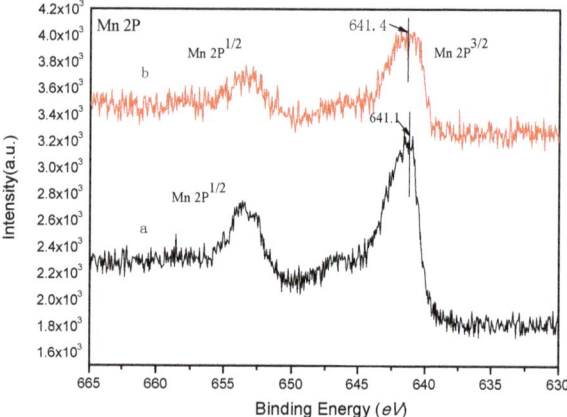

Figure 4. XPS spectra of Mn 2p for (**a**) salenMn and (**b**) GO-salenMn.

2.4. SEM and EDS Analysis

SEM images of GO, GO-NH$_2$, and GO-salenMn are shown in Figure 5a–c. As seen in Figure 5a, as a specific feature of GO, a large number of overlapping and curled slice layers existed, which indicated that GO had been successfully prepared. Figure 5b,c shows that the lamellar structure was damaged to a certain degree, which may have occurred after the amine and salenMn catalyst were introduced on the surface of GO, and the inter-layer space subsequently increased. Under the influence of ultrasonic dispersion, slices of GO would strip, and the same conclusion could be drawn from XRD. To further confirm that chiral salenMnCl complexes were successfully immobilized on carrier GO, which was modified by MPTMS, EDS analysis of GO-salenMn was carried out (Figure 5d–f). The results showed that, compared with GO (Figure 5d), characteristic spectral lines of Si and N appeared in GO-NH$_2$ (Figure 5e), indicating that Si-containing amine functional groups were successfully introduced in GO. Compared with GO-NH$_2$ (Figure 5e), GO-salenMn (Figure 5f) exhibited obvious characteristic spectral lines of Mn, which also proved that MPTMS-modified GO successfully achieved axial immobilization with chiral salenMnCl complexes. In addition, according to Figure 6a–d, EDS layers of GO-salenMn contained elements such as C, Oi, Mn, and Si, which further proved the heterogeneity of chiral salenMnCl.

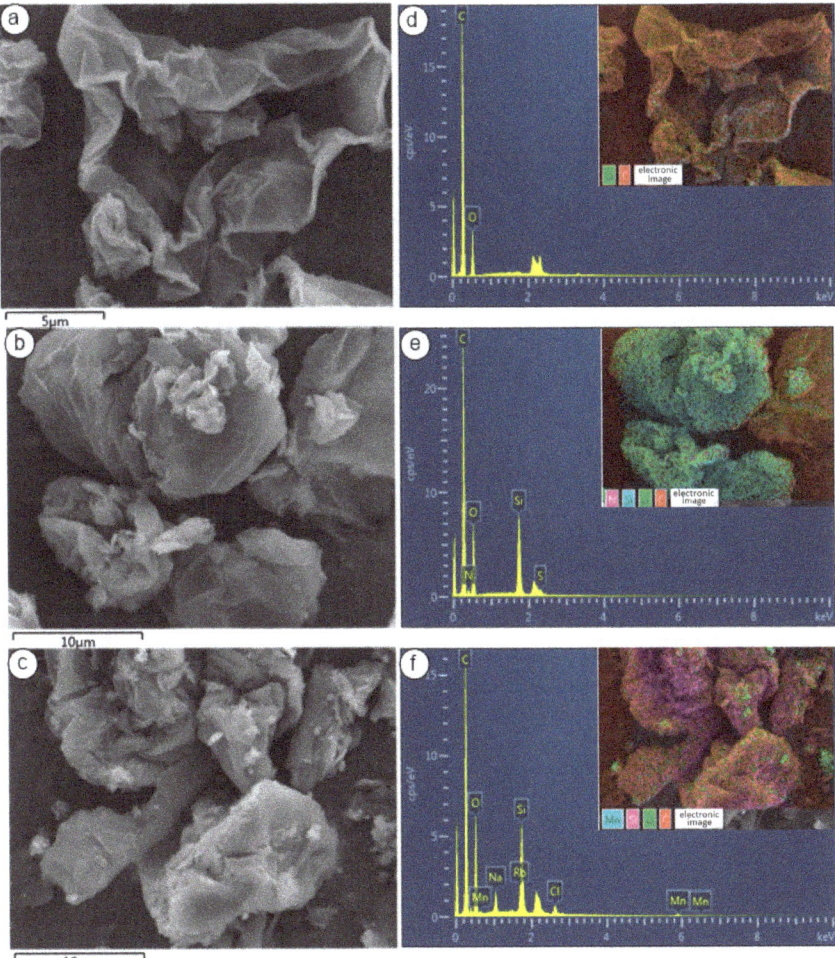

Figure 5. SEM photograph of (**a**) GO, (**b**) GO-NH$_2$, and (**c**) GO-salenMn, and the measured EDS image of (**d**) GO, (**e**) GO-NH$_2$, and (**f**) GO-salenMn.

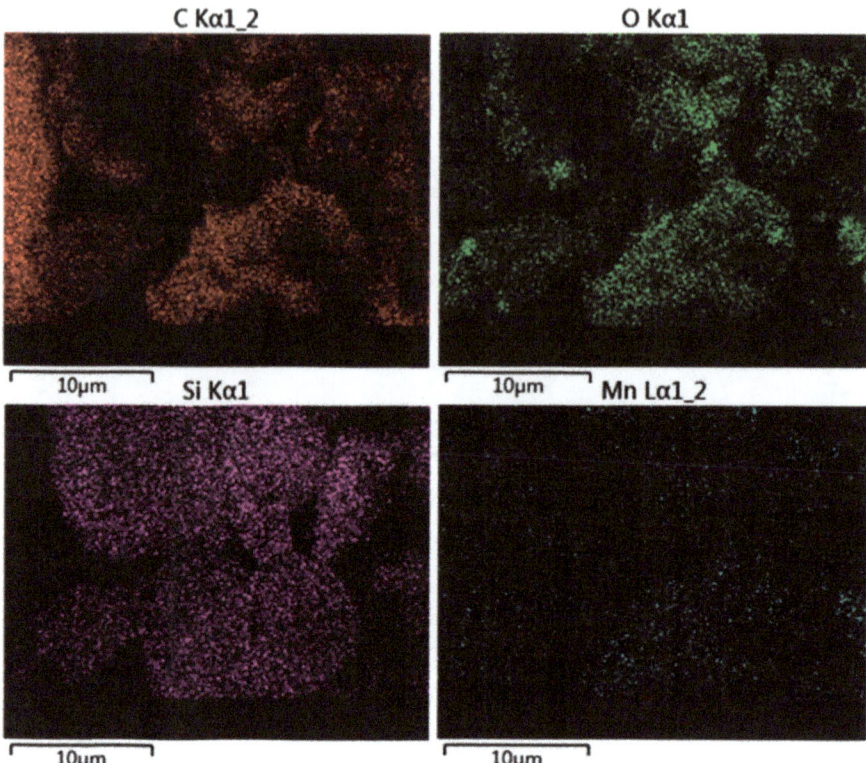

Figure 6. EDS elementary mapping of (**a**) C, (**b**) O, (**c**) Si, and (**d**) Mn in GO-salenMn.

2.5. XRD Analysis

The XRD results of GO, GO-NH$_2$, and GO-salenMn are shown in Figure 7. As seen in Figure 7a, a sharp diffraction peak at 2θ = 10.76° indicated the (001) characteristic diffraction of GO. A small diffraction peak appeared at 10°–26°, which was caused by the superposition of the laminate crystallization of GO with different thicknesses. This further proved that GO had been successfully prepared. Figure 7b,c shows no characteristic diffraction peak of GO, indicating that the original layer-cumulative structure of GO had been changed in the process of amine functionalization. In addition, after the homogeneous salenMn catalyst reacted with GO-NH$_2$, possibly due to the insertion of homogeneous salenMn catalyst between residual layers, the structure of GO-NH$_2$ was further destroyed, and the aggregation degree decreased. Combined with SEM analysis, these data show that GO-NH$_2$ and GO-salenMn became amorphous.

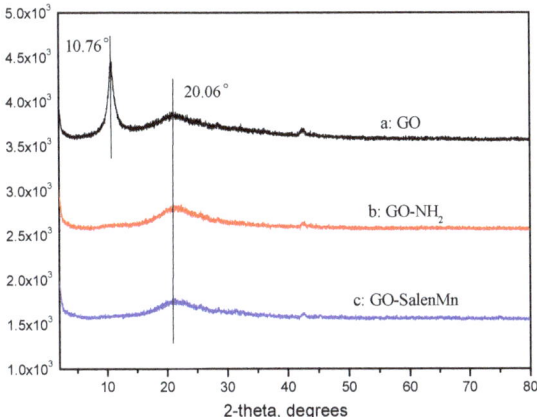

Figure 7. X-ray spectra of (**a**) GO, (**b**) GO-NH$_2$, and (**c**) GO-salenMn.

2.6. Asymmetric Epoxidation

Using *m*-CPBA/NMO as an oxidation system, the performance of salenMnCl and GO-salenMn catalysts for asymmetric epoxidation of α-methyl-styrene, styrene, and indene under the same conditions was compared. In addition, the performance of the catalytic reaction with or without the participation of axial additive NMO was studied in detail. The results are shown in Table 1. The results showed that the GO-salenMn catalyst could effectively catalyze epoxidation of alkenes (entries 2, 5, and 8). Compared with homogeneous salenMnCl, the GO-salenMn catalyst significantly increased the enantiomeric excess (*ee*) value of epoxides (entries 2, 5, and 8), The *ee* value of α-methyl-styrene epoxide increased from 52.0 to 83.2%. Similar results were observed in epoxidation of styrene and indene. The increase in chiral recognition was mainly attributed to the layered support effect [29]. In contrast, anther GO-salenMn catalyst [26], where salenMnCl was immobilized on amino-modified GO or imidazolium-based ionic liquid-functionalized GO with the methyl chloride group (–CH$_2$Cl) at the fifth position of the chiral salen ligand, gave a slightly lower *ee* value in the NaClO/PyNO oxidation system. The results showed that the combined effect of the immobilization mode and the oxidation system was beneficial to increase the *ee* value. Moreover, in our previous reports, the ZPS-PVPA-based catalyst-effectively catalyzed epoxidation of styrene and α-methyl-styrene (*ee*: 50 to 78% and 86% to >99%) with *m*-CPBA or NaClO. These results are significantly better than those achieved with the homogeneous chiral catalysts under the same reaction conditions (*ee*: 47% and 65%). Moreover, the immobilized catalysts could be reused at least 10 times without significant loss of activity and enantioselectivity. Furthermore, a point worth emphasizing is that the ZPS-PVPA-based catalyst resulted in remarkable increase of conversion and *ee* values in the absence of expensive NMO for the asymmetric epoxidation of olefins, which is exactly opposite to the literature reported earlier for both homogeneous and heterogeneous systems. This novel additive effect was mainly attributed to the support ZPS-PVPA and the axial phenoxyl linker group [10], α-methyl-styrene. Regrettably in this study, the *ee* values were sharply decreased when the epoxidation was carried out with GO-salenMn in the absence of axial ligand NMO (entries 3, 6, and 9). The results demonstrated that the NMO could coordinate with oxo-salenMn (V) and stabilize the generated intermediate oxo-salenMn (V) complex [30], so that the substrates and catalysts could fully react.

Table 1. Asymmetric epoxidation of different substrates by salenMnCl and GO-salenMn [a].

Entry	Substrate	Catalyst	Oxidant	Time (h)	Conv (%) [b]	Sele (%) [b]	ee (%) [b]	TOF[e] × 10^{-4} (s^{-1})
1		salenMnCl	m-CPBA/NMO	1	>99	>99	52.0 [c]	27.2
2		GO-salenMn	m-CPBA/NMO	2	>99	97.9	83.2 [c]	13.5
3		GO-salenMn	m-CPBA	2	65.9	76.5	8.6 [c]	7.0
4		salenMnCl	m-CPBA/NMO	1	>99	98.7	46.2 [c]	27.1
5		GO-salenMn	m-CPBA/NMO	2	92.5	93.1	67.8 [c]	12.0
6		GO-salenMn	m-CPBA	2	73.2	72.9	5.9 [c]	7.4
7		salenMnCl	m-CPBA/NMO	1	>99	>99	65.0 [d]	27.2
8		GO-salenMn	m-CPBA/NMO	2	>99	>99	83.2 [d]	13.6
9		GO-salenMn	m-CPBA	2	77.9	81.6	13.1 [d]	8.8

[a] Reactions were carried out in CH_2Cl_2 (3 mL) with alkene (0.5 mmol), m-CPBA (1.0 mmol), NMO (2.5 mmol, if necessary), nonane (internal standard, 0.5 mmol), and GO-salenMn (0.010 mmol, 2.0 mol%), based on elemental Mn at 0 °C. [b] Conversions, selectivity, and enantiomeric excess (ee) values were determined by GC with a chiral capillary column (HP19091G-B233, 30 m × 0.25 mm × 0.25 μm) by integration of product peaks against an internal quantitative standard (nonane), correcting for response factors. [c] Epoxide configuration R. [d] Epoxide configuration 1R, 2S. [e] Turnover frequency (TOF) is calculated by the expression of [product]/[catalyst] × time (s^{-1}).

2.7. Investigation of Reuse Performance of Supported Catalysts

The reuse performance of solid catalysts is an important standard to measure whether the catalytically active center has effectively and stably immobilized on the carrier. The reuse performance of GO-salenMn was studied in detail with α-methyl-styrene as the template reaction. The results are listed in Table 2 and show that the catalytic activity and enantioselectivity of the catalyst were not significantly changed after the catalyst was reused three times, and the conversion rate was still greater than 99%. With the increase in reuse times, enantioselectivity and chemical selectivity gradually decreased. However, even after the catalyst was reused 10 times, the ee value was still higher than 52%. However, compared with fresh catalysts, the yield dropped from 96.9 to 55.6%. A possible reason for deactivation of the GO-salenMn is the decomposition of salenMnCl under NaOH aqueous solution [31].

Table 2. The recycling of GO-salenMn in the epoxidation of α-methyl-styrene [a].

Run	Conv (%) [b]	Sele (%) [b]	ee (%) [b,c]	TOF [d] × 10^{-4} (s^{-1})
1	>99	97.9	83.2	13.5
2	>99	97.0	83.0	13.3
3	>99	96.3	82.4	13.3
4	98.2	95.2	82.2	13.0
5	96.9	94.3	81.7	12.7
6	95.3	92.2	81.0	12.2
7	92.1	86.9	80.3	11.1
8	89.3	79.2	78.7	9.8
9	86.2	73.2	76.2	8.8
10	80.6	69.0	71.9	7.7

[a,b,c] The reaction conditions are the same as those for Table 1. [d] Turnover frequency (TOF) is calculated by the expression of [product]/[catalyst] × time (s^{-1}).

3. Material and Methods

3.1. Materials

The chemicals (1R,2R)-(-)-1,2-diaminocyclohexane, α-methyl-styrene, indene, n-nonane, N-methylmorpholine N-oxide (NMO), m-chloroperbenzoic acid (m-CPBA), and γ-propyl mercaptotrimethoxysilane (MPTMS) were purchased from Alfa Aesar. The other commercially available chemicals that were procured were laboratory-grade reagents from local suppliers (Chuandong Chemical Group Co., Ltd. Chongqing, China). Styrene was passed through a pad of neutral alumina before use. Chiral salen ligand and chiral salenMnCl complex were synthesized according to previously published procedures [1].

3.2. Methods

Fourier transform infrared (FT–IR) spectra were obtained as potassium bromide pellets with a resolution of 4 cm^{-1} in the range of 400–4000 cm^{-1} using a Bruker RFS100/S spectrophotometer (Bruker, Karlsruhe, Germany). Atomic absorption spectroscopy (AAS) was used to determine the Mn content of the catalysts using a TAS-986G (Pgeneral, Beijing, China), where 0.02 g of GO-salenMn was calcined at 700 °C for 3 h, dissolved in 1:1 hydrochloric acid for 30 min, and the volume was adjusted. The loading of Mn^{2+} was measured by standard addition method. To explore the success of the GO-salenMn, high-resolution field emission scanning electron microscopy (FE-SEM, JSM 7800F, Tokyo, Japan) operating at 5 kV was used to analyze the topographic microstructure. Environmental SEM equipped with an energy dispersive (EDS) X-ray spectrometer was also used to investigate the composition and spatial distribution of elements in the samples. Scanning electron microscope (SEM) was carried out with a powder sample (100 mg) dispersed in alcohol (10 mL) and sonicated for 10 min. The dispersion was dripped onto the conductive tape, and the tape was blown dry with a blower and then tested. To carry out the EDS, the powder sample was put on a conductive adhesive cloth. The interlayer spacings were recorded on a LabXRD-6100 automated X-ray power diffractometer (XRD), using Cu Kα radiation and internal silicon powder standard with all samples (Shimadz, Kyoto, Japan). The patterns were generally measured between 2.00° and 80.00° and X-ray tube settings of 40 kV and 5 mA. Thermogravimetry–differential scanning calorimetry (TG–DSC) analyses (5 mg) were performed on a SBTQ600 thermal analyzer (TA, USA) with the heating rate of 10 °C·min^{-1} from 25 to 1000 °C under flowing N2 (100 mL·min^{-1}). X-ray photoelectron spectroscopy (XPS) data were obtained with an ESCALab250 instrument (MA, USA) electron spectrometer using 75 W Al Kα radiation. The base pressure was about 1×10^{-8} Pa. The conversions (with n-nonane as an internal standard) and the enantiomeric excess (*ee*) values were analyzed by gas chromatography (GC) with a Shimadzu GC-2014 instrument (Shimadzu, Japan) equipped with a chiral column (HP19091G-B233, 30 m × 0.25 mm × 0.25 μm) and flame ionization detector; injector 230 °C, detector 230 °C. The column temperature for *a*-methylstyrene, styrene, and indene was 80–180 °C. The retention times of the corresponding chiral epoxides were as follows: (a) α-methyl-styrene epoxide: the column temperature was 80 °C, t_S = 12.9 min, t_R = 13.0 min; (b) styrene epoxide: the column temperature was 80 °C, t_R = 14.7 min, t_S = 14.9 min; (c) indene epoxide: the column temperature was programmed from 80 to 180 °C, t_{SR} = 16.1 min, t_{RS} = 17.1 min.

4. Preparation of Catalysts

4.1. Synthesis of GO and GO-NH$_2$

The parent GO was prepared by the previously reported Hummers method [32,33]. Then, 0.50 g of GO was dispersed in 20 mL of anhydrous toluene. After ultrasonication for 2.0 h, a certain amount of APTES was added and stirred for 12 h under reflux [34] (Scheme 1). The solid was collected by filtration, washed with dichloromethane, and vacuum-dried at 60 °C for 10 h. The final catalyst was defined as amine-functionalized graphene oxide: GO-NH$_2$, yield: 93.6%.

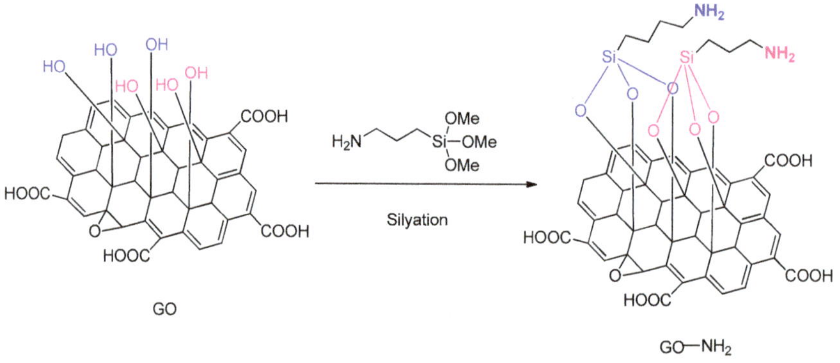

Scheme 1. Synthesis of GO-NH$_2$.

4.2. Synthesis of GO-salenMn Catalyst

In a 50 mL three-neck flask, 0.2 g amination carrier GO-NH$_2$ was pre-added to 10 mL toluene for undergoing ultrasonication for 30 min, followed by homogeneous chiral salenMnCl catalyst (543 mg, 1.0 mmol), Na$_2$CO$_3$ (120 mg, 3 mmol), and 20 mL toluene; the solution underwent reflux reaction for 24 h (Scheme 2). After the reaction stopped, the mixture was cooled to room temperature. Then, the solid mixture was soaked in methylene chloride, filtered by extraction, and centrifugally washed in methylene chloride and ethanol successively until the filtrate was clarified. The final filtrate was collected and tested by AAS, which showed no presence of Mn^{2+}. The brown solid catalyst powder was prepared by 60 °C vacuum drying to constant weight and the yield was 90.3%. AAS showed that the loading of the supported catalyst Mn^{2+} was 0.39 mmol/g.

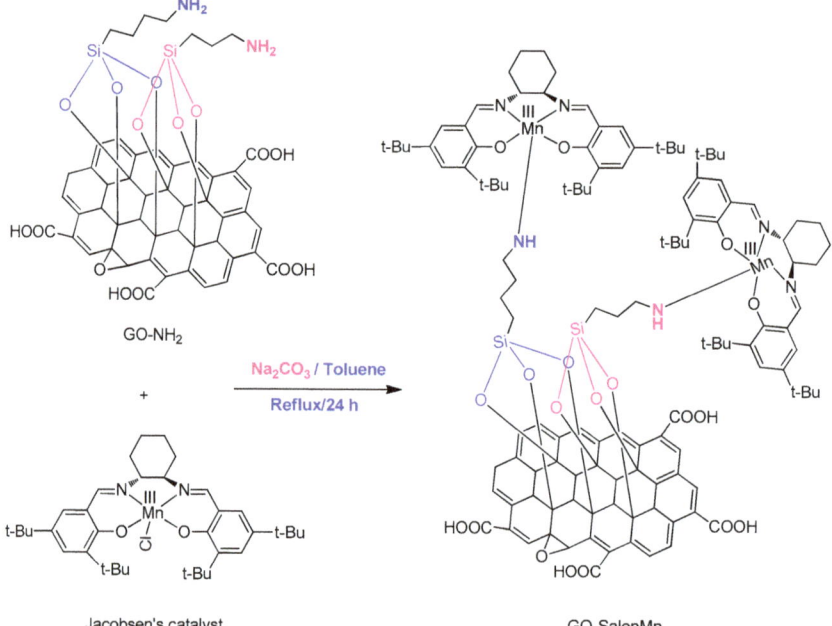

Scheme 2. Synthesis of the immobilized catalysts.

4.3. Asymmetric Epoxidation

At 0 °C, substrate (0.5 mmol), NMO (337.5 mg, 2.5 mmol, 5 equiv., if necessary), nonane (internal standard, 56.0 mL, 0.50 mmol), and a certain amount of GO-salenMn (0.010 mmol, 2.0%, manganese-based) were successively added to 3 mL CH_2Cl_2 solution. The oxidant *m*-CPBA (138 mg, 1.0 mmol, 2 equiv.) was added to the thermostatic reaction system in four portions within two minutes. After the reaction, NaOH aqueous solution (1 mol/L, 1.25 mL) was added to the system until the pH value was slightly higher than 7. After n-hexane was added, organic phase was extracted and dried by anhydrous sodium sulfate. Then column chromatography was conducted, and the chromatography liquid was rotated–evaporated and directly injected into the gas chromatograph to analyze the conversion rate and enantioselectivity. The racemic epoxides were prepared by epoxidation of the corresponding olefins by *m*-CPBA in CH_2Cl_2 and confirmed by NMR, and the GC was calibrated with the samples of *n*-nonane, olefins, and corresponding racemic epoxides.

4.4. Repeated Experiments of Supported Mn(Salen) Catalyst

In order to investigate the reuse performance of supported catalysts, α-methyl-styrene was used as a substrate. After the initial reaction, *n*-hexane was added directly to the reaction system and oscillated, and the organic phase at the top was separated after standing, and the supported catalyst at the bottom was removed and successively washed by distilled water, dichloromethane, and ethanol, and then dried for future use. Subsequent cycling with different amounts and product analysis was performed according to Section 4.3.

5. Conclusions

In summary, a simple and effective heterogeneous catalyst, GO-salenMn, was developed in this study and characterized in detail by FT–IR, TG–DSC, XPS, EDS, SEM, XRD, and AAS. The successful preparation of GO-salenMn was confirmed. GO-salenMn can efficiently catalyze α-methyl-styrene, styrene, and indene in *m*-CPBA/NMO as the oxidation system, and in particular, epoxides can be obtained with higher enantioselectivity than those obtained by homogeneous salenMnCl catalyst. However, enantioselectivity significantly decreased without the participation of axial promoter NMO. After the supported catalyst was reused 10 times, the *ee* value of α-methyl-styrene epoxides was still higher than that obtained by the homogeneous salenMnCl catalyst under the same conditions. However, after the catalyst was reused six times, the selectivity of the catalyst significantly decreased. The cause of catalyst deactivation still requires further investigation.

Author Contributions: X.Z. designed the experiments, wrote the paper, and provided the funds; F.W., T.H., S.R., Q.C., and X.D. performed the experiments; Z.T. and C.H. provided analytical testing.

Funding: This work was financially supported by the Children's Research Institute of National Center for the Schooling Development Programme and Chongqing University of Education (grant no. CSDP19FS01111), the Natural Science Foundation of Chongqing (no. cstc2018jcyjAX0110), the Science and Technology Research Program of Chongqing Municipal Education Commission (no. KJ201801607), the Key Laboratory for Green Chemical Technology of Chongqing University of Education (no. 2016xjpt08), and the "Qizhi" Creative Space Student Entrepreneurship Incubation Project of Chongqing University of Education (2019).

Conflicts of Interest: The authors declare no conflict of interest.

References

1. Zhang, W.; Loebach, J.L.; Wilson, S.R.; Jacobsen, E.N. Enantioselective epoxidation of unfunctionalized olefins catalyzed by salen manganese complexes. *J. Am. Chem. Soc.* **1990**, *112*, 2801–2803. [CrossRef]
2. Luo, Y.F.; Zou, X.C.; Fu, X.K.; Jia, Z.Y.; Huang, X.M. The advance in asymmetric epoxidation of olefins catalyzed by chiral Mn(salen). *Sci. China Chem.* **2011**, *41*, 433–450.
3. Zhang, H.D.; Zhang, Y.M.; Li, C. Asymmetric epoxidation of unfunctionalized olefins catalyzed by Mn(salen) axially immobilized onto insoluble polymers. *Tetrahedron Asymmetry* **2005**, *16*, 2417–2423. [CrossRef]

4. Aneta, N.; Agnieszka, W.; Michael, W. Recent advances in the catalytic oxidation of alkene and alkane substrates using immobilized manganese complexes with nitrogen containing ligands. *Coord. Chem. Rev.* **2019**, *382*, 181–216.
5. Reger, T.S.; Janda, K.D. Polymer-Supported (Salen)Mn Catalysts for Asymmetric Epoxidation: A Comparison Between Soluble and Insoluble Matrices. *J. Am. Chem. Soc.* **2000**, *122*, 6929–6934. [CrossRef]
6. Peng, M.; Chen, Y.J.; Tan, R.; Zheng, W.G.; Yin, D.H. A highly efficient and recyclable catalyst-dendrimer supported chiral salen Mn(III) complexes for asymmetric epoxidation. *RSC Adv.* **2013**, *3*, 20684–20692. [CrossRef]
7. Afsaneh, F.; Hassan, H.M. Highly efficient asymmetric epoxidation of olefins with a chiral manganese-porphyrin covalently bound to mesoporous SBA-15: Support effect. *J. Catal.* **2017**, *352*, 229–238.
8. Zou, X.C.; Shi, K.Y.; Li, J.; Wang, Y.; Deng, C.F.; Wang, C.; Ren, Y.R.; Tan, J. Research Progress on Epoxidation of Olefins Catalyzed by Mn(II, III, V) in Different Valence States. *Chin. J. Org. Chem.* **2016**, *36*, 1765–1778. [CrossRef]
9. Gong, B.W.; Fu, X.K.; Chen, J.X.; Li, Y.D.; Zou, X.C.; Tu, X.B.; Ding, P.P.; Ma, L.P. Synthesis of a new type of immobilized chiral salen Mn(III) complex as effective catalysts for asymmetric epoxidation of unfunctionalized olefins. *J. Catal.* **2009**, *262*, 9–17. [CrossRef]
10. Zou, X.C.; Fu, X.K.; Li, Y.D.; Tu, X.B.; Fu, S.D.; Luo, Y.F.; Wu, X.J. Highly enantioselective epoxidation of unfunctionalized olefins catalyzed by chiral Jacobsen's catalyst immobilized on phenoxyl modified Zirconium poly (styrene-phenylvinylphosphonate)-phosphate. *Adv. Synth. Catal.* **2010**, *352*, 163–170. [CrossRef]
11. Zou, X.C.; Wang, C.; Wang, Y.; Shi, K.Y.; Wang, Z.M.; Li, D.W.; Fu, X.K. Chiral MnIII (Salen) Covalently Bonded on Modified ZPS-PVPA and ZPS-IPPA as Efficient Catalysts for Enantioselective Epoxidation of Unfunctionalized Olefins. *Polymers* **2017**, *9*, 108. [CrossRef]
12. Zou, X.C.; Shi, K.Y.; Wang, C. Chiral MnIII(Salen) supported on tunable phenoxyl group modified zirconium poly (styrene-phenylvinylphosphonate)-phosphate as an efficient catalyst for epoxidation of unfunctionalized olefins. *Chin. J. Catal.* **2014**, *35*, 1446. [CrossRef]
13. Huang, J.; Liu, S.R.; Ma, Y.; Cai, J.L. Chiral salen Mn (III) immobilized on ZnPS-PVPA through alkoxyl-triazole for superior performance catalyst in asymmetric epoxidation of unfunctionalized olefins. *J. Organomet. Chem.* **2019**, *886*, 27–33. [CrossRef]
14. Huang, X.M.; Fu, X.K.; Jia, Z.Y.; Miao, Q.; Wang, G.M. Chiral salen Mn(III) complexes immobilized onto crystalline aluminium oligo-styrenyl phosphonate-hydrogen phosphate (AlSPP) for heterogeneous asymmetric epoxidation. *Catal. Sci. Technol.* **2013**, *3*, 415–424. [CrossRef]
15. Huang, J.; Tang, M.; Li, X.; Zhong, G.Z.; Li, C.M. Novel layered crystalline organic polymer-inorganic hybrid material comprising calcium phosphate with unique architectures for superior performance catalyst support. *Dalton Trans.* **2014**, *43*, 17500–17508. [CrossRef] [PubMed]
16. Li, Z.F.; Yang, C.L.; Cui, J.X.; Ma, Y.Y.; Kan, Q.B.; Guan, J.Q. Recent Advancements in Graphene-Based Supports of Metal Complexes/Oxides for Epoxidation of Alkenes. *Chem. Asian J.* **2018**, *13*, 3790–3799. [CrossRef]
17. Siegfried, E.; Andreas, H. Chemistry with graphene and graphene oxide-challenges for synthetic chemists. *Angew. Chem. Int. Ed.* **2014**, *53*, 7720–7738.
18. Scheuermann, G.M.; Rumi, L.; Steurer, P.; Bannwarth, W.; Mülhaupt, R. Palladium nanoparticles on graphite oxide and its functionalized graphene derivatives as highly active catalysts for the Suzuki-Miyaura coupling reaction. *J. Am. Chem. Soc.* **2009**, *131*, 8262–8270. [CrossRef]
19. Shao, L.D.; Huang, X.; Teschner, D.; Zhang, W. Gold Supported on Graphene Oxide: An Active and Selective Catalyst for Phenylacetylene Hydrogenations at Low Temperatures. *ACS Catal.* **2014**, *4*, 2369–2373. [CrossRef]
20. Li, X.; Yu, J.G.; Wageh, S.; Ghamdi, A.A.A.; Xie, J. Graphene in Photocatalysis: A Review. *Small* **2016**, *12*, 6640–6696. [CrossRef]
21. Liang, Y.Y.; Wang, H.L.; Casalongue, H.S.; Chen, Z.; Dai, H.J. TiO_2 nanocrystals grown on graphene as advanced photocatalytic hybrid materials. *Nano Res.* **2010**, *3*, 701–705. [CrossRef]
22. Yun, M.; Ahmed, M.S.; Jeon, S. Thiolated graphene oxide-supported palladium cobalt alloyed nanoparticles as high performance electrocatalyst for oxygen reduction reaction. *J. Power Sources* **2015**, *293*, 380–387. [CrossRef]

23. Yuan, D.C.; Chen, B.B. Synthesis and Characterization of Graphene Oxide Supported Schiff Base Palladium Catalyst and Its Catalytic Performance to Suzuki Reaction. *Chin. J. Org. Chem.* **2014**, *34*, 1630–1638. [CrossRef]
24. Tan, R.; Li, C.Y.; Luo, J.Q.; Kong, Y.; Zheng, W.G.; Yin, D.H. An effective heterogeneous L-proline catalyst for the direct asymmetric aldol reaction using graphene oxide as support. *J. Catal.* **2013**, *298*, 138–147. [CrossRef]
25. Shi, H.; Yu, C.; He, J. On the Structure of Layered Double Hydroxides Intercalated with Titanium Tartrate Complex for Catalytic Asymmetric Sulfoxidation. *J. Phys. Chem. C* **2010**, *114*, 17819–17828. [CrossRef]
26. Zheng, W.G.; Tan, R.; Yin, S.F.; Zhang, Y.Y.; Zhao, G.W.; Chen, Y.J.; Yin, D.H. Ionic liquid-functionalized graphene oxide as an efficient support for the chiral salen MnIII complex in asymmetric epoxidation of unfunctionalized olefins. *Catal. Sci. Technol.* **2015**, *5*, 2092–2102. [CrossRef]
27. Domenech, A.; Formentin, P.; Garcia, H.; Sabater, M.J. Combined electrochemical and EPR studies of manganese Schiff base complexes encapsulated within the cavities of zeolite Y. *Eur. J. Inorg. Chem.* **2000**, *2000*, 1339–1344. [CrossRef]
28. Zou, X.C.; Wang, Y.; Wang, C.; Shi, K.Y.; Ren, Y.R.; Zhao, X. Chiral MnIII (Salen) Immobilized on Organic Polymer/Inorganic Zirconium Hydrogen Phosphate Functionalized with 3-Aminopropyltrimethoxysilane as an Efficient and Recyclable Catalyst for Enantioselective Epoxidation of Styrene. *Polymers* **2019**, *11*, 212. [CrossRef]
29. Caplan, N.A.; Hancock, F.E.; Bulman, P.P.C.; Hutchings, G.J. Heterogeneous Enantioselective Catalyzed Carbonyl- and Imino-Ene Reactions using Copper Bis(Oxazoline) Zeolite, Y. *Angew. Chem. Int. Ed.* **2004**, *43*, 1685–1688. [CrossRef]
30. Venkataramanana, N.S.; Rajagopal, S. Effect of added donor ligands on the selective oxygenation of organic sulfides by oxo(salen)chromium(V) complexes. *Tetrahedron* **2006**, *62*, 5645–5651. [CrossRef]
31. Ma, X.B.; Wang, Y.H.; Wang, W.; Cao, J. Synthesis and characterization of mesoporous zirconium phosphonates: A novel supported cinchona alkaloid catalysts in asymmetric catalysis. *Catal. Commun.* **2010**, *11*, 401–407. [CrossRef]
32. Jia, H.P.; Dreyer, D.R.; Bielawski, C.W. Graphite Oxide as an Auto-Tandem Oxidation–Hydration–Aldol Coupling Catalyst. *Adv. Synth. Catal.* **2011**, *53*, 528–532. [CrossRef]
33. Zou, Z.G.; Yu, H.J.; Long, F.; Fan, Y.H. Preparation of Graphene Oxide by Ultrasound-Assisted Hummers Method. *Chin. J. Inorg. Chem.* **2011**, *27*, 1753–1757.
34. Zhang, F.; Jiang, H.Y.; Li, X.Y.; Wu, X.T.; Li, H.X. Amine-Functionalized GO as an Active and Reusable Acid–Base Bifunctional Catalyst for One-Pot Cascade Reactions. *ACS Catal.* **2014**, *4*, 394–401. [CrossRef]

© 2019 by the authors. Licensee MDPI, Basel, Switzerland. This article is an open access article distributed under the terms and conditions of the Creative Commons Attribution (CC BY) license (http://creativecommons.org/licenses/by/4.0/).

Communication

A Four Coordinated Iron(II)-Digermyl Complex as an Effective Precursor for the Catalytic Dehydrogenation of Ammonia Borane

Yoshinao Kobayashi [1] and Yusuke Sunada [1,2,*]

1. Department of Applied Chemistry, School of Engineering, The University of Tokyo, 4-6-1 Komaba Meguro-ku, Tokyo 153-8580, Japan; yk504@iis.u-tokyo.ac.jp
2. Institute of Industrial Science, The University of Tokyo, 4-6-1 Komaba Meguro-ku, Tokyo 153-8580, Japan
* Correspondence: sunada@iis.u-tokyo.ac.jp; Tel.: +81-3-5452-6361

Received: 11 December 2019; Accepted: 24 December 2019; Published: 25 December 2019

Abstract: A coordinatively unsaturated iron(II)-digermyl complex, $Fe[Ge(SiMe_3)_3]_2(THF)_2$ (**1**), was synthesized in one step by the reaction of $FeBr_2$ with 2 equiv of $KGe(SiMe_3)_3$. Complex **1** shows catalytic activity comparable to that of its silicon analogue in reduction reactions. In addition, **1** acts as an effective precursor for the catalytic dehydrogenation of ammonia borane. Catalytically active species can also be generated in situ by simple mixing of the easy-to-handle precursors $FeBr_2$, $Ge(SiMe_3)_4$, KO^tBu, and phenanthroline.

Keywords: iron catalyst; organogermyl ligand; reduction; dehydrogenation

1. Introduction

Group 14 ligands stabilize coordinatively unsaturated transition-metal centers due to their strong σ-donating properties and high *trans*-influence [1–5]. Among these, organosilyl ligands especially are widely used as supporting ligands. The creation of coordinatively unsaturated complexes is one of the most straightforward ways to develop efficient catalyst systems, especially in the field of base-metal catalysis [6–11], as base-metal compounds are usually obtained as coordinatively saturated species that exhibit low reactivity. The introduction of organosilyl ligands thus represents a potentially effective strategy to develop base-metal catalysts. We have recently reported that the four-coordinated iron(II)-disilyl complex $Fe[Si(SiMe_3)_3]_2(THF)_2$ (**2**) exhibits a coordinatively unsaturated character and can thus act as an effective catalyst for reduction reactions [12]. Notably, **2** can be synthesized easily via the reaction of a commercially available iron salt with silyl anions.

Recently, the use of heavier-group-14 elements other than silicon has also been identified as a potentially effective strategy to develop highly active catalysts. For instance, Iwasawa and Takaya have examined the catalytic activity of Pd complexes that bear PSiP or PGeP pincer-type ligands in the hydrocarboxylation of allenes, and found that the latter exhibit superior catalytic performance relative to the former [13–15]. However, examples of iron complexes that contain heavier-group-14 ligands, such as organogermyl ligands, remain scarce due to a lack of synthetic methods [16–20]. Most iron–organogermyl complexes have been obtained from the reactions of an iron precursor with organogermanes that bear Ge–H moieties; however, coordinatively saturated species are often formed via this route. Based on the synthetic route to Complex **2**, we hypothesized that a structurally similar coordinatively unsaturated iron(II)-digermyl complex could be accessed easily by using organogermyl anions instead of organosilyl anions. Herein, we report the corresponding synthesis of the germanium analogue of **2** and describe the catalytic performance of the resulting complex. Iron–digermyl Complex **1** was found to act as a good catalyst precursor for reduction reactions and the dehydrogenation of ammonia borane ($NH_3·BH_3$).

2. Results and Discussion

2.1. Synthesis of Iron–Digermyl Complex 1

Based on a slight modification of the method reported by Marschner et al. [21], the potassium salt of the tris(trimethylsilyl)germyl anion, KGe(SiMe$_3$)$_3$, was easily generated by treating Ge(SiMe$_3$)$_4$ with 1.05 equiv of KO^{t}Bu in THF. A THF solution of KGe(SiMe$_3$)$_3$ was then treated for 1 h with 0.5 equiv of iron(II) bromide at room temperature to furnish iron(II) digermyl complex Fe[Ge(SiMe$_3$)$_3$]$_2$(THF)$_2$ (1) in 74% yield as red-purple crystals (Scheme 1).

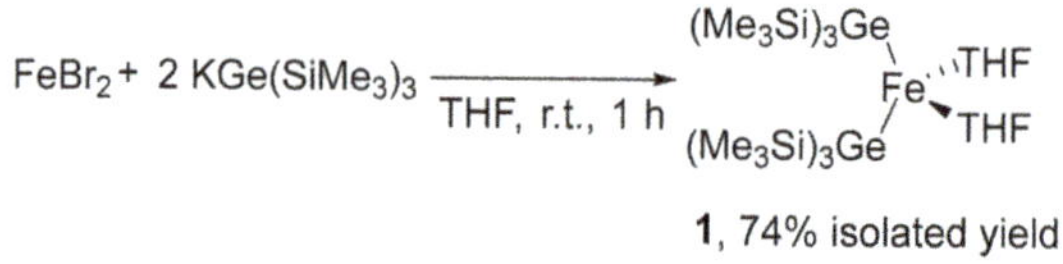

FeBr$_2$ + 2 KGe(SiMe$_3$)$_3$ $\xrightarrow{\text{THF, r.t., 1 h}}$ (Me$_3$Si)$_3$Ge–Fe(THF)(THF)–Ge(SiMe$_3$)$_3$

1, 74% isolated yield

Scheme 1. Synthesis of Iron–Digermyl Complex **1**.

Similar to its silicon analogue **2**, **1** quickly decomposes upon exposure to air or moisture. The molecular structure of **1** was determined by single-crystal X-ray diffraction analysis, and an ORTEP drawing of the molecular structure of **1** is depicted in Figure 1. Complex **1** adopts a distorted tetrahedral coordination geometry around its iron center, presumably due to the steric repulsion between the two Ge(SiMe$_3$)$_3$ ligands. The Ge–Fe–Ge angle of **1** (134.77(4)°) is comparable to the analogous Si–Fe–Si angle in the silicon analogue **2** (135.23(3)°). The Fe–Ge bond is slightly elongated (2.5589(8) Å) compared to those of previously reported iron–organogermyl complexes [17–19,22–26].

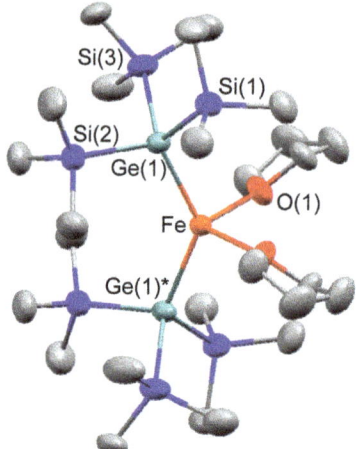

Figure 1. Molecular structure of **1** with 50% probability ellipsoids. All hydrogen atoms were omitted for clarity.

In the ^{1}H NMR spectrum of **1**, a broad singlet assignable to the SiMe$_3$ group appeared at 8.67 ppm, and two broad signals corresponding to two molecules of THF coordinated to the iron center were observed at 24.02 and 29.42 ppm, respectively. These spectral features indicate that **1** is paramagnetic; the magnetic moment of **1** estimated using the Evans method [27–29] is consistent with a high-spin $S = 2$ ground state ($\mu_{eff} = 4.85$). Despite its high sensitivity toward air and moisture, the elemental analysis of **1** was satisfactory.

2.2. Catalytic Performance of 1 toward Reduction Reactions

In our previous paper, 2 was found to exhibit good catalytic performance in reduction reactions, namely, the hydrosilylation of ketones and the reductive silylation of dinitrogen. A key aspect of these catalytic reactions is the fact that the organosilyl ligands promote the creation of a reaction site on the iron center owing to their high σ-electron donating ability and their strong *trans*-influence. In other words, coordinatively unsaturated catalytically active species are effectively generated in situ via the dissociation of THF. As 2 is a structural analogue of 1, we tested the catalytic performance of 1 in the two reduction reactions mentioned above.

Initially, we examined the 1-catalyzed hydrosilylation of ketones with Ph_2SiH_2. 4-Acetylbiphenyl and 4-phenyl-2-butanone were selected as representative aromatic and aliphatic ketones (Scheme 2). The hydrosilylation of 4-acetylbiphenyl with Ph_2SiH_2 in the presence of 0.1 mol % of 1 in dioxane at room temperature proceeded smoothly to afford the corresponding silyl-ether in quantitative yield within 4 h. Similarly, complete conversion of cyclohexanone was confirmed within 4 h when the reaction was conducted using Ph_2SiH_2 and 0.1 mol % of 1 at room temperature. It should be noted here that the catalysis mediated by the silicon analogue 2 was carried out under the reaction conditions shown in Scheme 2, indicating that the catalytic activity of 1 toward the hydrosilylation of ketones is comparable to that of 2.

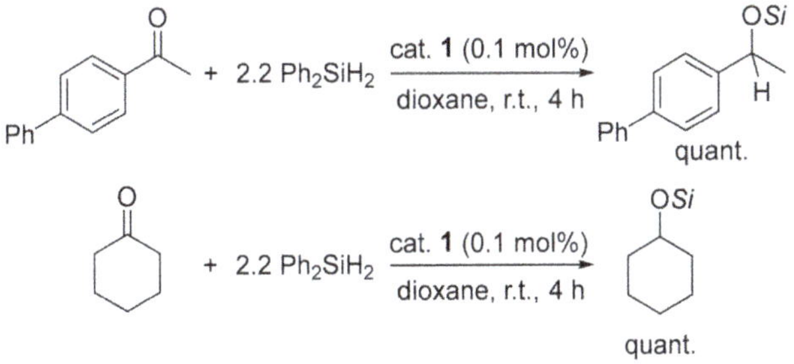

Scheme 2. Catalytic hydrosilylation of ketones catalyzed by 1.

Next, we examined the reductive silylation of dinitrogen catalyzed by 1. The reductive silylation of atmospheric dinitrogen was achieved using an excess of $SiMe_3Cl$ and KC_8 at room temperature in the presence of 0.2 mol % of 1. During this reaction, the formation of 11.4 equiv of $N(SiMe_3)_3$ per Fe atom was observed in THF, concomitant with the generation of byproducts including $Me_3Si–SiMe_3$ and mono- and bis-silylated ring-opened THF, i.e., $Me_3SiO(CH_2)_3CH_2R$ (R = H or $SiMe_3$). However, the TON of the reaction catalyzed by 2 reached 22.9; thus, the catalytic performance of 1 seems to have decreased slightly due to the replacement of silicon by germanium as the coordinating atom.

2.3. Catalytic Dehydrogenation of Ammonia Borane Catalysed by 1

These results indicate that catalytic performance of 1 in reduction reactions is decent. In addition to these reduction reactions, we found that 1 also promotes a catalytic oxidation reaction, i.e., the dehydrogenation of $NH_3 \cdot BH_3$, when the appropriate reagents are used. $NH_3 \cdot BH_3$ has been identified as an appropriate solid hydrogen-storage material due to its high hydrogen content, and accordingly, the catalytic production of hydrogen via the dehydrogenation of $NH_3 \cdot BH_3$ has attracted great attention. Recently, numerous transition-metal-mediated reactions have been developed based on the design and synthesis of appropriate catalysts [30–32]. Among these, iron-based catalysts represent some of the most promising candidates for the development of practical systems owing to their low cost, earth-abundance, and low toxicity [33–42]. Thus, great efforts have been devoted to developing iron-mediated reactions for the dehydrogenation of ammonia borane derivatives. For instance, Baker reported that the iron catalyst $(dcpe)Fe(N(Ph)CH_2CH_2(Ph)N)$ [dcpe = 1,2-bis(dicyclohexylphosphino)ethane], which contains a chelating diamide ligand, exhibits high catalytic performance and could complete the reaction within 15 min even at room temperature [33]. Several examples of iron catalysts that contain pincer-type ligands have also been reported to show good catalytic performance for this reaction [43]. For instance, the iron catalyst (PNP)Fe(H)(CO) [PNP = N(CH$_2$CH$_2$P^{i}Pr)$_2$], which contains a PNP pincer ligand, was reported by Schneider et al. and shows exceptionally high catalytic activity at room temperature [34]. Complete consumption of NH$_3$·BH$_3$ was confirmed at catalyst loadings as low as 0.5 mol %. A theoretical study revealed that the cooperative function between the iron center and the nitrogen atom of the PNP ligand plays a crucial role in the activation of NH$_3$·BH$_3$ in this catalytic reaction. Despite the large number of iron-catalyzed dehydrogenation reactions that have been developed, almost all require the synthesis of iron catalysts that bear bespoke and often difficult-to-synthesize auxiliary ligands or that are difficult to isolate due to their sensitivity toward air and/or moisture. An exceptional example was introduced by Manners and co-workers, who reported that the commercially available dinuclear iron catalyst [CpFe(CO)$_2$]$_2$ (Cp = η^5-C$_5$H$_5^-$) can effectively catalyze the dehydrogenation of Me$_2$NH·BH$_3$ under irradiation from a medium-pressure mercury lamp [35,36]. The [CpFe(CO)$_2$]$_2$-catalyzed dehydrogenation of NH$_3$·BH$_3$ under irradiation from an LED lamp was achieved by Waterman et al. [37]. However, in these catalytic reactions, continuous irradiation is required.

We discovered that NH$_3$·BH$_3$ underwent dehydrogenation at 55 °C in the presence of a catalytic amount of **1** and the appropriate auxiliary ligand and additive. Initially, we treated NH$_3$·BH$_3$ in THF (0.1 M) at 55 °C with 5 mol % of **1** in the presence of ^{t}BuOH (5 mol %) and a series of auxiliary ligands. As shown in Table 1, phenanthroline, bathophenanthroline, a diimine, an N-heterocyclic carbene (NHC), N,N,N',N'-tetramethylethylenediamine (TMEDA), and pyridine were examined. We found that the addition of phenanthroline resulted in the highest conversion of NH$_3$·BH$_3$ (Entry 2), while the more electron-withdrawing ligand bathophenanthroline was less effective. The reaction also proceeded with lower conversion in the presence of the π-accepting diimine ligand. In contrast, the combination of **1** and the σ-donating nitrogen-based ligands TMEDA and pyridine resulted in negligible catalytic activity (Entries 8 and 9), while the addition of the NHC ^{i}Pr$_2$IM^{Me} furnished the dehydrogenated product in 32% (Entry 7). The reaction did not occur in the absence of an additional auxiliary ligand (Entry 1). The optimal ratio of **1**:^{t}BuOH was found to be 1:1, and the conversion of NH$_3$·BH$_3$ decreased when the amount of ^{t}BuOH was increased (**1**:^{t}BuOH = 1:2; Entry 3). The conversion increased slightly when the reaction time was extended to 72 h (Entry 4). ^{11}B{^{1}H} and ^{11}B NMR spectroscopic measurements revealed that NH$_3$·BH$_3$ was converted into **3**, **4** and **5** in a ca. 3:5:8 molar ratio in each case [44].

Table 1. Dehydrogenation of ammonia borane catalyzed by combination of **1**, ^{t}BuOH, and ligands [a].

$$H_3N\text{-}BH_3 \xrightarrow[\text{THF (0.1 M),}~55~°\text{C, time}]{\substack{\text{cat. 1 (5 mol\%)} \\ ^t\text{BuOH (5 mol\%)} \\ \text{ligand (5 mol\%)}}} \mathbf{3} + \mathbf{4} + \mathbf{5} + H_2$$

Entry	Ligand	Time (h)	Conv. (%) [b]
1	none	24	~0
2	phenanthroline	24	50
3 [c]	phenanthroline	24	37
4	phenanthroline	72	70
5	bathophenanthroline	24	9
6	Diamine [d]	24	24
7	^{i}Pr$_2$IM^{Me} [e]	24	32
8	TMEDA	24	~0
9	pyridine	24	5

[a] All reactions were carried out with ammonia borane (0.5 mmol), catalytic amount of **1** (5 mol %), ^{t}BuOH (5 mol %), and ligand (5 mol %) at 55 °C in THF (5 mL). [b] Conversion of ammonia borane was determined by ^{11}B NMR. [c] reaction was performed with 10 mol % of ^{t}BuOH. [d] diimine = 2,3-bis(2,4,6-trimethylphenylimino)butane. [e] ^{i}Pr$_2$IM^{Me} = 1,3-diisopropyl-4,5-dimethyl-imidazol-2-ylidene.

After optimization of the auxiliary ligand, further screening of the other additives, except for ^{t}BuOH, was performed. As noted above, NH$_3$·BH$_3$ has been reported to be effectively activated in a cooperative fashion over the Fe-N bond in several iron catalysts. We therefore examined the potential generation of catalytically active species with Fe-N bonds by using amine additives (Table 2; Entries 4–8). As expected, the yield of the dehydrogenated product increased in the reactions shown in Entries 4–7. Among these, ^{t}BuNH$_2$ was identified as the most effective additive, which furnished 76% conversion of NH$_3$·BH$_3$ into **3**, **4**, and **5**. In contrast, the addition of 2,6-diisopropylaniline decreased the conversion, presumably due to the increased steric hinderance around the Fe-N bond (Entry 8). The dehydrogenation of NH$_3$·BH$_3$ also occurred in the presence of an aliphatic carboxylic acid, though the products were obtained in lower yield. It should be noted that the use of the iron complex **2** instead of **1** resulted in a slight increase in the conversion under the same reaction conditions (Entry 3).

Table 2. Dehydrogenation of ammonia borane catalyzed by combination of **1**, phenanthroline, and additives [a].

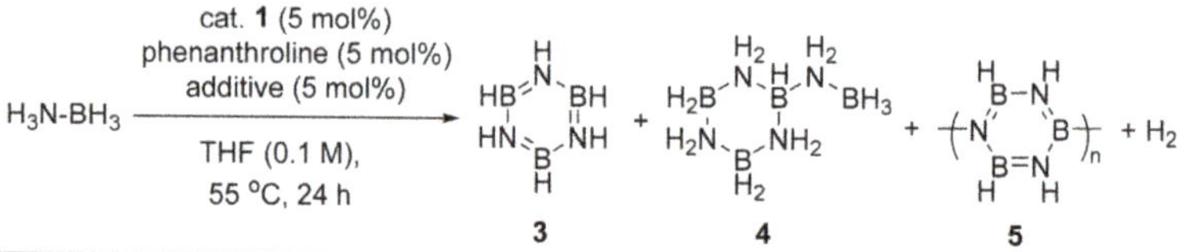

H$_3$N-BH$_3$ $\xrightarrow{\text{cat. 1 (5 mol\%), phenanthroline (5 mol\%), additive (5 mol\%), THF (0.1 M), 55 °C, 24 h}}$ **3** + **4** + **5** + H$_2$

Entry	Additive	Conv. (%) [b]
1	none	5
2	^{t}BuOH	50
3 [c]	^{t}BuOH	72
4	^{t}BuNH$_2$	76
5	diethylamine	64
6	diphenyamine	61
7	2,4,6-trimethylaniline	64
8	2,6-diisopropylaniline	30
9	2-ethylhexanoic acid	39

[a] All reactions were carried out with ammonia borane (0.5 mmol), a catalytic amount of **1** (5 mol %), phenanthrolne (5 mol %), and additives (5 mol %) at 55 °C in THF (5 mL) for 24 h. [b] Conversion of ammonia borane was determined by ^{11}B NMR. [c] Complex **2** was used instead of **1**.

The organogermyl anion KGe(SiMe$_3$)$_3$ can be quantitatively generated in situ by the reaction of Ge(SiMe$_3$)$_4$ and KO^{t}Bu in THF. Thus, we hypothesized that the catalytically active species for the dehydrogenation of NH$_3$·BH$_3$ could be easily generated in situ by simply mixing FeBr$_2$, Ge(SiMe$_3$)$_4$, KO^{t}Bu, and phenanthroline. In fact, NH$_3$·BH$_3$ could be dehydrogenated using this catalyst system, and the conversion reached 26% (Scheme 3). Although the conversion was lower than that obtained by the reaction catalyzed by isolated **1**, it should be noted that all the reagents used in the present catalysis are easily available and easy to handle. Thus, the catalysis shown in Scheme 3 should allow for a practical and convenient catalytic dehydrogenation of NH$_3$·BH$_3$ via a base-metal catalyst.

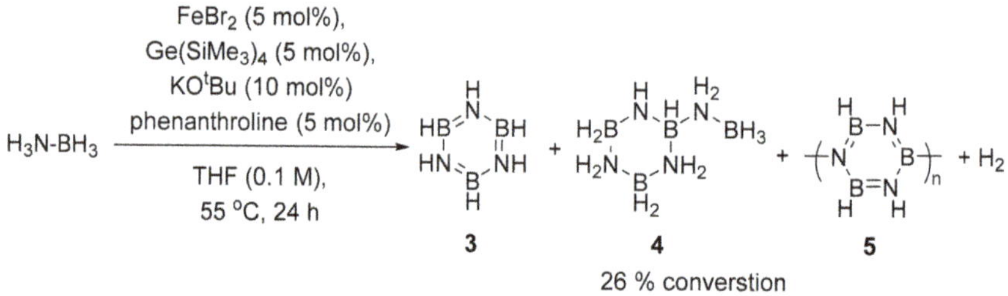

H$_3$N-BH$_3$ $\xrightarrow{\text{FeBr}_2 \text{ (5 mol\%), Ge(SiMe}_3\text{)}_4 \text{ (5 mol\%), KO}^t\text{Bu (10 mol\%), phenanthroline (5 mol\%), THF (0.1 M), 55 °C, 24 h}}$ **3** + **4** + **5** + H$_2$

26 % converstion

Scheme 3. In situ generation of catalytically active species.

To gain some insight into the reaction mechanism of the dehydrogenation of ammonia borane, a stoichiometric reaction between **1** and ^{t}BuOH was performed in THF at room temperature. Although the isolated yield was relatively low (~5%), the dinuclear iron complex **6**, which contains a bridging O^{t}Bu group, was formed in this reaction (Scheme 4). Complex **6** was also obtained from the reaction of FeBr$_2$, KGe(SiMe$_3$)$_3$, and KO^{t}Bu in a 1:1:1 ratio in THF (details are shown in the supporting information). The molecular structure was determined via single-crystal X-ray diffraction analysis, and the structural parameters are summarized in the supporting information. It is interesting to note that one of the two iron centers adopts three-coordinated trigonal planar geometry. These results suggests that the catalytically active species may be the mononuclear iron species "(phenanthroline)[Ge(SiMe$_3$)$_3$]Fe(O^{t}Bu)," which was generated in situ via the cleavage of the Fe–O bond of **6**, and the coordination of phenanthroline. The higher σ-electron-donating properties and the stronger *trans*-influence of the organosilyl ligand compared to the organogermyl ligand may allow for a more effective cleavage of the bridging Fe–O bond, leading to the somewhat superior catalytic activity of **2** compared to **1**. However, it should be emphasized here that the structurally similar silyl analogue of **6** could not be accessed via a reaction analogous to that shown in Scheme 4 starting from **2**. The decomposition of the product might be ascribed to the high oxo-affinity of silicon. Thus, the use of the organogermyl ligand contributed to the elucidation of the reaction mechanism by stabilizing the intermediary species.

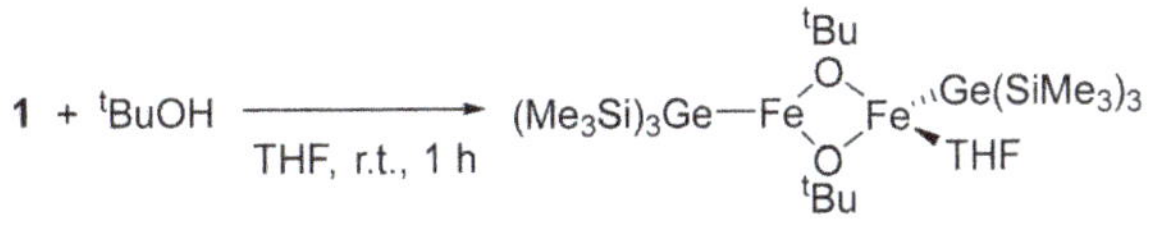

$$1 + {}^tBuOH \xrightarrow{\text{THF, r.t., 1 h}} (Me_3Si)_3Ge-Fe(\mu\text{-O}{}^tBu)_2Fe-Ge(SiMe_3)_3 \cdot THF$$

6, ~ 5% isolated yield

Scheme 4. Reaction of **1** and 1 equiv of ^{t}BuOH to afford the dinuclear complex **3**.

3. Materials and Methods

Manipulation of air and moisture sensitive compounds was carried out under a dry nitrogen atmosphere using Schlenk tube techniques associated with a high-vacuum line or in the glove box which was filled with dry nitrogen. All solvents were purchased from Kanto Chemical Co. Inc. (Tokyo, Japan), and were dried over activated molecular sieves. ^{1}H, ^{11}B, and ^{11}B{^{1}H} NMR spectra were recorded on a JEOL Lambda 400 spectrometer at ambient temperature unless otherwise noted. ^{1}H NMR chemical shifts (δ values) were given in ppm relative to the solvent signal, whereas ^{11}B NMR chemical shifts were given relative to the standard resonances (NaBH$_4$). Elemental analyses were performed by the Thermo Scientific FLASH 2000 Organic Elemental Analyzer. The starting compounds, Ge(SiMe$_3$)$_4$ and potassium tris(trimethylsilyl)germanide, KGe(SiMe$_3$)$_3$ [21], were synthesized by a method reported in the literature. All reagents were purchased from FUJIFILM Wako Chemicals (Osaka, Japan), Tokyo Chemical Industries Co., Ltd. (Tokyo, Japan) or Sigma-Aldrich (St. Louis, USA), and were used without further purification.

3.1. Preparation of Potassium Tris(trimethylsilyl)germanide

Potassium salt of germyl anion, KGe(SiMe$_3$)$_3$, was prepared from a reaction of Ge(SiMe$_3$)$_4$ (3.142 g, 8.6 mmol) with KO^{t}Bu (1.009 g, 9.0 mmol) in THF (40 mL) at room temperature. The solvent was removed in vacuo, and the remaining crude product was then dried at 40 °C overnight to afford KGe(SiMe$_3$)$_3$ as pale yellow powder in 58% yield (1.64 g). Formation of KGe(SiMe$_3$)$_3$ was confirmed by ^{1}H NMR spectra.

3.2. Synthesis of (THF)$_2$Fe[Ge(SiMe$_3$)$_3$]$_2$ (**1**)

In a 50 mL schlenk tube, FeBr$_2$ (86.3 mg, 0.4 mmol) was suspended in THF (10 mL), and THF (5 mL) solution of KGe(SiMe$_3$)$_3$ (278 mg, 0.84 mmol) was then slowly added to this solution at room temperature. The solution was stirred at room temperature for 1 h. In the course of this reaction, the color of the solution turned to dark red purple, and the solution was then centrifuged to remove the insoluble materials. The mother liquid was evaporated in vacuo, and the obtained solid was dissolved in pentane (20 mL). The solution was again centrifuged to remove the insoluble materials. The supernatant was collected, and THF (1 mL) was added. The solvent was concentrated to ca. 3 mL, and the remaining solution was cooled at −30 °C. Complex **1** was obtained as red-purple crystals in 74% yield (0.232 g). ^{1}H NMR (400 MHz, r.t., C$_6$D$_6$): δ = 8.67 (brs, 54H, SiMe$_3$), 24.04 (brs, 8H, THF), 29.43 (brs, 8H, THF). Magnetic susceptibility (Evans): μ_{eff} = 4.85 (in C$_6$D$_6$, 20.0 °C). Anal. Calcd. for C$_{26}$H$_{70}$O$_2$Ge$_2$Si$_6$Fe$_1$: C, 39.81; H, 8.99. Found: C, 40.16; H, 8.93.

3.3. General Procedure for the Reduction of Carbonyl Compounds Catalyzed by 1

Catalyst **1** (2.0 mg, 0.003 mmol, 0.1 mol %) was placed in a 10 mL flask, and the substrate, Ph_2SiH_2 (1.28 mg, 6.6 mmol), 4-acetylbiphenyl or 4-phenyl-2-butanone (3 mmol), and dioxane (0.5 mL) were then added, in this order. The resulting mixture was stirred at room temperature for 4 h, and the complete consumption of the substrate and the yield of the product were then confirmed by ^1H NMR spectrum with anisole as an internal standard.

3.4. General Procedure for the Catalytic Silylation of N_2 Catalyzed by 1

The following procedure is analogous to those reported in previous papers [12]. A 10 mL flask was charged with KC_8 (270 mg, 2.0 mmol) and THF (5 mL), and Complex **1** (0.004 mmol, 0.2 mol %) and Me_3SiCl (0.25 mL, 2.0 mmol) was then added to this suspension. The remaining suspension was vigorously stirred at room temperature for 24 h under successive supply of N_2 (1 atm). After the reaction, cyclododecane (60 mg, 0.36 mmol) was added as an internal standard, and the yield of the formed $N(SiMe_3)_3$ was then determined by GC/GC–MS.

3.5. General Procedure for the Dehydrogeantion of Ammonia Borane Catalyzed by 1

Catalyst **1** (20 mg, 0.025 mmol, 5 mol %) was dissolved in THF (2 mL) in a 4 mL reaction vessel, and the additives shown in Tables S1 or S2 in the supporting information were then added to this solution. The obtained solution was stirred for 15 min at room temperature, and ligands shown in Tables S1 or S2 were then added. After stirring this solution for 15 min at room temperature, the obtained solution was added to THF (2 mL) solution of ammonia borane (15 mg, 0.5 mmol). The resulting mixture was stirred at 55 °C for the time indicated in Tables S1 and S2. Conversion of ammonia borane was confirmed by ^{11}B and $^{11}B\{^1H\}$ NMR spectrum.

3.6. Synthesis of Dinuclear Iron Complex, $[Ge(SiMe_3)_3]Fe(\mu-O^tBu)_2Fe(THF)[Ge(SiMe_3)_3]$ (6)

In a 50 mL schlenk tube, Complex **1** (335 mg, 0.43 mmol) was dissolved in THF (2 mL), and ^tBuOH (40 μL, 0.43 mmol) was added to this solution at room temperature. The solution was stirred at room temperature for 1 h. In the course of this reaction, the color of the solution turned from dark red purple to brown, and the solvent was then removed in vacuo. The crude product was dissolved in pentane (1 mL), and the remaining solution was cooled at −20 °C. Complex **6** was obtained as brown crystals in ~5% yield. The molecular structure of **6** was exclusively determined by X-ray diffraction analysis.

Complex **6** was alternatively obtained from the reaction of $FeBr_2$, $KGe(SiMe_3)_3$, and KO^tBu. Iron dibromide (216 mg, 1.0 mmol) was dissolved in THF (2 mL), and a KO^tBu (112 mg, 1.0 mmol) and THF (3 mL) solution of $KGe(SiMe_3)_3$, prepared in situ from the reaction of $Ge(SiMe_3)_4$ (385 mg, 1.05 mmol) and KO^tBu (123 mg, 1.1 mmol), was added to this solution at room temperature. The solution was stirred at room temperature for 1 h. The solution was centrifuged to remove insoluble materials, and the mother liquid was dried under vacuum. The remaining crude product was dissolved in pentane (5 mL), and centrifuged to remove a small amount of insoluble materials, and the mother liquid was then concentrated to ca. 1 mL. After cooling this solution at −20 °C, Complex **6** was obtained as brown crystals. The cell parameters of the obtained crystals were consistent with those found in crystals of **6**, which was synthesized in the procedure described above.

3.7. X-ray Data Collection and Reduction

X-ray crystallography for Complexes **1** and **6** was performed on a Rigaku Saturn CCD area detector with graphite monochromated Mo–Kα radiation (λ = 0.71075 Å). The data were collected at 183(2) K using a ω scan in the θ range of $3.01 \leq \theta \leq 27.45$ deg for **1** and $3.28 \leq \theta \leq 27.48$ deg for **6**. The data obtained were processed using Crystal-Clear (Rigaku) on a Pentium computer, and were corrected for Lorentz and polarization effects. The structures were solved by direct methods [45], and expanded using Fourier techniques. Hydrogen atoms were refined using the riding model.

The final cycle of full-matrix least-squares refinement on F^2 was based on 5750 observed reflections and 188 variable parameters for **1** and 11,471 observed reflections and 388 variable parameters for **6**. Neutral atom scattering factors were taken from International Tables for Crystallography (IT), Vol. C, Table 6.1.1.4 [46]. Anomalous dispersion effects were included in Fcalc [47]; the values for $\Delta f'$ and $\Delta f''$ were those of Creagh and McAuley [48]. The values for the mass attenuation coefficients are those of Creagh and Hubbell [49]. All calculations were performed using the CrystalStructure [50] crystallographic software package except for refinement, which was performed using SHELXL Version 2017/1 [51]. Details of the final refinement as well as the bond lengths and angle are summarized in Tables S3 and S4, and the numbering scheme employed is shown in Figures S3 and S4, which were drawn with ORTEP at 50% probability ellipsoids. CCDC numbers 1971242 (**1**) and 1971243 (**6**) contain the supplementary crystallographic data for this paper. These data can be obtained free of charge from the Cambridge Crystallographic Data Centre via www.ccdc.cam.ac.uk/data_request/cif.

4. Conclusions

In summary, we have described a facile synthesis of the four-coordinated iron–digermyl complex **1** via the reaction of $FeBr_2$ and in-situ-generated $KGe(SiMe_3)_3$. Complex **1** shows high catalytic performance in the hydrosilylation of ketones and in the reductive silylation of dinitrogen. Additionally, **1** acts as a good precursor for the catalytic dehydrogenation of ammonia borane ($NH_3 \cdot BH_3$) in the presence of $^tBuNH_2$ and phenanthroline. It should be emphasized that the catalytic dehydrogenation of $NH_3 \cdot BH_3$ was realized using a catalyst system consisting of stable and easy-to-handle precursors. The use of the organogermyl ligand was found to contribute to both the generation of coordinatively unsaturated reactive complexes and the isolation of a key intermediary species involved in the catalysis. Efforts to develop efficient base-metal catalysts based on the introduction of heavier-group-14 ligands are currently in progress in our laboratory.

Supplementary Materials: The following are available online at http://www.mdpi.com/2073-4344/10/1/29/s1: CIF and cif-checked files, detailed crystallographic data, detailed results obtained by the dehydrogenation of ammonia borane catalyzed by **1** in the presence of various ligands and additives, the actual NMR charts of Complex **1**, and the crude product obtained from the hydrogenation ammonia borane catalyzed by **1**.

Author Contributions: Y.S. conceived and designed the experiments; Y.K performed the experiments; Y.K. and Y.S. analyzed the data; Y.S. wrote the paper. All authors have read and agreed to the published version of the manuscript.

Funding: This research was funded by a Grant in Aid for Scientific Research (B) (No. 16H04120) and a Grant in Aid for Scientific Research on Innovative Areas "Precise Formation of a Catalyst Having a Specified Field for Use in Extremely Difficult Substrate Conversion Reactions" (No. 18H04240) from the Ministry of Education, Culture, Sports, Science and Technology, Japan. This work was also supported by Iketani Science and Technology Foundation, and the special fund of the Institute of Industrial Science, The University of Tokyo.

Conflicts of Interest: The authors declare that there is no conflict of interest.

References

1. Corey, J.Y. Reactions of Hydrosilanes with Transition Metal Complexes. *Chem. Rev.* **2016**, *116*, 11291–11435. [CrossRef] [PubMed]
2. Corey, J.Y. Reactions of Hydrosilanes with Transition Metal Complexes and Characterization of the Products. *Chem. Rev.* **2011**, *111*, 863–1071. [CrossRef] [PubMed]
3. Corey, J.Y.; Braddock-Wilking, J. Reactions of Hydrosilanes with Transition-Metal Complexes: Formation of Stable Transition-Metal Silyl Compound. *Chem. Rev.* **1999**, *99*, 175–292. [CrossRef] [PubMed]
4. Tanabe, M.; Osakada, K. Sila-and Germametallacycles of Late Transition Metals. *Organometallics* **2010**, *29*, 4702–4710. [CrossRef]
5. Tilley, T.D. Transition-Metal Silyl Derivatives. In *The Chemistry of Organic Silicon Compounds*; Patai, S., Rappoport, Z.J., Eds.; Wiley and Sons: New York, NY, USA, 1989; pp. 1415–1477.
6. Marek, I.; Rappoport, Z. *The Chemistry of Organoiron Compounds*; Patai, S., Rappoport, Z.J., Eds.; Wiley and Sons: New York, NY, USA, 2014.
7. Plietker, B. (Ed.) *Iron Catalysis in Organic Chemistry*; WILEY-VCH: Weinheim, Germany, 2008.
8. Nakamura, E.; Yoshikai, N. Low-Valent Iron-Catalyzed C–C Bond Formation–Addition, Substitution, and C–H Bond Activation. *J. Org. Chem.* **2010**, *75*, 6061–6067. [CrossRef]
9. Bauer, I.; Knölker, H.-J. Iron Catalysis in Organic Synthesis. *Chem. Rev.* **2015**, *115*, 3170–3387. [CrossRef]
10. Sun, C.-L.; Li, B.-J.; Shi, Z.-J.; Direct, C.-H. Transformation via Iron Catalysis. *Chem. Rev.* **2011**, *111*, 1293–1314. [CrossRef]
11. Bolm, C.; Legros, J.; LePaih, J.; Zani, L. Iron-Catalyzed Reactions in Organic Synthesis. *Chem. Rev.* **2004**, *104*, 6217–6254. [CrossRef]
12. Arata, S.; Sunada, Y. An isolable iron(ii) bis(supersilyl) complex as an effective catalyst for reduction reactions. *Dalton Trans.* **2019**, *48*, 2891–2895. [CrossRef]
13. Zhu, C.; Takaya, J.; Iwasawa, N. Use of Formate Salts as a Hydride and a CO_2 Source in PGeP-Palladium Complex-CatalyzedHydrocarboxylation of Allenes. *Org. Lett.* **2015**, *17*, 1814–1817. [CrossRef]
14. Takaya, J.; Nakamura, S.; Iwasawa, N. Synthesis, Structure, and Catalytic Activity of Palladium Complexes Bearing a Tridentate PXP-Pincer Ligand of Heavier Group 14 Element (X = Ge, Sn). *Chem. Lett.* **2012**, *41*, 967–969. [CrossRef]
15. Takaya, J.; Iwasawa, N. Synthesis, Structure, and Reactivity of a Mononuclear η^2-(Ge–H)palladium(0) Complex Bearing a PGeP-Pincer-Type Germyl Ligand: Reactivity Differences between Silicon and Germanium. *Eur. J. Inorg. Chem.* **2018**, *2018*, 5012–5018. [CrossRef]
16. Barrau, J.; Hamida, N.B.; Agrebi, A.; Satge, J. Bis(dimethylgermyl)alkane-iron tetracarbonyls: Synthesis, photolysis, and reactivity. *Organometallics* **1989**, *8*, 1585–1593. [CrossRef]
17. Itazaki, M.; Kamitani, M.; Hashimoto, Y.; Nakazawa, H. Synthesis, Characterization, and Crystal Structure of Germyl(phosphine)iron Complexes, $Cp(CO)Fe(PPh_3)(GeR_3)$ (R = Et, ^{n}Bu, Ph), Prepared from $Cp(CO)Fe(PPh_3)(Me)$ and $HGeR_3$. *Phosphorus Sulfur Silicon Relat. Elem.* **2010**, *185*, 1054–1060. [CrossRef]
18. Itazaki, M.; Kamitani, M.; Nakazawa, H. Trans-selective hydrogermylation of alkynes promoted by methyliron and bis(germyl)hydridoiron complexes as a catalyst precursor. *Chem. Commun.* **2011**, *47*, 7854–7856. [CrossRef] [PubMed]
19. Itazaki, M.; Kamitani, M.; Ueda, K.; Nakazawa, H. Syntheses and Ligand Exchange Reaction of Iron(IV) Complexes with Two Different Group 14 Element Ligands, $Cp(CO)FeH(EEt_3)(E'Et_3)$ (E, E' = Si, Ge, Sn). *Organometallics* **2009**, *28*, 3601–3603. [CrossRef]
20. Anema, S.G.; Mackay, K.M.; Nicholson, B.K.; Tiel, M.V. Synthesis of iron carbonyl clusters with trigonal-bipyramidal E_2Fe_3 cores (E = Ge, Si). Crystal structures of $(\mu_3\text{-GeEt})_2Fe_3(CO)_9$, $[\mu_3\text{-Ge}\{Fe(CO)_2Cp\}]_2Fe_3(CO)_9$, and $[\mu_3\text{-Si}\{Fe(CO)_2Cp\}]_2Fe_3(CO)_9$. *Organometallics* **1990**, *9*, 2436–2442. [CrossRef]
21. Fischer, J.; Baumgartner, J.; Marschner, C. Silylgermylpotassium Compounds. *Organometallics* **2005**, *24*, 1263–1268. [CrossRef]
22. Okazaki, M.; Kimura, H.; Komuro, T.; Okada, H.; Tobita, H. Synthesis, Structure, and Properties of Three-and Six-Membered Metallacycles Composed of Iron, Germanium, and Sulfur Atoms. *Chem. Lett.* **2007**, *36*, 990–991. [CrossRef]

23. Tobita, H.; Ishiyama, K.; Kawano, Y.; Inomata, S.; Ogino, H. Synthesis of Cationic Germyleneiron Complexes and X-ray Structure of [Cp*(CO)$_2$FeGeMe$_2$·DMAP]BPh$_4$·CH$_3$CN (Cp* = C$_5$Me$_5$, DMAP = 4-(Dimethylamino)pyridine). *Organometallics* **1998**, *17*, 789–794. [CrossRef]
24. Meier-Brocks, F.; Weiss, E. Tetraphenylzirkonacyclopentadien-derivate als synthone für tetraphenylthiophenmonoxid und substituierte germanole. *J. Organomet. Chem.* **1993**, *453*, 33–45. [CrossRef]
25. Xie, W.; Wang, B.; Dai, X.; Xu, S.; Zhou, X. Novel Rearrangement Reactions. 3. Thermal Rearrangement of the Diruthenium Complex (Me$_2$SiSiMe$_2$)[(η^5-C$_5$H$_4$)Ru(CO)]$_2$(μ-CO)$_2$. *Organometallics* **1998**, *17*, 5406–5410. [CrossRef]
26. Wang, B.; Zhu, B.; Zhang, J.; Xu, S.; Zhou, X.; Weng, L. Reactions of Doubly Bridged Bis(cyclopentadienes) with Iron Pentacarbonyl. *Organometallics* **2003**, *22*, 5543–5555. [CrossRef]
27. Schubert, E.M. Utilizing the Evans method with a superconducting NMR spectrometer in the undergraduate laboratory. *J. Chem. Educ.* **1992**, *69*, 62. [CrossRef]
28. Piguet, C. Paramagnetic Susceptibility by NMR: The "Solvent Correction" Removed for Large Paramagnetic Molecules. *J. Chem. Educ.* **1997**, *74*, 815–816. [CrossRef]
29. Bain, G.A.; Berry, J.F. Diamagnetic Corrections and Pascal's Constants. *J. Chem. Educ.* **2008**, *85*, 532–536. [CrossRef]
30. Rossin, A.; Peruzzini, M. Ammonia–Borane and Amine–Borane Dehydrogenation Mediated by Complex Metal Hydrides. *Chem. Rev.* **2016**, *116*, 8848–8872. [CrossRef]
31. Balaraman, E.; Nandakumar, A.; Jaiswal, G.; Sahoo, M.K. Iron-catalyzed dehydrogenation reactions and their applications in sustainable energy and catalysis. *Catal. Sci. Technol.* **2017**, *7*, 3177–3195. [CrossRef]
32. Coles, N.T.; Webster, R.L. Iron Catalyzed Dehydrocoupling of Amine-and Phosphine-Boranes. *Isr. J. Chem.* **2017**, *57*, 1070–1081. [CrossRef]
33. Baker, R.T.; Gordon, J.C.; Hamilton, C.W.; Henson, N.J.; Lin, P.-H.; Maguire, S.; Murugesu, M.; Scott, B.L.; Smythe, N.C. Iron Complex-Catalyzed Ammonia–Borane Dehydrogenation. A Potential Route toward B–N-Containing Polymer Motifs Using Earth-Abundant Metal Catalysts. *J. Am. Chem. Soc.* **2012**, *134*, 5598–5609. [CrossRef]
34. Glüer, A.; Förster, M.; Celinski, V.R.; Schmedt auf der Günne, J.; Holthausen, M.C.; Schneider, S. Highly Active Iron Catalyst for Ammonia Borane Dehydrocoupling at Room Temperature. *ACS Catal.* **2015**, *5*, 7214–7217. [CrossRef]
35. Vance, J.R.; Robertson, A.P.M.; Lee, K.; Manners, I. Photoactivated, Iron-Catalyzed Dehydrocoupling of Amine–Borane Adducts: Formation of Boron–Nitrogen Oligomers and Polymers. *Chem. Eur. J.* **2011**, *17*, 4099–4103. [CrossRef] [PubMed]
36. Vance, J.R.; Schäfer, A.; Robertson, A.P.M.; Lee, K.; Turner, J.; Whittell, G.R.; Manners, I. Iron-Catalyzed Dehydrocoupling/Dehydrogenation of Amine–Boranes. *J. Am. Chem. Soc.* **2014**, *136*, 3048–3064. [CrossRef] [PubMed]
37. Pagano, J.K.; Bange, C.A.; Farmiloe, S.E.; Waterman, R. Visible Light Photocatalysis Using a Commercially Available Iron Compound. *Organometallics* **2017**, *26*, 3891–3895. [CrossRef]
38. Bhattacharya, P.; Krause, J.A.; Guan, H. Mechanistic Studies of Ammonia Borane Dehydrogenation Catalyzed by Iron Pincer Complexes. *J. Am. Chem. Soc.* **2014**, *136*, 11153–11161. [CrossRef]
39. Kuroki, A.; Ushiyama, H.; Yamashita, K. Theoretical studies on ammonia borane dehydrogenation catalyzed by iron pincer complexes. *Comput. Theor. Chem.* **2016**, *1090*, 214–217. [CrossRef]
40. Gorgas, N.; Stöger, B.; Veiros, L.F.; Kirchner, K. Access to FeII Bis(σ-B–H) Aminoborane Complexes through Protonation of a Borohydride Complex and Dehydrogenation of Amine-Boranes. *Angew. Chem. Int. Ed.* **2019**, *58*, 13874–13879. [CrossRef]
41. Coles, N.T.; Mahon, M.F.; Webster, R.L. Phosphine-and Amine-Borane Dehydrocoupling Using a Three-Coordinate Iron(II) β-Diketiminate Precatalyst. *Organometallics* **2017**, *36*, 2262–2268. [CrossRef]
42. Bäcker, A.; Li, Y.; Fritz, M.; Grätz, M.; Ke, Z.; Langer, R. Redox-Active, Boron-Based Ligands in Iron Complexes with Inverted Hydride Reactivity in Dehydrogenation Catalysis. *ACS Catal.* **2019**, *9*, 7300–7309. [CrossRef]
43. Alig, L.; Fritz, M.; Schneider, S. First-Row Transition Metal (De)Hydrogenation Catalysis Based On Functional Pincer Ligands. *Chem. Rev.* **2019**, *119*, 2681–2751. [CrossRef]

44. Todisco, S.; Luconi, L.; Giambastiani, G.; Rossin, A.; Peruzzini, M.; Golub, I.E.; Filippov, O.A.; Belkova, N.V.; Shubina, E.S. Ammonia Borane Dehydrogenation Catalyzed by (κ^4-EP$_3$)Co(H) [EP$_3$ = E(CH$_2$CH$_2$PPh$_2$)$_3$; E = N, P] and H$_2$ Evolution from Their Interaction with NH Acids. *Inorg. Chem.* **2017**, *56*, 4296–4307. [CrossRef] [PubMed]
45. Burla, M.C.; Caliandro, R.; Camalli, M.; Carrozzini, B.; Cascarano, G.L.; De Caro, L.; Giacovazzo, C.; Polidori, G.; Siliqi, D.; Spagna, R. IL MILIONE: a suite of computer programs for crystal structure solution of proteins. *J. Appl. Cryst.* **2007**, *40*, 609–613. [CrossRef]
46. Creagh, D.C.; McAuley, W.J. *International Tables for Crystallography*; Table 6.1.1.4; Wilson, A.J.C., Ed.; Kluwer Academic Publishers: Dordrecht, The Netherlands, 1992; p. 572.
47. Ibers, J.A.; Hamilton, W.C. Dispersion corrections and crystal structure refinements. *Acta Cryst.* **1964**, *17*, 781–782. [CrossRef]
48. Creagh, D.C.; McAuley, W.J. *International Tables for Crystallography*; Table 4.2.6.8; Wilson, A.J.C., Ed.; Kluwer Academic Publishers: Dordrecht, The Netherlands, 1992; pp. 219–222.
49. Creagh, D.C.; Hubbell, J.H. *International Tables for Crystallography*; Table 4.2.4.3; Wilson, A.J.C., Ed.; Kluwer Academic Publishers: Dordrecht, The Netherlands, 1992; pp. 200–206.
50. *CrystalStructure 4.2.5: Crystal Structure Analysis Package*; Rigaku Corporation: Tokyo 196-8666, Japan, 2000–2017.
51. Sheldrick, G.M. SHELXL Version 2017/1: A short history of *SHELX*. *Acta Cryst.* **2008**, *64*, 112–122. [CrossRef] [PubMed]

© 2019 by the authors. Licensee MDPI, Basel, Switzerland. This article is an open access article distributed under the terms and conditions of the Creative Commons Attribution (CC BY) license (http://creativecommons.org/licenses/by/4.0/).

Article

Improved NOx Reduction Using C_3H_8 and H_2 with Ag/Al_2O_3 Catalysts Promoted with Pt and WOx

Naomi N. González Hernández [1], José Luis Contreras [1,*], Marcos Pinto [1], Beatriz Zeifert [2], Jorge L. Flores Moreno [1], Gustavo A. Fuentes [3], María E. Hernández-Terán [3], Tamara Vázquez [2], José Salmones [2] and José M. Jurado [1]

- [1] Energy Department, CBI, Universidad Autónoma Metropolitana-Azcapotzalco. Av. Sn. Pablo 180, Col. Reynosa, México City C.P.02200, Mexico; iq_naomi@hotmail.com (N.N.G.H.); socram_pinto@hotmail.com (M.P.); jflores@correo.azc.uam.mx (J.L.F.M.) josemanueljuradoflores@outlook.com (J.M.J.)
- [2] Instituto Politécnico Nacional, ESIQIE. U.P. López Mateos Zacatenco, Mexico City C.P. 07738, Mexico; bzeifert@yahoo.com (B.Z.); avazquezrdz@hotmail.com (T.V.); jose_salmones@yahoo.com.mx (J.S.)
- [3] Universidad Autónoma Metropolitana-Iztapalapa, CBI-IPH, Mexico City C.P. 09340, Mexico; gfuentes@xanum.uam.mx (G.A.F.); meht4@hotmail.com (M.E.H.-T.)
- * Correspondence: jlcl@azc.uam.mx; Tel.: +52-55591911047

Received: 29 June 2020; Accepted: 14 October 2020; Published: 19 October 2020

Abstract: The addition of Pt (0.1 wt%Pt) to the 2 wt%Ag/Al_2O_3-WOx catalyst improved the C_3H_8–Selective Catalytic Reduction (SCR) of NO assisted by H_2 and widened the range of the operation window. During H_2–C_3H_8–SCR of NO, the bimetallic Pt–Ag catalyst showed two maxima in conversion: 80% (at 130 °C) and 91% (between 260 and 350 °C). This PtAg bimetallic catalyst showed that it could combine the catalytic properties of Pt at low temperature, with the properties of Ag/Al_2O_3 at high temperature. These PtAg catalysts were composed of Ag^+, $Ag_n^{\delta+}$ clusters, and PtAg nanoparticles. The catalysts were characterized by Temperature Programmed Reduction (TPR), Ultraviolet Visible Spectroscopy (UV-Vis), Scanning Electron Microscopy (SEM)/ Energy Dispersed X-ray Spectroscopy (EDS), x-ray Diffraction (XRD) and N_2 physisorption. The PtAg bimetallic catalysts were able to chemisorb H_2. The dispersion of Pt in the bimetallic catalysts was the largest for the catalyst with the lowest Pt/Ag atomic ratio. Through SEM, mainly spherical clusters smaller than 10 nm were observed in the PtAg catalyst. There were about 32% of particles with size equal or below 10 nm. The PtAg bimetallic catalysts produced NO_2 in the intermediate temperature range as well as some N_2O. The yield to N_2O was proportional to the Pt/Ag atomic ratio and reached 8.5% N_2O. WOx stabilizes Al_2O_3 at temperatures ≥650 °C, and also stabilizes Pt when it is reduced in H_2 at high temperature (800 °C).

Keywords: H_2–C_3H_8–SCR–NO; PtAg/Al_2O_3; bimetallic clusters; Al_2O_3–WOx

1. Introduction

The conversion of nitrogen oxide emissions (NOx) from diesel machines can be carried out through selective catalytic reduction (SCR) using reductants such as hydrocarbons, alcohols, NH_3, H_2, and supported metal catalysts [1,2]. The exhaust atmosphere is oxidizing, which hinders the use of three-way technology. Hydrocarbons, present in small amounts in emissions, can be used as a reducing agent to react competitively with O_2 or NOx, producing N_2, CO_2, and H_2O and traces of N_2O.

Among the NOx reduction technologies today, urea-based SCR is used in new heavy trucks and some types of cars that run with diesel-type engines. The requirement to add a urea solution into the gas emissions is inconvenient for bus operators and for passenger cars. The possibility of using hydrocarbons to carry out the SCR of NOx (HC–SCR) or alcohols to decrease the contaminants continuously is still attractive [3]. For the SCR of NO using hydrocarbons (HC), catalysts based on Pt, Cu, Ir, Rh, and Ag have been studied [1,2]. The latter metal supported on γ-Al_2O_3 is very interesting because, during SCR using C_3H_6 or C_8H_{18}, the main product is N_2 and not N_2O [2]. It has also been shown to have good stability in the presence of water vapor [4,5] and some tolerance to SO_2 [6].

Hydrocarbons have been studied as reducers, but alcohols such as methanol, ethanol, and butanol have also been considered [7–10]. The SCR of NO from Ag/Al_2O_3 catalysts in the presence of O_2 depends on the concentration and structure of Ag moieties on the surface (i.e., Ag^+ cations and Ag_n (n = 8) nanoclusters). Ag^0 nanoparticles have been reported to catalyze the total oxidation of hydrocarbons or alcohols to CO_2 and H_2O [8]. In studies using reductants such as C_3H_6 and impregnation of Ag precursor in Al_2O_3, it has been found that there is an optimum concentration in Al_2O_3 between 1 and 3 wt% [3,11]. Studies on the nature and role of active Ag species during NOx reduction in the presence of C_3H_6 and water have been done, and proposals for reaction mechanisms in the presence of hydrocarbons using spectroscopic techniques have been made [12–24].

There are some drawbacks of the Ag/Al_2O_3 catalyst. The main one is that it is active in a narrow range of high temperatures and has low activity below 400 °C in the case of SCR with light hydrocarbons [7,8]. This is an issue because the exhaust gases of lean-burn diesel engines have low temperatures during standard driving conditions.

This problem can be addressed because the operating temperature range has been expanded with the addition of H_2 [24–30], which results in the presence of two NO reduction zones: one at low-temperature (80–180 °C) and the other at high temperature (180 to 480 °C). In the search for other reducers, research has also been done using H_2 and NH_3 together [31,32] and on the coaddition of NH_3 and ethanol [33], obtaining good reduction results at low temperatures, although there may be NH_3 emissions.

The study of the effect of the hydrocarbon's molecular weight has been carried out using propene and octane [3]. There is one work about the use of gasoline and ethanol as reducing agents [34]. The results showed the presence of NH_3, and both NO and ethanol began to react at low temperatures (200–300 °C) on Ag/Al_2O_3. Studies with other metals such as In/Ag/Al_2O_3–TiO_2 using CO showed an improvement in yield to N_2 when In was added to Ag [35].

The combined effect of CO and C_3H_8 was analyzed, and the temperature window was expanded, but a 5 wt%Ag/Al_2O_3 catalyst was required [36]. In this case, the addition of noble metals such as Pt to Ag is useful in obtaining high conversions of NO at low temperatures by using octane as a reduction agent [37]. The catalyst composition showing the highest activity for NOx reduction was a 2 wt%Ag/Al_2O_3 doped with 500 ppm Pt. This catalyst showed great capacity for adsorption and partial oxidation of the hydrocarbon, where the Pt has a predominant role.

There have been studies of intermediate storage of NO (passive NOx trap) at low temperatures and regeneration of the adsorbed compound (NO) at high temperatures when the Ag/Al_2O_3 catalyst was active to reduce NOx. A Pt/Ba/Al_2O_3 catalyst proposed by Tamm et al. [38] in the presence of H_2 demonstrated the importance of this to increase the amount of NO stored on the catalyst between 100 and 200 °C. Other noble metals such as the addition of Pd to the Ag/Al_2O_3 catalyst [39] showed that the catalytic activity of the catalyst promoted with Pd was higher than the activity of the Ag single metal catalyst in the oxidation of CO and hydrocarbon as well as in reducing NOx.

In another work [40], the Pd–Ag/Al_2O_3 catalyst showed higher activity than the Ag/Al_2O_3 catalyst at temperatures of 300 to 450 °C. It was found that Pd catalyzed the formation of enolic species, which were converted from C_3H_6. The superficial enolic species were quite reactive toward NO_3- and NO_2 to form superficial species of –NCO. In the case of the addition of Rh to Ag, theoretical studies have been done that mention the higher capacity of Rh than Ag for NO reduction reactions [41]. Another study of NO reduction where C_3H_6, Pt, Rh, and Ag/Al_2O_3 were investigated showed that Ag was the most active at higher temperatures, while Pt and Rh were at lower temperatures (200–250 °C) [42].

As has been observed, the SCR of NOx with hydrocarbons (HC–SCR–NOx) has been of great interest until now, because in the presence of an oxidizing atmosphere, as diesel engines work, it is possible to reduce NOx to N_2 [24,40]. In the case of Pt catalysts supported in WO_3/ZrO_2 and the presence of H_2 without Ag for the reduction of NO [43], the authors found high activity at temperatures below 200 °C and high selectivity to N_2 (90%). Additionally, the catalyst showed outstanding hydrothermal stability as well as SO_2 resistance. Along these same lines, the exceptional stability of WOx has been studied in Al_2O_3 and Pt/Al_2O_3 [44,45]. It was adopted for the present study, especially to preserve the thermal stability of Pt species', Ag^0, Ag^+ cations, and Ag_n nanoclusters.

As previously mentioned, the present study contributes to improving the Ag/Al_2O_3 catalyst by adding minimal amounts of Pt and WOx in the presence of H_2 and C_3H_8, which allows for improvement in the SCR of NO. The $PtAg/Al_2O_3$-WOx catalysts were prepared in powder form with the optimal amount of WOx that allows for a high metallic dispersion of Pt and Ag to be obtained, since it is resistant to deactivation by sintering, stabilizing the porous structure of Al_2O_3. The study combined H_2 and C_3H_8 reducers and the presence of small amounts of Pt, which allows for a higher conversion of NO at low temperatures.

2. Results and Discussion

Seven Ag and Pt catalysts supported on γ-alumina (with or without WOx) were prepared by the incipient wetness impregnation method with $AgNO_3$ and H_2PtCl_6 aqueous solutions. The preparation method is reported in greater detail in the Materials and Methods section. The characterization section is presented first and the catalytic evaluation section later.

2.1. Characterization

2.1.1. Textural Properties

The synthesized γ-Al_2O_3 (A) presented type IV isotherms, according to the International Union of Pure and Applied Chemistry (IUPAC), the Brunauer Emmett and Teller (BET) area was higher than some commercial alumina with an average unimodal pore diameter of 54 Å (Table 1). The impregnation of Pt as well as Ag and WOx did not significantly modify the area and the other properties.

Table 1. Composition of Pt, Ag, and $PtAg/Al_2O_3$ catalysts with 2 wt% Ag and 0.5 wt% W (WOx) prepared in powder. BET area, texture, Pt/Ag atomic ratio, H_2 consumption by TPR, and Pt dispersion.

Catalyst Name	Pt (%)	Ag (%)	Pt/Ag Atomic Ratio	BET Area (m²/g)	Pore Vol. (cm³/g)	Pore Diam. (Å)	H_2 Consumption (µmol/g_c)	Pt Dispersion (%)
A	0	0	0	267	0.36	54	0	0
0.4Pt/A	0.4	0	-	256	0.36	55	45.1	61
2Ag/AW	0	2	-	264	0.38	56	67	0
0.4Pt/AW	0.4	0	-	230	0.39	68	43	57
0.1PtAg/AW	0.1	2	0.027	226	0.37	66	2.5	60
0.25PtAg/AW	0.25	2	0.069	218	0.36	67	26	46
0.4PtAg/AW	0.4	2	0.110	225	0.37	66	41	38
1PtAg/AW	1	2	0.270	228	0.37	65	112	21

2.1.2. X-Ray Diffraction (XRD)

The powder x-ray diffraction pattern of all samples showed typical reflections of the g-alumina phase (Figure 1a), with peaks at 2θ = 37°, 46°, and 67°. The presence of another phase was not observed, and the materials were amorphous [46]. According to Aguado et al. [47], these three main peaks correspond to the reflections (311), (400), and (440).

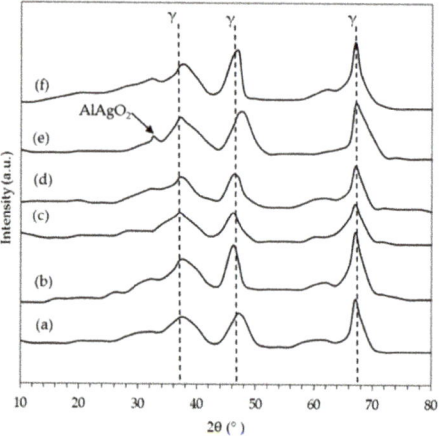

Figure 1. X-ray diffraction (XRD) patterns of the catalysts: (**a**) 0.4Pt/AW; (**b**) 2Ag/AW; (**c**) 0.1PtAg/AW; (**d**) 0.25PtAg/AW; (**e**) 0.4PtAg/AW; (**f**) 1PtAg/AW.

Figure 1a–d,f show similar diffractograms of the samples calcined at 500 °C and reduced to 450 °C and no Pt or Ag signals were observed due to their low concentration. However, it was reported that a catalyst of 5 wt%Ag/Al$_2$O$_3$ showed reflections of metallic Ag [48].

Sample 0.4PtAg/AW (Figure 1e) showed a reflection at 2θ = 38.7°, probably attributed to AlAg$_2$O, however, the other reflections of this compound (2θ = 50.5° and 66.8°) were not noted or revealed the presence of supported Ag$_2$O particles with a size greater than 5 nm. The broad peaks of these samples showed a stable amorphous and meta structure. The metallic compound Ag$_2$O has been reported in other studies [49] using Ag concentrations higher than 5 wt%.

2.1.3. Temperature Programmed Reduction (TPR)

In the case of 2Ag/AW silver catalyst reduction (Figure 2a), two peaks located at 100 °C and 340 °C was observed, corresponding to the reduction of AgO and Ag$_2$O clusters. This result has already been reported using a 2 wt%Ag/Al$_2$O$_3$ catalyst by Bethke and Kung [11] and Maria E. Hernández-Terán [50]. In this last peak, the inflection point or maximum was not observed. These small reduction peaks could be caused by part of the Ag compound already decomposed to metallic silver during calcination [50,51].

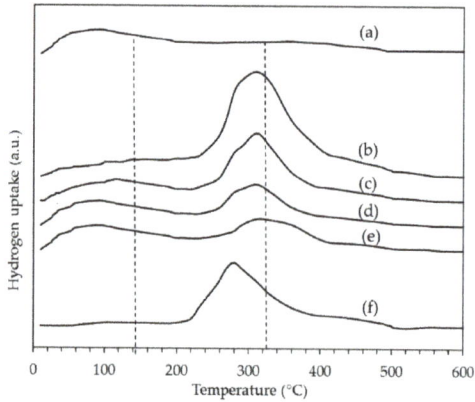

Figure 2. TPR of the catalysts: (**a**) 2Ag/AW; (**b**) 1PtAg/AW; (**c**) 0.4PtAg/AW; (**d**) 0.25PtAg/AW; (**e**) 0.1PtAg/AW; (**f**) 0.4Pt/AW catalyst.

In the case of the 0.4Pt/AW catalyst (Figure 2f), the Pt reduction peak could be observed due to the Pt-oxychloride complexes (PtOxCly) located at 280 °C [52]. The H_2 consumption for the reduction of (PtOxCly) corresponds to the reduction from Pt^{+4} to Pt^0. It is known that the temperature of the reduction peak depends on the precursor of Pt [52]. If a chlorine-free Pt precursor such as $Pt(NH_3)_4(NO_3)_2$ is used, the temperature is close to 70 °C (reduction of PtO_2), whereas if H_2PtCl_6 is used, the temperature is 290 °C.

In the case of the catalyst with the highest concentration of Pt (1PtAg/AW), two peaks located at 100 °C and 315 °C were observed (Figure 2b). Again, the first corresponded to the reduction of AgO, while the second corresponded to the co-reduction of the two metals' oxides, as has been reported in the literature [50,53]. The maximum peak temperature of this catalyst was 35 °C higher than the peak of the 0.4Pt/AW catalyst (Figure 2f).

This kind of peak has been mentioned in the literature; for example, the Pt–Ag/SiO_2 catalyst has been reported as an alloy when Ag is impregnated on the Pt/SiO_2 catalyst [54]. It has been found that in such alloys, the Pt and Ag can be secreted by high-temperature oxidation.

The catalysts with lower Pt concentrations (Figure 2c–e) also showed two peaks at temperatures of 100 °C and 305 °C. This last peak was again found at 35 °C higher than the maximum of the 0.4Pt/AW catalyst peak (Figure 2f). The ratio of μmoles of H_2 consumed per g of catalyst (g_c) for the Pt peaks is shown in Table 1. It was observed that the bimetallic catalysts of PtAg consumed H_2 as a function of the concentration of Pt.

In the case of the reduction peak at 100 °C (reduction of AgO), for the 2Ag/AW catalyst (Figure 2a), it was 68 μmol H_2/g_c. This value was almost constant for the other values of the reduction of AgO of the bimetallic catalysts since the Ag content was constant (2 wt%Ag), and only in the 1PtAg/AW catalyst did it decrease slightly (Figure 2b).

The ratio of moles of H_2 consumed per g of catalyst (g_c) for Pt is shown in Table 1. The bimetallic catalysts of PtAg consumed H_2 as a function of the concentration of Pt, as reported in the literature [53]. In general, reducing the Pt oxides was completed, while in the case of Ag oxides, it was not completed. Only a part of Ag oxides seemed to be susceptible to reduction, because another part was already in a metallic state after calcination, as has been reported in the literature [50,54].

2.1.4. H_2 Chemisorption

The chemisorption of H_2 was carried out mainly at the Pt sites. H_2 chemisorption of part of the Ag was not observed; even in the case of bimetallic catalysts, it is evident that the consumption of H_2 is proportional to the concentration of Pt (Table 1). For the Pt dispersion calculation, the stoichiometry H/Pt was 1, as reported in the literature [55], the dilution state that Ag exerts on Pt atoms is evident, as reported in the literature [53].

In the case of the catalyst with a high Pt content (1PtAg/AW), a large part of the Pt was reduced to metal; however, only a fraction of it remained on the surface of the bimetallic PtAg particles, so it showed a low dispersion (21%). On the contrary, in the case of the catalyst with a low Pt content (0.1PtAg/AW) with a dispersion of 60%, there was a better ratio of surface Pt atoms to Pt atoms in the bulk of the bimetallic. For the 0.25PtAg/AW and 0.4Pt/AW catalysts, there were intermediate Pt dispersions (46 and 38%) that could explain the NO conversion profiles in terms of the activation of C_3H_8 and H_2.

Some authors have mentioned the presence of a "Pt–Ag alloy" [54]. However, this assertion is doubtful because they did not check if there was a solid solution of Pt and Ag, which is why in our case, we only speak of a bimetallic PtAg system that could have Ag particles decorated with Pt, (or bimetallic core-shell structures), since in all our catalysts, the Ag was always higher in concentration.

On the other hand, there have been studies where the Ag chemisorbs O_2, and it has been used to determine its dispersion [56]. The authors validated the stoichiometry of chemisorption of O_2 (O_2/Ag = 2) by comparing the average particle size using the bright-field TEM, high angle annular dark-field (HAADF), and O_2 chemisorption techniques. The active Ag dispersion values they found were: 57.6% (for 1.28 wt%Ag), 51% (for 1.91 wt%Ag), 44.8% (for 2.88 wt%Ag), and 51.2% (for the 6 wt%Ag). The particle sizes were 2.63 nm, 2.62 nm, 3 nm, and 2.63 nm, respectively, at the percentages of Ag, as above-mentioned.

2.1.5. SEM of the Catalysts

The calcined 0.4PtAg/AW catalyst showed spherical particles (Figure 3a) that could be related to the presence of Ag_2O [48]. The distribution of particle diameters indicates the predominance (31.64%) of particles of 10 nm, followed by those of 25 and 30 nm (total 42%) (Figure 3b). Finally, those of 50 nm represented 9.9%, and the particles of larger sizes decreased. These results approximate the results reported by Richter et al. [48]. The presence of nanoparticles of different sizes that can vary depending on the type of support has been mentioned. The diameters they found were between 2 to 40 nm with a load of 5 wt%Ag, but predominantly between 5 and 10 nm.

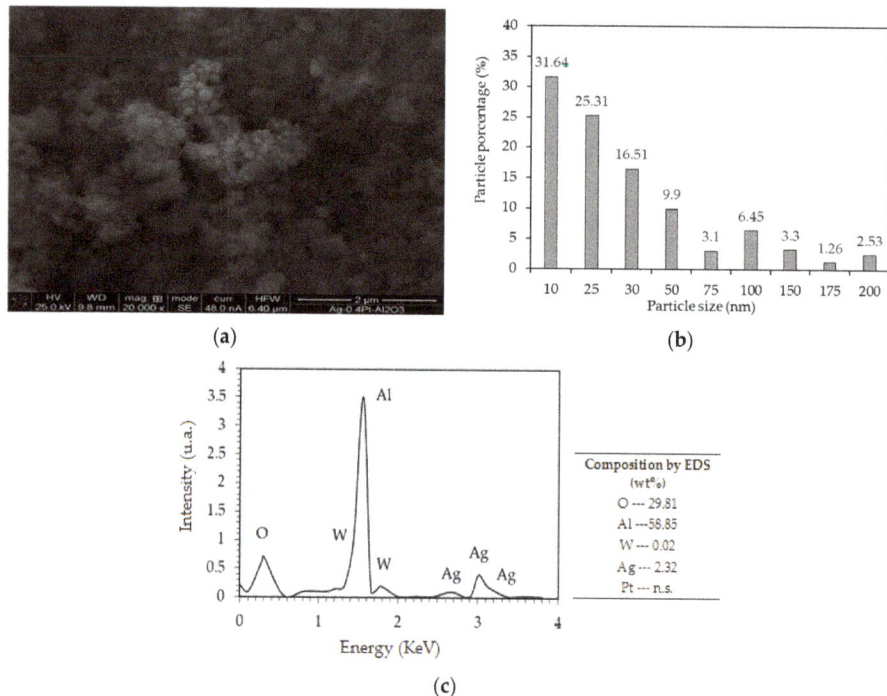

Figure 3. Morphological and chemical composition of the catalyst 0.4PtAg/AW calcined at 500 °C. (a) SEM micrograph; (b) The distribution of particle diameters; (c) EDS analysis of the spherical cumulus zone where Pt cannot be analyzed due to its low concentration.

The EDS analysis of the observed area is shown in Figure 3c, where we can observe the presence of Ag and W in amounts approximately at the nominal ones. Richter et al. [48] also identified Ag_2O by convergent beam diffraction, where Ag_2O was indexed, and the primary signal was attributed to it when they were analyzed by temperature-programmed reduction (TPR). The authors found that HAADF was more suitable because the contours of the Ag particles were better distinguished. We also found this problem (Figure 3a) because the particles of Ag showed a dark silhouette that could be confused with part of the alumina support.

Due to the TPR studies, the presence of AgO and Ag_2O is possible, however it was not possible to confirm them by other more advanced techniques. Arve et al. [56] found that both Ag metal and

Ag_2O phases were present in their catalysts, and did not find AgO and cubic Ag_2O. The authors concluded that in small particles, Ag_2O is the predominant phase, while metallic Ag is more likely in large particles.

2.1.6. UV–Vis Spectroscopy

2Ag/AW catalyst

The existence of different Ag oxidation states was demonstrated by ex situ UV–Vis analysis when the 2Ag/AW catalyst was calcined at 500 °C (Figure 4a). In this case, ionic Ag (Ag^+) species were observed showing absorption peaks in the range between 200 and 230 nm [17,49]. The spectrum of the 2Ag/AW sample with a band at 220–235 nm was attributed to the $4d^{10}$ to $4d^9 5s^1$ electronic transitions due to highly dispersed Ag^+ ions [16,57]. A similar band was observed for the Ag/Al_2O_3 and Ag^+/H-ZSM-5 catalysts [9,12,30].

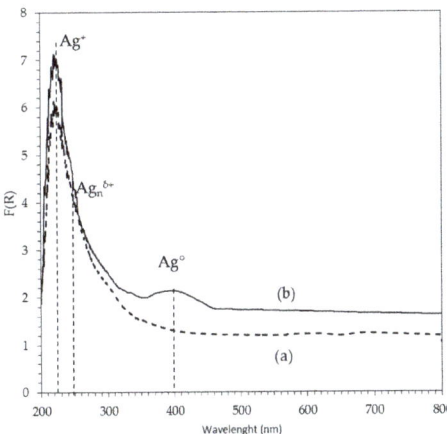

Figure 4. Ex situ UV–Vis spectra of the catalysts: (**a**) Calcined 2Ag/AW; (**b**) 2Ag/AW after H_2 reduction in a flow of H_2 (30 cm³/min) at 500 °C for 2 h (spectrum taken immediately after reduction in H_2).

The absorption in the range of 240–288 nm is commonly ascribed to silver nanoclusters $Ag_n^{\delta+}$ (n < 8) with a variety of cluster sizes and different oxidation states. After H_2 reduction, an absorption band at 340 and 423 nm (Figure 4b) was assigned to larger silver nanoclusters (n > 8) and metallic Ag nanoparticles [16].

0.4Pt/AW Catalyst

The spectrum of the 0.4Pt/AW catalyst calcined at 500 °C showed a band with a maximum at 215–240 nm (Figure 5a). This band was very close to the band found by Lietz et al. [58] at 217 nm for a Pt catalyst prepared by impregnation with H_2PtCl_6 in Al_2O_3. The authors attributed this signal to a charge transfer band due to the presence of a compound of the type $[PtCl_5OH]^{2-}$, which compares well with the literature data for octahedral Pt^{4+}.

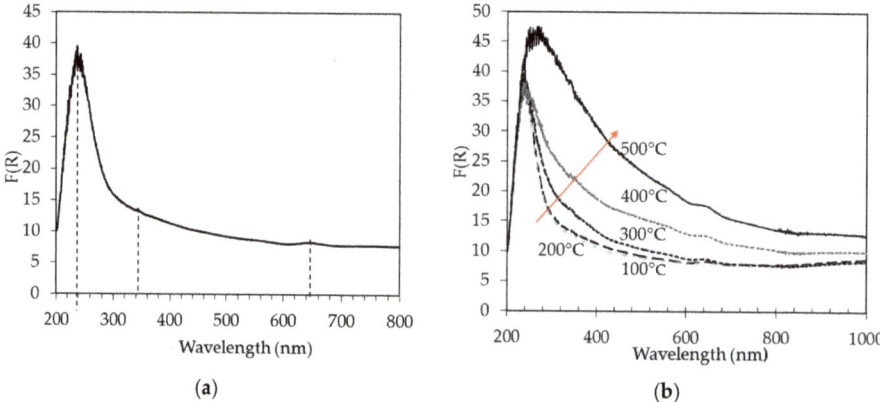

Figure 5. In situ UV–Vis spectrum of: (**a**) 0.4Pt/AW catalyst calcined at 500 °C; (**b**) 0.4Pt/AW catalyst during the reduction process with 5 vol.% H_2/N_2, 0.5 cm^3/s.

We found two bands located at 360 nm and 640 nm (Figure 5a). These bands were close with the bands reported by Lietz et al. [58] for a Pt/Al$_2$O$_3$ catalyst calcined at 500 °C. Their bands were located at 340, 450, and 550 nm, which were associated with the [PtOxCly]$_s$ complexes. We did not find the band located at 450 nm.

The UV–Vis spectra of the 0.4Pt/AW catalyst during the "in situ" reduction with H_2 (Figure 5b) showed that with increasing reduction temperature, an increase in the values of the function F(R) corresponding to the band at 320 nm were related to the formation of metallic Pt [58]. During this "in situ" reduction with H_2 from 100 °C to 500 °C of the 0.4Pt/AW catalyst, an increase in the F(R) function was observed with respect to the F(R) function of the same calcined catalyst shown in Figure 5a, over a whole spectrum wavelength range from 250 to 1000 nm.

This increase in absorbance due to the reduction of Pt oxychlorocomplexes to metallic particles is responsible for the color change to dark gray of the catalysts. This is related to the so-called color centers [59] and to the appearance of a spectroscopic signal of greater intensity that is due to a greater electronic conduction on the surface of the solid that is associated with the formation of Pt crystallites formed during this reduction process.

PtAg/AW Catalysts

For calcined PtAg/AW catalysts, a band at 220 nm corresponding to Ag$^+$ was observed (Figure 6), as previously reported (Figure 4). This band could also represent the Pt band located at a wavelength of 215–240 nm, which can be attributed to [PtCl$_5$OH]$^{2-}$ related to octahedral Pt^{+4}, as mentioned above. This band was noticeable in the two catalysts with high Pt concentration; 1PtAg/AW and 0.4PtAg/AW (Figure 6a,b), however it did not appear in the two catalysts with low Pt concentration; 0.25PtAg/AW and 0.1PtAg/AW (Figure 6c,d).

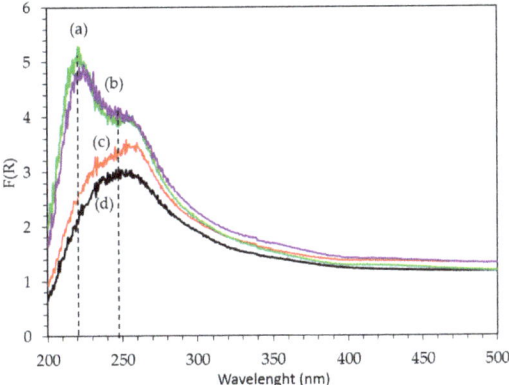

Figure 6. Ex situ UV–Vis spectra of the PtAg/AW catalysts calcined at 500 °C when the Pt concentration changed from 0.1 to 1 wt% on Al_2O_3–WOx catalysts. (**a**) 1PtAg/AW; (**b**) 0.4PtAg/AW; (**c**) 0.25PtAg/AW; and (**d**) 0.1PtAg/AW.

The broadband with a maximum at 255 nm could correspond to the signal of the Ag metal clusters, $(Ag_n^{\delta+})$, as already mentioned [30]. In this band, the spectroscopic contribution of the signal due to Ag appeared to be higher than the small-signal at 360 nm, as shown by the 0.4Pt/AW catalyst (Figure 5).

2.2. Catalytic Activity

2.2.1. SCR of NO on Pt and Ag Catalysts

In Figure 7a, the conversion of the 2Ag/AW catalyst started at 370 °C and reached 80% at around 450 °C, followed by a sharp decrease. This reaction temperature window has been previously reported [3,20,30] to be narrow and dependent on the Al_2O_3 preparation method [3].

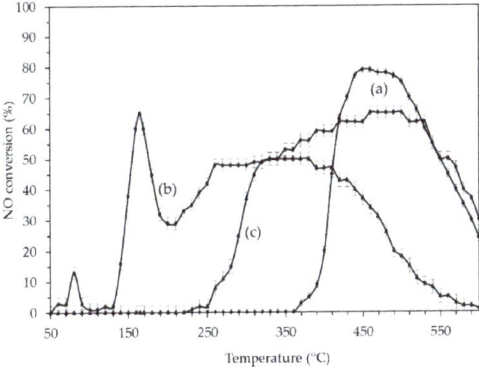

Figure 7. C_3H_8–SCR–NOx from (**a**) 2Ag/AW catalyst in the absence of H_2; (**b**) 2Ag/AW catalyst in the presence of H_2; (**c**) 0.4Pt/AW catalyst in the presence of H_2. Inlet gas composition: 500 ppm NO, 625 ppm C_3H_8, 200 ppm CO, 660 ppm H_2, 2 vol.% O_2, N_2 balance; GHSV = 128,000 h^{-1}.

The impregnation method is better than the sol-gel method when the Ag concentration is 2 wt%; if the Ag concentration increases to 5 or 8 wt%, the lyophilized sol-gel method is better. In other words, the method of preparation and the concentration of Ag could provide better catalysts. However, it has been mentioned that the optimal amount of Ag is defined between 1 to 3 wt% [9–11,15,16].

The drawbacks of the Ag/Al$_2$O$_3$ catalyst behavior when using hydrocarbons as a reducing agent are that the operating temperature window to reduce NO is narrow as well as its low activity below 400 °C. These characteristics do not favor the reduction of NOx emitted by the diesel engines since the temperatures of these emissions are low. Fortunately, when small amounts of H$_2$ are added to the emissions using the Ag/Al$_2$O$_3$ catalyst, advantages are obtained both in conversion and in the operating window [26].

We could verify this effect when we added small amounts of H$_2$ to our 2Ag/AW catalyst in the flue gas stream (Figure 7b). It was observed that the temperature window in which the catalyst showed activity widened from 130 °C to 500 °C. Additionally, a range of activity appeared at low temperatures (130–200 °C) with medium conversions. The interval from 200 to 470 °C showed better conversions (63%), in agreement with the literature [25–28,30,50].

The reaction interval of 130–200 °C has been the subject of debate on the participation of H$_2$ and the nature of the active species capable of favoring the reaction at low temperatures [27,30]. Studies using Diffuse Reflectance Infrared Fourier Transform Spectroscopy (DRIFTS) of NO and temperature-programmed desorption found two groups of surface NOx species: a less thermally stable group of low temperature (LT) species and a more thermally stable group of high-temperature species (HT). The existence of LT species was attributable to the decomposition of the superficial NOx species formed in the active sites where there is elimination by the addition of H$_2$ or thermal decomposition related to higher oxidation of NO and NOx [27].

The 0.4Pt/AW catalyst in the presence of H$_2$ (Figure 7c) showed a maximum conversion close to 50% at a temperature of 300 °C, starting at 250 °C, and this result coincided with that reported by Lanza et al. [42]. These authors investigated Pt and Rh and found high conversions at low temperature (200–250 °C), showing high selectivity to NO$_2$. On the other hand, when they investigated Ag, a higher temperature was required but with high selectivity to N$_2$. In this study, it was found that the presence of H$_2$ triggered the conversion of NO.

In the case of the SCR of NO on the catalyst with the addition of H$_2$ (2Ag/AW + H$_2$), specifically with 660 ppm H$_2$, a decrease in the light-off temperature down to 150 °C was observed (Figure 7b). This remarkable decrease (from 400 to 150 °C) was similar to that reported by Satokawa et al. [60], and we observed two reaction zones. In the low temperature range (100–180 °C), H$_2$ allows the reactants (or reaction intermediates) to be activated, significantly reducing the activation energy (Ea) of the entire global reaction [48].

According to some authors [48], H$_2$ contributes to reducing oxidized silver species such as Ag$_2$O to Ag^0 on which the nitrate adducts, as above-mentioned, will be adsorbed. The same authors suggest that nano-sized Ag$_2$O clusters can be reversibly reduced and reoxidized in the presence of H$_2$. These authors also found that the presence of H$_2$O did not change the Ea. They also found that more adducts or nitrate species were found adsorbed in the presence of H$_2$, and attributed this to a dissociative activation of O$_2$ in the gas phase on the Ag^0 particles.

Based on the studies by Azis et al. [27] using DRIFTS and TPD, it appears that there is a formation of stable superficial NOx species that are related to the promoter effect of H$_2$ on the Ag/Al$_2$O$_3$ catalyst.

2.2.2. SCR of NO with C$_3$H$_8$ on PtAg Catalysts

The SCR of NO with C$_3$H$_8$ on the catalyst 2Ag/AW (Figure 8d) in the absence of H$_2$ showed a volcano-like profile that has already been reported [2], where the starting temperature (light-off) was 400 °C with a maximum at 470 °C. The possible presence of monodentate nitrate species at high temperatures is feasible [27]. In contrast, the other bidentate and bridged species that are less stable thermally may not be present.

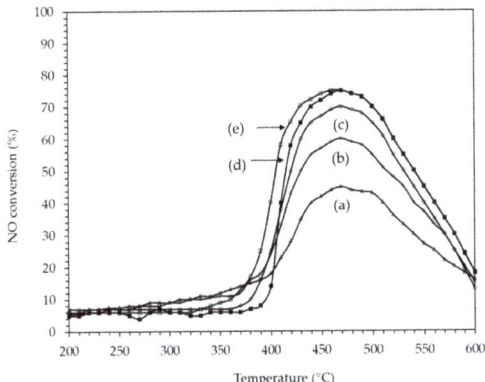

Figure 8. C_3H_8–SCR–NOx from (**a**) 1PtAg/AW catalyst; (**b**) 0.4PtAg/AW catalyst; (**c**) 0.25PtAg/AW catalyst; (**d**) 2Ag/AW catalyst; (**e**) 0.1PtAg/AW catalyst. Inlet gas composition: 500 ppm NO, 625 ppm C_3H_8, 200 ppm CO, 2vol.% O_2, N_2 balance; GHSV = 128,000 h^{-1}.

In general, the addition of Pt to the Ag/AW catalyst did not contribute significantly to the NO conversion at temperatures below 360 °C. In the case of the 1PtAg/AW catalyst (Figure 8a), it only showed a 9% conversion at less than 300 °C. Despite the high Pt content, the NO conversion at 470 °C was the lowest (45%). This behavior indicates that the bimetallic PtAg particles at a Pt/Ag atomic ratio of 0.27 do not provide the best active sites for NO reduction reactions. On the other hand, the Pt dispersion for this catalyst (21%) was the lowest (Table 1). The dilution effect of Pt by Ag seemed evident.

As the Pt concentration decreased in the 0.4PtAg/AW catalyst (Figure 8b) and 0.25PtAg/AW catalyst (Figure 8c), the NO conversion increased to 470 °C, approaching the catalyst conversion of Ag without Pt (Figure 8d). Only when the Pt concentration was less than the 0.1PtAg/AW catalyst (Figure 8e) for a Pt/Ag atomic ratio of 0.027 was the NO conversion slightly higher (75.3%) at 470 °C.

This behavior was similar to that found by Wang et al. [61] for the Pd–Ag bimetallic system. The authors found a 0.01 wt%Pd-5 wt%Ag/Al_2O_3 catalyst with higher activity than the 5 wt%Ag/Al_2O_3 catalyst. The authors attributed this increase in NO conversion to the presence of enolic species of the type [$H_2C = CH - O - M$], which comes from the partial oxidation of C_3H_6 and are highly reactive with NOx adsorbed forming –NCO and –CN. The authors proposed a new reaction mechanism different to that proposed by Burch et al. [2], which helps us explain our results.

It was observed in our catalysts that the 1PtAg/AW catalyst with a high Pt content did not show a high conversion of NO, nor a high dispersion of Pt (Table 1), which suggests that a large part of the total Pt was diluted or covered by Ag atoms, despite showing high H_2 consumption values by TPR in Figure 2. It is probable that the addition of Pt (and Pd) to the Ag particles can produce changes of an electronic and superficial type that favor the formation of enolic structures, as demonstrated by Wang et al. [61].

2.2.3. H_2 Assisted SCR of NO on PtAg Catalysts

The conversion values of NO versus the reaction temperature in the bimetallic PtAg catalysts were dependent on the Pt concentration (Figure 9). The 0.1PtAg/AW catalyst (Figure 9a) with the lowest concentration of Pt showed two high conversion regions (110–180 °C and 180–500 °C), having a maximum conversion of 80% (at 120 °C) and 90% (at 250 °C). The conversion-temperature profile was similar to that shown by the 2Ag/AW catalyst in the presence of H_2 (Figure 7b).

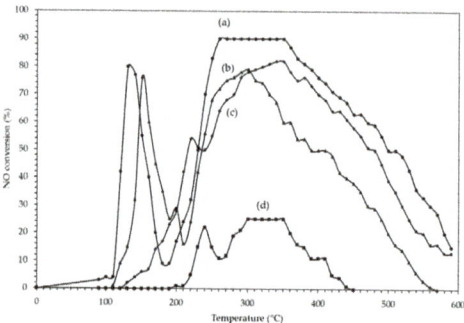

Figure 9. Bimetallic catalyst of PtAg for the C_3H_8–SCR–NO: (**a**) 0.1PtAg/AW; (**b**) 0.25PtAg/AW; (**c**) 0.4PtAg/AW; (**d**) 1PtAg/AW in the presence of H_2. Inlet gas composition: 500 ppm NO, 625 ppm C_3H_8, 200 ppm CO, 660 ppm H_2, 2 wt% O_2, N_2 balance; GHSV = 128,000 h^{-1}.

The addition of Pt to the 2Ag/AW catalyst showed better conversions and a full temperature window and was also more active at low temperatures (between 100 to 180 °C). These advantages have also been reported in the literature by Gunnarsson et al. [37]. These authors used lower Pt concentrations and reported as optimal a catalyst with 2 wt%Ag plus 500 ppm Pt, which showed the highest activity at low temperature.

The authors attributed this behavior to an ability to adsorb hydrocarbons and partially oxidize them on the surface of bimetallic Pt-Ag particles. The presence of Pt could produce a lower barrier for dissociative hydrocarbon adsorption as well as a change in oxidation potential, which in turn, could be attributed to the Pt doping.

The NO conversions of the catalyst 0.25PtAg/AW (Figure 9b) were 78% (at 150 °C) and 78% (at 300 °C), while when we increased the Pt load in the 0.4PtAg/AW catalyst (Figure 9c), there was perhaps a drop in the NO conversion to 54% (at 220 °C) and 80% (at 330 °C). Finally, when we increased the Pt concentration to 1% with the 1PtAg/AW catalyst (Figure 9d), we observed a low conversion of 22% (at 240 °C) and 25% (at 300 °C).

Additionally, in the low temperature region (100–180 °C), the catalyst with 0.1PtAg/AW showed a maximum conversion of 82% versus the 64% conversion of the 2Ag/AW catalyst (a difference of 18%) (Figure 10a). In the high temperature region (250 to 360 °C), the Pt catalyst showed a maximum conversion of 90% compared to a maximum conversion of 65% without Pt. With these results, it was demonstrated that the addition of Pt in low concentrations improved the activity of the 2Ag/AW catalyst.

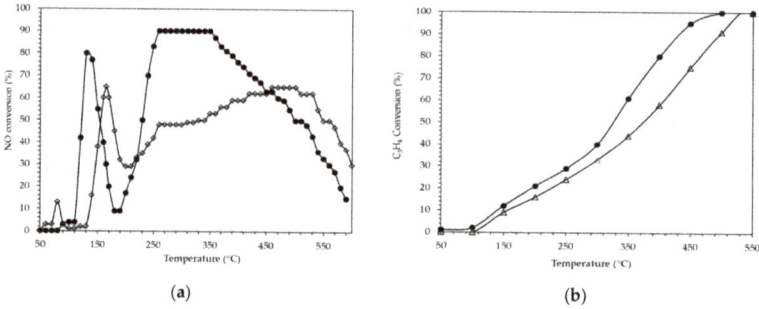

Figure 10. NO and C_3H_8 conversion in the reaction C_3H_8–SCR–NO. (**a**) NO conversion with (●) 0.1PtAg/AW and (◊) 2Ag/AW; (**b**) C_3H_8 Conversion with (●) 0.1PtAg/AW and (Δ) 2Ag/AW catalysts. Inlet gas composition: 500 ppm NO, 625 ppm C_3H_8, 200 ppm CO, 660 ppm H_2, 2 vol.% O_2, N_2 balance; GHSV = 128,000 h^{-1}.

During the combustion of C_3H_8 in the presence of Ag or (AgPt), we observed the effect of Pt in the presence of Ag clusters (Figure 10b). The C_3H_8 conversion for the 0.1PtAg/AW catalyst was found to be higher than the conversion of the catalyst containing only Ag (2Ag/AW). This experimental fact was also reported by Gunnarsson et al. [37]. The presence of Pt in the Ag/Al$_2$O$_3$ catalyst modified the C_3H_8 conversion, (Figure 10b) presenting two regions, with a turning point of the C_3H_8 conversion at 300 °C, which was related to the NOx reduction activity where two conversion regions were also observed.

The main factor that allows for the increase in NOx reduction at low temperatures is related to the increase in C_3H_8 chemisorption due to the presence of small amounts of Pt on the Ag particles and its corresponding oxidation from NO to NO_2.

That is, the Pt–Ag catalyst appears to be able to convert more of the C_3H_8 species available on its surface. This phenomenon could be related to a modification of the metallic Ag particles, in terms of both the surface structure and the bulk of their crystal lattice.

Thus, for example, Bordley and El Sayed, [62] prepared Pt–Ag catalysts for electrochemical reactions where the formation of Pt–Ag nano-boxes was reported. They found that the bimetallic particles showed an expanded atomic mesh compared to the Pt atomic mesh. This resulted in the Pt–Ag particles with the least amount of Pt showing improved catalytic activity in the O_2 reduction reaction due to a higher binding energy of Pt, which in turn favored an advantageous change in the decrease in the adsorption energy of the intermediates containing oxygen on the surface of these alloyed particles.

According to Gunnarsson et al. [37] the addition of Pt to its Ag/Al$_2$O$_3$ catalysts produced a greater adsorption of hydrocarbon and with this, an increase in the reduction of NOx at low temperatures. The initial step in the activation of hydrocarbons (such as C_3H_8) is known to be the chemisorption and dissociation of a hydrogen atom on the surface of Pt [57].

The dissociation energy of different molecules (or hydrocarbons) present on the surfaces of Ag and Pt is always lower for Pt [49].

During the reaction experiments, the conversion of CO increased from low temperatures to about 375 °C (Figure 11). It was observed that the 0.1PtAg/AW catalyst showed slightly higher conversion than the 2Ag/AW catalyst. These results are similar to those obtained by Shang et al. [36] in terms of CO conversion versus temperature. The authors showed a steep 95% conversion at 200 °C for a 5% Ag/Al$_2$O$_3$ catalyst. The combustion of both the CO fed to the reactor, and the CO from the combustion of C_3H_8 in our case, was carried out in the temperature range from 150 °C to less than 400 °C. Additionally, our results of CO coincided with the study by Gunnarsson et al. [37], in which they found a very low concentration of CO (<5 ppm) at temperatures of 350 °C at the outlet of their reactor for their PtAg/Al$_2$O$_3$ catalyst.

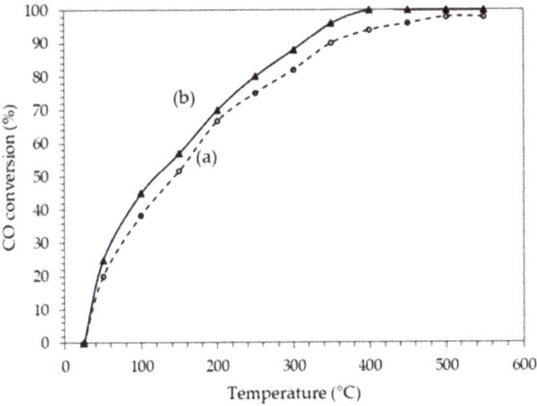

Figure 11. CO conversion of catalysts: (a) 2Ag/AW and (b) 0.1PtAg/AW. Inlet gas composition: 500 ppm NO, 625 ppm C_3H_8, 200 ppm CO, 660 ppm H_2, 2 vol.% O_2, N_2 balance; GHSV = 128,000 h^{-1}.

The emissions of the nitrogen compounds as a function of the reaction temperature for the 0.1PtAg/AW catalyst are shown in Figure 12. It can be observed that NO_2 formed in the interval between 150 to 400 °C (with a volcano-like profile), as has been reported by other authors in the case of Ag/Al_2O_3 catalysts [3,30,50] or also $PtAg/Al_2O_3$ [37]. In the latter case, the authors reported the presence of NO_2 between 250 to 350 °C.

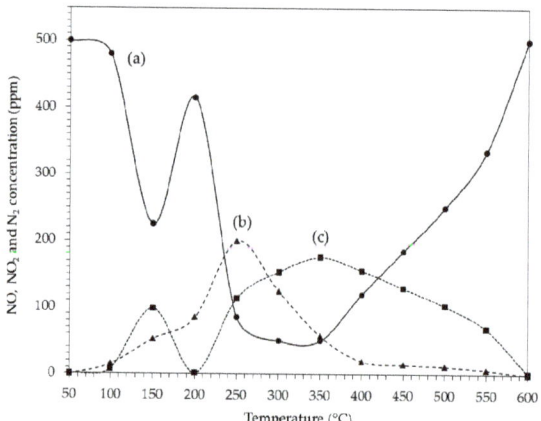

Figure 12. Composition of (a) NO; (b) NO_2 and (c) N_2 (calculated) in the microreactor as a function of temperature during the C_3H_8–SCR–NO with the catalyst 0.1PtAg/AW in the presence of H_2. Inlet gas composition: 500 ppm NO, 625 ppm C_3H_8, 200 ppm CO, 660 ppm H_2, 2vol.% O_2, N_2 balance; GHSV = 128,000 h^{-1}.

Between 200 and 350 °C, the NO_2 that did not react for the formation of nitrates was being desorbed, as has been reported in the literature [26]. The oxidation of NO with O_2 to NO_2 is lower than the overall conversion that occurs in the presence of hydrocarbon and NO to N_2 [62].

Based on the mass balance of nitrogen compounds at the inlet and outlet of the micro reactor, and Equation (1) proposed by Richter et al. [48], the concentration of N_2 at the outlet of the reactor (Figure 12) can be calculated as follows:

$$[N_2] = \frac{1}{2} [[NO]_o - [NO] - [NO_2] - 2[N_2O]] \tag{1}$$

where [N_2] is the calculated concentration of N_2 (mol/L); [NO]$_o$ is the initial concentration of NO (mol/L); [NO] is the present concentration of NO (mol/L); [NO_2] is the present concentration of NO_2 (mol/L); and [N_2O] is the present concentration of N_2O (mol/L).

The decrease in NO_2 at 250 °C is related to the partial oxidation of C_3H_8, as can be seen in Figure 10b. The formation of N_2 showed a maximum at 350 °C and then decreased as a result of the parallel and series reactions that were being carried out.

These bimetallic catalysts produced N_2O at several temperatures. The results of the formation of N_2O in our study are shown in Table 2, and expressed in Y_{N2O} yield, which is defined as $Y_{N2O} = 2 \times [N_2O]/[NO]_o \times 100$. These yields ($Y_{N2O}$) are reported for two temperatures: 200 °C and 400 °C.

Table 2. N_2O yield (%) of the PtAg/AW catalysts in C_3H_8–SCR of NO with H_2.

Catalyst	Pt/W Atomic Ratio	Y_{N2O}(%) at T(°C)	
		200	400
Ag/Al$_2$O$_3$	0	7	5
0.1PtAg/AW	0.027	8.5	6.5
0.25PtAg/AW	0.069	9	6.5
0.4PtAg/AW	0.11	12	7
1PtAg/AW	0.27	14	6

The results at 200 °C showed that the 2Ag/AW catalyst produced a 7% yield to N_2O while the 0.1PtAg/AW catalyst showed a 8.5% yield at the same temperature. For the 0.25Pt/AW catalyst, the N_2O yield was 9%, which was very similar to that observed for the 0.1PtAg/AW catalyst. In the case of 0.4PtAg/AW and 1PtAg/AW catalysts, the yields were 12.5 and 14%, respectively.

We observed that the formation of N_2O was greater at 200 °C than at 400 °C, as has been found by Shaieb et al. [25]. It was also observed that the formation of the Pt–Ag bimetallic was so strong, probably in the form of the Pt–Ag alloy [54], that Ag decreased the selectivity of Pt to produce N_2O, as observed by Gunnarsson et al. [37] when studying catalysts of 2%Ag–0.05%Pt/Al$_2$O$_3$ in this reaction.

In the case of the 2%Ag/Al$_2$O$_3$ catalyst (without Pt), other authors such as Meunier et al. [14] reported an 8% yield of N_2O at 450 °C; Richter et al. [48] reported a 2.25% N_2O yield at 267 °C; and Shaieb et al. [25] reported a 6% N_2O at 200 °C. Hernandez and Fuentes [30] reported 1% N_2O, Kannisto et al. [3] reported 1.6% at 250 °C, and finally Iglesias-Juez et al. [18] reported 12.5%.

2.2.4. Effect of H_2O on H_2–C_3H_8–SCR of NO

The effect of water in this reaction was studied with the catalyst 2 wt%Ag/γ-Al$_2$O$_3$ [50] by adding 6%vol. H_2O. It was found that the addition of H_2O decreased the NO conversion (9% on average) at 150 °C compared to the NO conversion without H_2O, but increased it (8.4% on average) in the temperature range of 200 to 360 °C. The opposite happened with the C_3H_8 conversion. In this case, the addition of H_2O improved the conversion (6% on average) compared with the C_3H_8 conversion without H_2O in the range of 250 to 450 °C.

The presence of water vapor could decrease the concentration of carbonaceous deposits that block adsorption sites on the catalyst surface [2], which could largely explain the increase in C_3H_8 conversion. The authors studied the SCR of NO with 10 vol.%H_2O on an In$_2$O$_3$/Ga$_2$O$_3$/Al$_2$O$_3$ catalyst. The authors mentioned that the presence of water vapor partially inhibited the non-selective combustion of C_3H_8 with O_2. As a result, more hydrocarbons could be available for the SCR reaction resulting in increased activity and selectivity for this reaction.

This behavior has already been reported when small hydrocarbons such as C_3H_8 react. It has been mentioned that one possible reason is the lower enthalpy of adsorption of the short alkanes compared to the large chain alkanes.

3. Materials and Methods

3.1. Preparation of Catalysts

The synthesis of Al$_2$O$_3$ (A) was carried out by the precipitation of a 0.44 mg/mL solution of Al(NO$_3$)$_3$·9H$_2$O (Fermont, Mexico) to which a solution of NH$_4$OH at 30 vol.% was added dropwise (JT Baker, USA) under stirring, until a boehmite suspension with a pH of 9–10 was obtained, which was left to stand for 12 h. The solid was filtered, dried at 110 °C for 24 h, and then calcined at 500 °C for 6 h.

The catalyst containing 0.4 wt%Pt/Al$_2$O$_3$ (0.4Pt/A) was prepared by the incipient wetness impregnation method using 52.8 mL of an H$_2$PtCl$_6$ solution (Aldrich, USA) with a concentration of 0.38 mgPt/mL on 5 g of Al$_2$O$_3$. The impregnation started with a pH of 2.5 at 60 °C for 2 h, then dried at 110 °C for 12 h, and finally calcined at 500 °C for 6 h.

The synthesis of the Al$_2$O$_3$ support promoted with WOx(AW) was carried out with the same procedure as the Al$_2$O$_3$ (A) synthesis by adding the required amount of (NH$_4$)$_{12}$W$_{12}$O$_{40}$·5H$_2$O (Aldrich, USA) to obtain a nominal content of 0.5 wt%W during precipitation.

The 0.4 wt%Pt/Al$_2$O$_3$–WOx (0.4Pt/AW) catalyst was prepared using the same H$_2$PtCl$_6$ incipient wetness impregnation method used in the 0.4Pt/Al$_2$O$_3$ catalyst.

The 2 wt%Ag/Al$_2$O$_3$–WOx (2Ag/AW) catalyst was also obtained by impregnation by the incipient wetness method using 25 mL of a AgNO$_3$ solution (Aldrich, USA) containing 4 mgAg/mL on 5 g of Al$_2$O$_3$–WOx (AW) at 60 °C for 2 h. The solid was dried at 110 °C for 12 h and finally calcined in air at 500 °C for 6 h.

The Pt–Ag/Al$_2$O$_3$–WOx (PtAg/AW) bimetallic catalysts were prepared with the same incipient wetness impregnation method used for the preparation of 0.4Pt/AW and 2Ag/AW monometallic catalysts; however, they were impregnated sequentially, starting with the impregnation of Pt and then the Ag to achieve strong contact between the two metals [53]. During the impregnation, 13.15, 32.89, 52.8, and 131.56 mL of H$_2$PtCl$_6$ solution in 5 g of Al$_2$O$_3$-WOx (pH = 2.5) were used to obtain solids with a Pt content of 0.1, 0.25, 0.4, and 1 wt% Pt, respectively. The solids were dried at 110 °C for 12 h and calcined at 500 °C for 6 h. Subsequently, 5 g of the calcined solids were impregnated with 25 mL of a solution of AgNO$_3$ (4 mgAg/mL) (Aldrich, USA) to obtain a concentration of 2 wt%Ag. This solution was soaked at 60 °C for 2 h, dried at 110 °C for 12 h, and calcined at 500 °C for 6 h. All catalysts were reduced in H$_2$ flow (30 cm^3/min) at 500 °C for 2 h.

3.2. Catalyst Characterization

The catalysts were characterized by adsorption-desorption of N$_2$. The measurement of the isotherms was carried out on ASAP-2460 Version 2.01 (Micromeritics, Norcross, GA, USA) equipment. The samples received pretreatment in a vacuum (1 × 10^{-4} Torr) at 300 °C for 14 h; after that, physisorption with N$_2$ was performed at −196 °C (77 K). The BET and BJH methods were used to determine the specific area, diameter, and pore volume.

The crystalline phases were obtained in a Bruker diffractometer (D8FOCUS) (Bruker, Karlsruhe, Germany) operated at 35 kV and 25 mA using Cu Kα radiation (λ = 0.154 nm) at a goniometer speed of 2°/min with a sweep of 10° $\leq 2\theta \leq$ 100°. A special detector called "Lynx Eyes" was used. The identification of the different crystalline phases was compared with the data from the corresponding JCPDS diffraction cards.

Temperature programmed reduction (TPR) profiles of the calcined Pt/Al$_2$O$_3$ samples were obtained under H$_2$ flow (10 vol.%H$_2$/Ar) by using a commercial thermodesorption apparatus (multipulse RIG model, from ISRI) equipped with a thermal conductivity detector (TCD). Samples of 30 mg and a gas flow rate of 25 cm^3/min were used in the experiments. The TPR profiles were registered by heating the sample from 25 to 600 °C at a rate of 10 °C/min, and a TCD monitored the rate of H$_2$ consumption. The amount of H$_2$ consumed was obtained by the deconvolution and integration of the TPR peaks using the Peak Fit program. The calibration was done by measuring the change in weight due to a reduction in H$_2$ of 2 mg of CuO using an electrobalance Cahn-RG. The TPR signal of CuO was made and correlated with the stoichiometric H$_2$ consumption.

Chemisorption measurements of H$_2$ were performed using a conventional volumetric glass apparatus (base pressure 1 × 10^{-5} Torr). The amount of chemisorbed H$_2$ was determined from adsorption isotherms measured at room temperature (25 °C). In a typical experiment, the catalysts (0.5 g) were reduced in H$_2$ at 500 °C for 1 h, then evacuated at the same temperature for 2 h and cooled down under vacuum to 25 °C. After that, the first adsorption isotherm was measured. The catalyst was then evacuated to 1 × 10^{-5} Torr for 30 min at 25 °C to remove the physisorbed species and back-sorption isotherm. The linear parts of the isotherms were extrapolated to zero pressure. The subtraction of the two isotherms gave the amount of H$_2$ strongly chemisorbed on metal particles. These values were then used to calculate the Pt dispersion (H/Pt ratio). In preliminary experiments, it was found that chemisorptions of hydrogen on the Al_2O_3 support were negligible at 25 °C. The uncertainty of the reported uptakes was ±0.45 µmol H_2/g_{cat}.

The materials' microstructure images were taken by scanning electron microscopy (SEM) with field emission and high resolution in a Joel microscope (model JFM-6701-F) (JEOL Ltd., Tokyo, Japan) using secondary electrons. The qualitative and quantitative chemical analyses and their corresponding images were obtained by attaching an x-ray energy dispersion spectroscopy (EDS) probe to the microscope.

Ex situ UV–Visible spectra of the powder samples calcined at 500 °C were obtained with a UV–Vis spectrophotometer (GBC model Cintra 20) (GBC Scientific Equipment, Braeside, Australia) with a wavelength of 200 to 800 nm under ambient conditions.

In situ UV–Visible spectra of the powder 0.4Pt/AW catalyst calcined at 500 °C were collected using an Agilent Cary 5000 spectrometer (Agilent Technologies Inc., Santa Clara, CA, USA) equipped with a Harrick Praying Mantis. During the experiment, 50 mg of the sample was packed in the sample holder. The spectra were recorded in the range of 200–1000 nm with a resolution of 0.1 nm each 100 °C from 25 to 500 °C in a gas mixture of 5 vol.%H_2/N_2, with a flow of 0.5 cm^3/s.

3.3. Catalytic Evaluation

The catalytic activity for C_3H_8–SCR was carried out in a quartz microreactor with a diameter of 10 mm, which was operated under kinetic conditions where the phenomena of mass transfer by internal and external diffusion were minimized, for which the value of the Weisz and Prater criterion was calculated [63–65] (Supplementary Material SI.2).

The effect of the reducing agent, H_2, was investigated at a gas hourly space velocity (GHSV) of 128,000 h^{-1}. The flow used was 400 cm^3/min and a sample of 150 mg was placed (100 US mesh, 0.148 mm). Prior to each test, the samples were pretreated in a flow of 20 cm^3/min with a mixture containing 13.2 vol.% H_2 with N_2 at 500 °C for 2 h.

The temperature of reaction increased from 50 to 600 °C at 5 °C/min. The feed mixture to the microreactor was: 500 ppm of NO, 625 ppm of C_3H_8, 200 ppm of CO, 660 ppm of H_2, 2 vol.%O_2, and N_2 (balance). The pure gases were chromatographic grade O_2, H_2, and N_2 (99.90% Infra S.A.) and a prepared mixture of NO, CO, and propane (Praxair, USA). The concentration of the gases in the feed to the microreactor varied according to the reaction experiment (C_3H_8–SCR or H_2–C_3H_8–SCR). The gas flows were controlled by mass flow controllers (AALBORG, Repeatability: ±0.25% of full scale) and the inlet and outlet gas composition was analyzed as described below.

The C_3H_8 was analyzed in a gas chromatograph (Gow-Mac, 550, GOW-MAC Instrument Company, Bethlehem, PA, USA) with a flame ionization detector and six feet packed column using a stationary phase, tri-cresyl phosphate in bentonite-34. The analysis of NO and NO_2 was performed by chemiluminescence with a Rosemount Analytical analyzer (Model 951A NO/NOx Analyzer, Rosemount Analytical Inc., Anaheim, CA, USA). The N_2O was analyzed in another gas chromatograph (Gow-Mac, 580) with a thermal conductivity detector using He as the carrier gas (52 cm^3/min) with a 10 feet packed column of Porapak Q at a temperature of 50 °C (note that this column can also analyze air, NO, NO_2, N_2O, CO_2 and H_2O). Bridge current: 150 mA, Response time: 0.5 s, noise: 10 µVmax. (within operating parameters), drift: 40 µV/hour max. The repeatability of the experiments carried out was 0.71, which means that it represents a moderate repeatability in the measurement of experiments according to David G.C. [66], and the absolute error of ±5 ppm N_2O. The average retention times were: 0.62 min (Air), 0.75 min (NO), 1.1 min (NO_2), 2.1 min (CO_2), 2.75 min (N_2O), and 8.25 min (H_2O). The analysis of CO and H_2 was made in a Gow-Mac, 580 gas chromatograph with a TCD detector using He as the carrier gas and a 13× molecular sieve packed column (1/8 in × 8 ft).

The experiments of the combustion of C_3H_8 (see Supplementary Material SI.1 and Appendix A) were made in a flow microreactor connected in-line to a gas chromatograph (Gow-Mac, 550) with a flame ionization detector and six feet packed column using a stationary phase, tri-cresyl phosphate in bentonite-34. The propane composition was 999 ppm in dry air (Linde). The total pressure within the reaction system remained constant at 590 Torr and the gas flow used in all of the experiments was

300 cm^3/min. The amounts of catalyst evaluated were 20 mg. The samples were evaluated by scanning temperatures from room temperature to 500 °C at a fixed time of 90 min. The deactivation tests were carried out at the same final temperature and for 180 min, keeping all other variables constant.

The equations to calculate the conversion of NO(X) and the yield to a product (Yi) are as follows:

$$X = ([NO]_o - [NO])/[NO]_o \times 100 \qquad (2)$$

$$Y_{N2O} = 2 \times [N_2O]/[NO]_o \times 100 \qquad (3)$$

4. Conclusions

The addition of 0.1 wt%Pt to the 2 wt%Ag/Al$_2$O$_3$–WOx catalyst improved the C$_3$H$_8$–SCR of NO assisted by H$_2$ and widened the range of conversions with respect to the reaction temperature.

Bimetallic PtAg particles were formed, having a strong contact between the metals, and had the capacity of adsorbing H$_2$. Pt dispersion was more significant in the particles with a lower concentration of Pt, and the Ag monometallic catalyst did not show H$_2$ chemisorption.

After reduction with H$_2$, Ag and PtAg particles were obtained in all the bimetallic catalysts. It appears that the bimetallic PtAg particles have adsorption properties that explain the differences with the Ag/Al$_2$O$_3$ catalysts.

Utilizing ex situ UV–Vis spectroscopy, species such as Ag$^+$ and Ag$_n^{\delta+}$ were found as well as Ag^0 nanoparticles after SCR of NO with C$_3$H$_8$ and H$_2$ in a wide temperature range. By SEM, mainly spherical clusters of small particles of less than 10 nm were found in the calcined Pt catalyst, which was probably related to the presence of Ag$_2$O. The distribution of particle diameters indicated the predominance (31.64%) of particles of 10 nm or less.

The C$_3$H$_8$–SCR of NO from the PtAg/Al$_2$O$_3$ bimetallic catalysts promoted with WOx was considerably improved by adding H$_2$ to the combustion gases due to the formation of Ag clusters, (or PtAg clusters), enolic species, and the decrease in nitrate self-poisoning, which is a stage before the formation of N-containing species.

The addition of 0.5 wt%W to the Al$_2$O$_3$ and the Pt/Al$_2$O$_3$-WOx catalysts stabilized them in the propane combustion reaction.

Supplementary Materials: The following are available online at http://www.mdpi.com/2073-4344/10/10/1212/s1, Figure S1: C$_3$H$_8$ combustion in air over the (●) 0.4Pt/A; (□) 0.4Pt/AW catalysts reduced to 500 and 800°C. (a) Effect of reduction temperature on catalyst 0.4Pt/A without W; (b) Effect of reduction temperature on catalyst 0.4Pt/AW with W; (c) Effect of WOx during catalyst deactivation 0.4Pt/A without W and 0.4Pt/AW with W reduced to 500°C and (d) Effect of WOx during deactivation of 0.4Pt/A catalysts without W and 0.4Pt/AW with W reduced to 800°C. Evaluation conditions: 300cm^3/min of air mixture flow plus 999 ppm of C$_3$H$_8$, catalyst weight: 20 mg, Table S1: Microreactor operating conditions to evaluate the SCR of NO at 250°C (average temperature).

Author Contributions: Conceptualization, J.L.C., G.A.F., N.N.G.H., M.E.H.-T., and M.P.; Methodology, J.L.C., N.N.G.H., and M.P.; Investigation, G.A.F., N.N.G.H., and M.P.; Resources, J.L.C., B.Z., J.L.F.M., T.V., and J.M.J.; Data curation, J.L.C.; Writing-preparing the initial draft, N.N.G.H.; Critical writing-review and editing, J.L.C., J.S., and B.Z.; Visualization, N.N.G.H.; Supervision, J.L.C.; Project administration, J.L.C.; Procurement of funds, J.L.C. All authors have read and agreed to the published version of the manuscript.

Funding: This research received no external funding.

Acknowledgments: Naomi N. González Hernández thanks CONACYT for the postgraduate scholarship awarded through program 001379. We thank the Catalysis Laboratory, UAM-A, and the Chemical Industry Process Laboratory, UAM-A., J.L. Contreras appreciates the support of the company Synthesis and Industrial Applications, S.A.

Conflicts of Interest: The authors declare no conflict of interest.

Appendix A. Stabilization of Al$_2$O$_3$ and Pt/Al$_2$O$_3$ Catalysts

The addition of tungsten oxides (WOx) to the Al$_2$O$_3$ support means that at low concentrations of W (0.5 wt%W), it is possible to thermally stabilize the Al$_2$O$_3$ structure at high temperatures (600 to 900 °C), as can be seen in Figure A1. The BET area of the Al$_2$O$_3$ samples calcined at 650, 800, and 950 °C was higher when more than 0.5 wt% of W was added. In other words, W acts as a structural promoter of Al_2O_3, allowing the Ag/Al_2O_3 catalyst to more extensively withstand the inevitable thermal sintering at high temperatures during reactions.

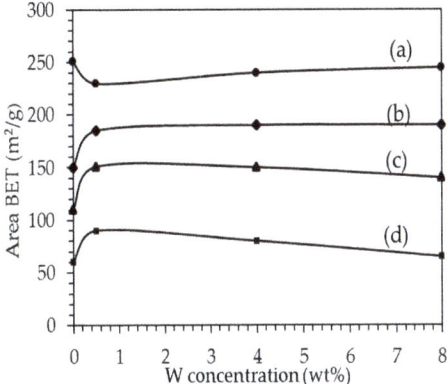

Figure A1. BET area against W concentration (in wt%W) named Al_2O_3–WOx, when the calcination temperature increased: (**a**) 500 °C; (**b**) 650 °C; (**c**) 800 °C; (**d**) 950 °C for 6 h.

On the other hand, we studied the effect of the W/Pt ratio on the Pt's dispersion for catalysts supported in Al_2O_3 [44,45] when they were subjected to reduction in H_2 at 500 °C and 800 °C (Figure A2). It was observed that the catalyst reduced to 800 °C without W (ratio W/Pt = 0) showed a dispersion of 42%, while the catalyst with a W/Pt ratio of 3.28 showed a dispersion of 60%.

However, as Figure A2 shows, as the W/Pt ratio increases, the dispersion of Pt decreases, and it can be observed that this trend is more pronounced in samples reduced to 500 °C than in samples reduced to 800 °C. This behavior suggests that the presence of WOx in the presence of PtOxCly could inhibit the formation of metallic Pt because a higher reduction temperature (800 °C) would be required to obtain a better dispersion of Pt.

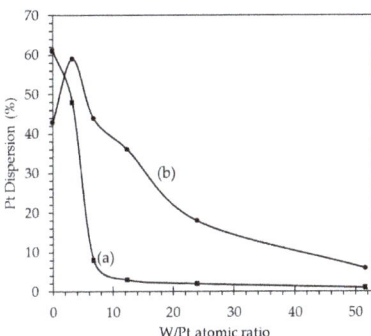

Figure A2. Dispersion of Pt in Pt/AW catalysts increasing the W/Pt atomic ratio when the reduction temperature in H_2 increased from (**a**) 500 to (**b**) 800 °C (for 4 h).

References

1. Twigg, M.V. Progress and future challenges in controlling automotive exhaust gas emissions. *Appl. Catal. B Environ.* **2007**, *70*, 2–15. [CrossRef]
2. Burch, R.; Breen, J.P.; Meunier, F.C. A review of the selective reduction of NOx with hydrocarbons under lean-burn conditions with non-zeolitic oxide and platinum group metal catalysts. *Appl. Catal. B Environ.* **2002**, *39*, 283–303. [CrossRef]
3. Kannisto, H.; Ingelsten, H.H.; Skoglundh, M. Ag–Al$_2$O$_3$ catalysts for lean NOx reduction—Influence of preparation method and reductant. *J. Mol. Catal. A Chem.* **2009**, *302*, 86–96. [CrossRef]
4. Nakatsuji, T.; Yasukawa, R.; Tabata, K.; Ueda, K.; Niwa, M. Catalytic reduction system of NOx in exhaust gases from diesel engines with secondary fuel injection. *Appl. Catal. B Environ.* **1998**, *17*, 333–345. [CrossRef]
5. Miyadera, T.; Yoshida, K. Alumina-supported catalysts for the selective reduction of nitric oxide by propene. *Chem. Lett.* **1993**, *9*, 1483–1486. [CrossRef]
6. Satokawa, S.; Yamaseki, K.; Uchida, H. Influence of low concentration of SO$_2$ for selective reduction of NO by C$_3$H$_8$ in lean-exhaust conditions on the activity of Ag/Al$_2$O$_3$ catalyst. *Appl. Catal. B Environ.* **2001**, *34*, 299–306. [CrossRef]
7. Kyriienko, P.; Popovych, N.; Soloviev, S.; Orlyk, S.; Dzwigaj, S. Remarkable activity of Ag/Al$_2$O$_3$/cordierite catalysts in SCR of NO with ethanol and butanol. *Appl. Catal. B Environ.* **2013**, *140–141*, 691–699. [CrossRef]
8. Popovych, N.O.; Soloviev, S.O.; Orlyk, S.M. Selective reduction of nitrogen oxides (NOx) with oxygenates and hydrocarbons over bifunctional, silver–alumina catalysts: A review. *Theor. Exp. Chem.* **2016**, *52*, 133–151. [CrossRef]
9. Männikkö, M.; Wang, X.; Skoglundh, M.; Härelind, H. Characterization of the active species in the silver/alumina system for lean NOx reduction with methanol. *Catal. Today* **2016**, *267*, 76–81. [CrossRef]
10. Deng, H.; Yu, Y.; He, H. Adsorption states of typical intermediates on Ag/Al$_2$O$_3$ catalyst employed in the selective catalytic reduction of NOx by ethanol. *Chin. J. Catal.* **2015**, *36*, 1312–1320. [CrossRef]
11. Bethke, K.A.; Kung, H.H. Supported Ag catalysts for the lean reduction of NO with C$_3$H$_6$. *J. Catal.* **1997**, *172*, 93–102. [CrossRef]
12. Hoost, T.E.; Kudla, R.J.; Collins, K.M.; Chattha, M.S. Characterization of Ag/γ-Al$_2$O$_3$ catalysts and their lean-NOx properties. *Appl. Catal. B Environ.* **1997**, *13*, 59–67. [CrossRef]
13. Kung, M.C.; Kung, H.H. Lean NOx catalysis over alumina-supported catalysts. *Top. Catal.* **2000**, *10*, 21–26. [CrossRef]
14. Meunier, F.C.; Breen, J.P.; Zuzaniuk, V.; Olsson, M.; Ross, J.R.H. Mechanistic aspects of the selective reduction of NO by propene over alumina and silver-alumina catalysts. *J. Catal.* **1999**, *187*, 493–505. [CrossRef]
15. Martínez-Arias, A.; Fernández-García, M.; Iglesias-Juez, A.; Anderson, J.A.; Conesa, J.C.; Soria, J. Study of the lean NOx reduction with C$_3$H$_6$ in the presence of water over silver/alumina catalysts prepared from inverse microemulsions. *Appl. Catal. B Environ.* **2000**, *28*, 29–41. [CrossRef]
16. Shimizu, K.I.; Shibata, J.; Yoshida, H.; Satsuma, A.; Hattori, T. Silver-alumina catalysts for selective reduction of NO by higher hydrocarbons: Structure of active sites and reaction mechanism. *Appl. Catal. B Environ.* **2001**, *30*, 151–162. [CrossRef]
17. Bogdanchikova, N.; Meunier, F.C.; Avalos-Borja, M.; Breen, J.P.; Pestryakov, A. On the nature of the silver phases of Ag/Al$_2$O$_3$ catalysts for reactions involving nitric oxide. *Appl. Catal. B Environ.* **2002**, *36*, 287–297. [CrossRef]
18. Iglesias-Juez, A.; Hungría, A.B.; Martínez-Arias, A.; Fuerte, A.; Fernández-García, M.; Anderson, J.A. Nature and catalytic role of active silver species in the lean NOx reduction with C$_3$H$_6$ in the presence of water. *J. Catal.* **2003**, *217*, 310–323. [CrossRef]
19. Arve, K.; Capek, L.; Klingstedt, F.; Eränen, K.; Lindfors, L.E.; Murzin, D.Y. Preparation and characterization of Ag/alumina catalysts for the removal of NOx emissions under oxygen rich conditions. *Top. Catal.* **2004**, *30*, 91–95. [CrossRef]
20. Mrabet, D.; Manh-Hiep, V.; Kaliaguine, S.; Trong-On, D. A new route to the shape-controlled synthesis of nano-sized γ-alumina and Ag/γ-alumina for selective catalytic reduction of NO in the presence of propene. *J. Colloid Interface Sci.* **2017**, *485*, 144–151. [CrossRef]
21. Keshavaraja, A.; She, X.; Flytzani-Stephanopoulos, M. Selective catalytic reduction of NO with methane over Ag-alumina catalysts. *Appl. Catal. B Environ.* **2000**, *27*, L1–L9. [CrossRef]

22. Shimizu, K.; Satsuma, A.; Hattori, T. Catalytic performance of Ag–Al$_2$O$_3$ catalyst for the selective catalytic reduction of NO by higher hydrocarbons. *Appl. Catal. B Environ.* **2000**, *25*, 239–247. [CrossRef]
23. She, X.; Flytzani-Stephanopoulos, M. The role of Ag–O–Al species in silver–alumina catalysts for the selective catalytic reduction of NOx with methane. *J. Catal.* **2006**, *237*, 79–93. [CrossRef]
24. Luo, Y.; Hao, J.; Hou, Z.; Fu, L.; Li, R.; Ning, P.; Zheng, X. Influence of preparation methods on selective catalytic reduction of nitric oxides by propene over silver–alumina catalyst. *Catal. Today* **2004**, *93–95*, 797–803. [CrossRef]
25. Chaieb, T.; Delannoy, L.; Costentin, G.; Louis, C.; Casale, S.; Chantry, R.L.; Li, Z.Y.; Thomas, C. Insights into the influence of the Ag loading on Al$_2$O$_3$ in the H$_2$-assisted C$_3$H$_6$-SCR of NOx. *Appl. Catal. B Environ.* **2014**, *156–157*, 192–201. [CrossRef]
26. Tamm, S.; Vallim, N.; Skoglundh, M.; Olsson, L. The influence of hydrogen on the stability of nitrates during H$_2$-assisted SCR over Ag/Al$_2$O$_3$ catalysts—A DRIFT study. *J. Catal.* **2013**, *307*, 153–161. [CrossRef]
27. Azis, M.M.; Härelind, H.; Creaser, D. On the role of H$_2$ to modify surface NOx species, over Ag–Al$_2$O$_3$ as lean NOx reduction catalyst: TPD and DRIFTS studies. *Catal. Sci. Technol.* **2015**, *5*, 296–309. [CrossRef]
28. Azis, M.M.; Härelind, H.; Creaser, D. Kinetic modeling of H$_2$-assisted C$_3$H$_6$ selective catalytic reduction of NO over silver alumina catalyst. *Chem. Eng. J.* **2015**, *278*, 394–406. [CrossRef]
29. Singh, P.; Yadav, D.; Thakur, P.; Pandey, J.; Prasad, R. Studies on H$_2$-Assisted Liquefied Petroleum Gas Reduction of NO over Ag/Al$_2$O$_3$ Catalyst. *Bull. Chem. React. Eng. Catal.* **2018**, *13*, 227–235.
30. Hernández-Terán, M.E.; Fuentes, G.A. Enhancement by H$_2$ of C$_3$H$_8$-SCR of NOx using Ag/γ-Al$_2$O$_3$. *Fuel* **2014**, *138*, 91–97. [CrossRef]
31. Ström, L.; Carlsson, P.-A.; Skoglundh, M.; Härelind, H. Surface Species and Metal Oxidation State during H$_2$-Assisted NH$_3$-SCR of NOx over Alumina-Supported Silver and Indium. *Catalysts* **2018**, *8*, 38. [CrossRef]
32. Xu, G.; Ma, J.; Wang, L.; Lv, Z.; Wang, S.; Yu, Y.; He, H. Mechanism of the H$_2$ Effect on NH$_3$-Selective Catalytic Reduction over Ag/Al$_2$O$_3$: Kinetic and Diffuse Reflectance Infrared Fourier Transform Spectroscopy Studies. *ACS Catal.* **2019**, *9*, 10489–10498. [CrossRef]
33. Barreau, M.; Tarot, M.-L.; Duprez, D.; Courtois, X.; Can, F. Remarkable enhancement of the selective catalytic reduction of NO at low temperature by collaborative effect of ethanol and NH$_3$ over silver supported catalyst. *Appl. Catal. B Environ.* **2018**, *220*, 19–30. [CrossRef]
34. Pihl, J.A.; Toops, T.J.; Fisher, G.B.; West, B.H. Selective catalytic reduction of nitric oxide with ethanol/gasoline blends over a silver/alumina catalyst. *Catal. Today* **2014**, *231*, 46–55. [CrossRef]
35. Wu, S.; Li, X.; Fang, X.; Sun, Y.; Sun, J.; Zhou, M.; Zang, S. NO reduction by CO over TiO$_2$-γ-Al$_2$O$_3$ supported In/Ag catalyst under lean burn conditions. *Chin. J. Catal.* **2016**, *37*, 2018–2024. [CrossRef]
36. Shang, Z.; Cao, J.; Wang, L.; Guo, Y.; Lu, G.; Guo, Y. The study of C$_3$H$_8$-SCR on Ag/Al$_2$O$_3$ catalysts with the presence of CO. *Catal. Today* **2017**, *281*, 605–660. [CrossRef]
37. Gunnarsson, F.; Kannisto, H.; Skoglundh, M.; Härelind, H. Improved low-temperature activity of silver-alumina for lean NOx reduction—Effects of Ag loading and low-level Pt doping. *Appl. Catal. B Environ.* **2014**, *152–153*, 218–225. [CrossRef]
38. Tamm, S.; Andonova, S.; Olsson, L. The Effect of Hydrogen on the Storage of NOx over Silver, Platinum and Barium Containing NSR Catalysts. *Catal. Lett.* **2014**, *144*, 1101–1112. [CrossRef]
39. Bonet, F.; Grugeon, S.; Herrera Urbina, R.H.; Tekaia-Elhsissen, K.; Tarascon, J.-M. In situ deposition of silver and palladium nanoparticles prepared by the polyol process and their performance as catalytic converters of automobile exhaust gases. *Solid State Sci.* **2002**, *4*, 665–670. [CrossRef]
40. He, H.; Wang, J.; Feng, Q.; Yu, Y.; Yoshida, K. Novel Pd, promoted Ag/Al$_2$O$_3$ catalyst for the selective reduction of NOx. *Appl. Catal. B Environ.* **2003**, *46*, 365–370. [CrossRef]
41. Inderwildi, O.R.; Jenkins, S.J.; King, D.A. When adding an unreactive metal enhances catalytic activity: NOx decomposition over silver–rhodium bimetallic surfaces. *Surf. Sci.* **2007**, *601*, L103–L108. [CrossRef]
42. Lanza, R.; Eriksson, E.; Pettersson, L.J. NOx selective catalytic reduction over supported metallic catalysts. *Catal. Today* **2009**, *147S*, S279–S284. [CrossRef]
43. Schott, F.J.P.; Balle, P.; Adler, J.; Kureti, S. Reduction of NOx by H$_2$ on Pt/WO$_3$/ZrO$_2$ catalysts in oxygen-rich exhaust. *Appl. Catal. B Environ.* **2009**, *87*, 18–29. [CrossRef]

44. Contreras, J.L.; Fuentes, G.A.; García, L.A.; Salmones, J.; Zeifert, B. WOx effect on the catalytic properties of Pt particles on Al$_2$O$_3$. *J. Alloy Compd.* **2009**, *483*, 450–452. [CrossRef]
45. Contreras, J.L.; Fuentes, G.A.; Zeifert, B.; García, L.A.; Salmones, J. Stabilization of Supported Platinum Nanoparticles on γ-Alumina Catalysts by Addition of Tungsten. *J. Alloy Compd.* **2009**, *483*, 371–373. [CrossRef]
46. Muñoz, H.P.; Delmás, R.D. Alumina synthesis from AlCl$_3$ acid solution and XRD characterization. *J. Per. Quim. Ing. Química* **2001**, *4*, 68–71.
47. Aguado, J.; Escola, J.M.; Castro, M.C. Influence of the thermal treatment upon the textural properties of sol–gel mesoporous γ-alumina synthesized with cationic surfactants. *Microporous Mesoporous Mater.* **2010**, *128*, 48–55. [CrossRef]
48. Richter, M.; Bentrup, U.; Eckelt, R.; Schneider, M.; Pohl, M.M.; Fricke, R. The effect of hydrogen on the selective catalytic reduction of NO in excess oxygen over Ag/Al$_2$O$_3$. *Appl. Catal. B Environ.* **2004**, *51*, 261–274. [CrossRef]
49. Vojvodic, A.; Calle-Vallejo, F.; Guo, W.; Wang, S.; Toftelund, A.; Studt, F.; Martínez, J.I.; Shen, J.; Man, J.; Rossmeisl, I.C.; et al. On the behavior of Brønsted-Evans-Polanyi relations for transition metal oxides. *J. Chem. Phys.* **2011**, *134*, 244–509. [CrossRef]
50. Hernández-Terán, M.E. Development of Ag/γ-Al$_2$O$_3$ and Ag/η-Al$_2$O$_3$ Catalytic Systems for H$_2$-Assisted NO C$_3$H$_8$-SCR under Oxidizing Operation for Emission Control Systems for Diesel Engines or Stationary Sources. Ph.D. Thesis, Universidad Autónoma Metropolitana-Iztapalapa, Ciudad de México, Mexico, 2020.
51. Zhang, Q.; Li, J.; Liu, X.; Zhu, Q. Synergetic effect of Pd and Ag dispersed on Al$_2$O$_3$ in the selective hydrogenation of acetylene. *Appl. Catal. A General* **2000**, *197*, 221–228. [CrossRef]
52. Lieske, H.; Lietz, G.; Spindler, H.; Volter, J. Reactions of Platinum in Oxygen- and Hydrogen-Treated Pt/Al$_2$O$_3$ Catalysts. Temperature-Programmed Reduction, Adsorption, and Redispersion of Platinum. *J. Catal.* **1983**, *81*, 8–16.
53. Gauthard, F.; Epron, F.; Barbier, J. Palladium and platinum-based catalysts in the catalytic reduction of nitrate in water: Effect of cooper, silver or gold addition. *J. Catal.* **2003**, *220*, 182–191. [CrossRef]
54. De Jong, K.P.; Bongenaar-Schlenter, B.E.; Meima, G.R.; Verkerk, R.C.; Lammers, M.J.J.; Geus, J.W. Investigations on silica-supported platinum-silver alloy particles by infrared spectra of adsorbed CO and N$_2$. *J. Catal.* **1983**, *81*, 67–76. [CrossRef]
55. Prasad, J.; Murthy, K.R.; Menon, P.G. The stoichiometry of hydrogen-oxygen titrations on supported platinum catalysts. *J. Catal.* **1978**, *52*, 515–520. [CrossRef]
56. Arve, K.; Svennerberg, K.; Klingstedt, F.; Eranen, K.; Wallenberg, L.R.; Bovin, J.O.; Capek, L.; Murzin, D.Y. Structure-Activity Relationship in HC-SCR of NOx by TEM, O$_2$-Chemisorption, and EDXS Study of Ag/Al$_2$O$_3$. *J. Phys. Chem. B* **2006**, *110*, 420–427. [CrossRef]
57. Bligaard, T.; Nørskov, J.K.; Dahl, S.; Matthiesen, J.; Christensen, C.H.; Sehested, J. The Brønsted–Evans–Polanyi relation and the volcano curve in heterogeneous catalysis. *J. Catal.* **2004**, *224*, 206–217. [CrossRef]
58. Lietz, G.; Lieske, H.; Spindler, H.; Hanke, W.; Völter, J. Reactions of Platinum in Oxygen- and Hydrogen-Treated Pt/γ-Al$_2$O$_3$ catalysts, II. Ultraviolet-Visible Studies, Sintering of Platinum, and Soluble Platinum. *J. Catal.* **1983**, *81*, 17–25. [CrossRef]
59. Barton, D.G.; Soled, S.L.; Meitzner, G.D.; Fuentes, G.A.; Iglesia, E. Structural and Catalytic Characterization of Solid Acids Based on Zirconia Modified by Tungsten Oxide. *J. Catal.* **1999**, *181*, 57–72. [CrossRef]
60. Satokawa, S. Enhancing the NO/C$_3$H$_8$/O$_2$ Reaction by Using H$_2$ over Ag/Al$_2$O$_3$ Catalysts under Lean-Exhaust Conditions. *Chem. Lett.* **2000**, *29*, 294–295. [CrossRef]
61. Wang, J.; He, H.; Feng, Q.; Yu, Y.; Yoshida, K. Selective catalytic reduction of NOx with C$_3$H$_6$ over Ag/Al$_2$O$_3$ catalyst with a small quantity of noble metal. *Catal. Today* **2004**, *93–95*, 783–789. [CrossRef]
62. Bordley, J.A.; El-Sayed, M.A. Enhanced Electrocatalytic Activity toward the Oxygen Reduction Reaction through Alloy Formation: Platinum-Silver Alloy Nano cages. *J. Phys. Chem.* **2016**, *120*, 14643–14651. [CrossRef]
63. Fogler, S.H. *Elements of Chemical Reaction Engineering*, 3rd ed.; Prentice Hall Inc.: Upper Saddle River, NJ, USA, 1999; pp. 738–758.
64. Smith, J.F. *Chemical Engineering Kinetics*, 3rd ed.; McGraw-Hill Inc.: New York, NY, USA, 1986; pp. 524–534.

65. Hirshfelder, J.O.; Curtis, C.F.; Bird, R.B. *Molecular Theory of Gases and Liquids*, 1st ed.; Wiley: Hoboken, NJ, USA, 1964; ISBN 0-471-40065-3.
66. David, G.C. Some comments on the repeatability of measurements. *Ringing Migr.* **1994**, *15*, 84–90.

Publisher's Note: MDPI stays neutral with regard to jurisdictional claims in published maps and institutional affiliations.

 © 2020 by the authors. Licensee MDPI, Basel, Switzerland. This article is an open access article distributed under the terms and conditions of the Creative Commons Attribution (CC BY) license (http://creativecommons.org/licenses/by/4.0/).

Article

Promoting Effect of the Core-Shell Structure of MnO$_2$@TiO$_2$ Nanorods on SO$_2$ Resistance in Hg^0 Removal Process

Xiaopeng Zhang, Xiangkai Han, Chengfeng Li, Xinxin Song, Hongda Zhu, Junjiang Bao, Ning Zhang and Gaohong He *

State Key Laboratory of Fine Chemicals, School of Chemical Engineering at Panjin, Dalian University of Technology, Panjin 124221, China; xiaopengzhang@dlut.edu.cn (X.Z.); hanxiangkai@mail.dlut.edu.cn (X.H.); li_cf@mail.dlut.edu.cn (C.L.); songxinxinhuagong@163.com (X.S.); dada@mail.dlut.edu.cn (H.Z.); baojj@dlut.edu.cn (J.B.); zhangning@dlut.edu.cn (N.Z.)
* Correspondence: hgaohong@dlut.edu.cn; Tel.: +86-427-2631916

Received: 5 December 2019; Accepted: 2 January 2020; Published: 3 January 2020

Abstract: Sorbent of αMnO$_2$ nanorods coating TiO$_2$ shell (denoted as αMnO$_2$-NR@TiO$_2$) was prepared to investigate the elemental mercury (Hg^0) removal performance in the presence of SO$_2$. Due the core-shell structure, αMnO$_2$-NR@TiO$_2$ has a better SO$_2$ resistance when compared to αMnO$_2$ nanorods (denoted as αMnO$_2$-NR). Kinetic studies have shown that both the sorption rates of αMnO$_2$-NR and αMnO$_2$-NR@TiO$_2$, which can be described by pseudo second-order models and SO$_2$ treatment, did not change the kinetic models for both the two catalysts. In contrast, X-ray photoelectron spectroscopy (XPS) results showed that, after reaction in the presence of SO$_2$, S concentration on αMnO$_2$-NR@TiO$_2$ surface is lower than on αMnO$_2$-NR surface, which demonstrated that TiO$_2$ shell could effectively inhibit the SO$_2$ diffusion onto MnO$_2$ surface. Thermogravimetry-differential thermosgravimetry (TG-DTG) results further pointed that SO$_2$ mainly react with TiO$_2$ forming Ti(SO$_4$)O in αMnO$_2$-NR@TiO$_2$, which will protect Mn from being deactivated by SO$_2$. These results were the reason for the better SO$_2$ resistance of αMnO$_2$-NR@TiO$_2$.

Keywords: core-shell structure; αMnO$_2$ nanorods; elemental mercury removal; SO$_2$ resistance

---

1. Introduction

The emission of mercury from coal-fired power plants has drawn wide public concern in modern society. Mercury emissions are a long-term threat to human health and the environment because of extreme toxicity, persistence, and bioaccumulation. Therefore, controlling mercury emitted from coal-fired power plants has practical significance. Mercury in coal combustion flue gas is mainly present in three forms: Elemental mercury (Hg^0), oxidized mercury (Hg^{2+}), and particulate-bound mercury (Hg^p). Particulate-bound mercury (Hg^p) can be removed by electrostatic precipitators (ESP) and fabric filters (FF), while oxidized mercury (Hg^{2+}) can be captured by wet flue gas desulfurization system (WFGD). However, existing air pollution control devices can hardly remove Hg^0 due to its high volatility and low solubility.

Hg^0 capture with specific adsorbents is a usual way to control Hg^0 emissions from coal-fired power plants [1]. Activated cabon (AC) has been widely used for the adsorption of Hg^0 in coal-fired flue gas [2,3]. However, a huge amount of AC needs to be injected into flue gas because of its low Hg^0 capture capacity, which leads to a high operating cost of this technology. Sulfur or halogen modification can enhance adsorption ability of AC [4,5]. However, the injected AC is usually captured together with fly ash by particulate control device, and the Hg^0 adsorbed on AC will influence the fly ash utilization [6]. Therefore, alternative economic sorbents with high Hg^0 removal efficiency are necessary.

Oxides, such as CuO_x [7,8], FeO_x [9,10], CeO_x [11,12] and MnO_x [13–15], with high redox properties, exhibit great potential for Hg^0 adsorption. Among these oxides, MnO_x is a commonly available and inexpensive material has received extensive attention due to the redox couples of Mn^{2+}/Mn^{3+} and Mn^{3+}/Mn^{4+} [16]. Electronic shift between the different valence states of Mn is active and leads to a high redox capacity. Stefano Cimino et al. [14] investigated the Hg^0 removal performance of Mn/TiO_2 and found that Hg^0 capture efficiency was about 57% at 70 °C. After modification by some other transition metal oxides, Mn-based materials, such as $Mn-FeO_x$ [15], $Mn-ZrO_x$ [17], $Mn-CeO_x$ [18], and $Mn-CuO_x$ [19] can remove Hg^0 better. Furthermore, it has been reported that the shape and crystallographic phases of Mn based sorbents have serious effects on Hg^0 removal performance. Xu et al. [20] synthesized three different crystallographic phases of MnO_2 and found that α-MnO_2 had the highest capacity due to its larger surface area and oxidizability. Chalkidis et al. [21] pointed out that MnO_2 nano-rods possessed good Hg^0 removal capacity owing to the higher surface adsorbed oxygen species.

However, Mn-based sorbents usually have a poor SO_2 resistance as SO_2 can easily react with Mn, thereby forming $MnSO_4$ and leading to a largely suppressed Hg^0 removal activity. Even a little amount of SO_2 will results in serious inhibited effects on Hg^0 removal process. Our previous work has indicated that Ce-Zr modified Mn sorbent will be totally deactivated in 1h after the introduction of 50 ppm SO_2 due to SO_2 poisoning Mn forming $MnSO_4$ [22]. TiO_2 is a traditional way to enhance the SO_2 resistance of MnO_x [23] as TiO_2 can inhibit the deposition of sulfates on sorbents surface [24]. But the Hg^0 removal activity of MnO_x/TiO_2 is unsatisfactory because the active component of Mn is still exposed in SO_2 atmosphere. Core-shell is a structure with active component core and supporting components shell. The shell can inhibit the interaction between SO_2 and sorbent surface and efficiently protect active component core [25]. Therefore, synthesizing a core-shell structure with MnO_x core and TiO_2 shell may obtain a better SO_2 resistance.

Inspired by this, αMnO_2 nanorods and αMnO_2 nanorods coating TiO_2 shell were synthesized in the present work to investigate the Hg^0 removal efficiency in the presence of SO_2. Thermo-gravimetric (TG) and X-ray photoelectron spectroscopy (XPS) were performed to determine the role of SO_2 in the Hg^0 oxidation and adsorption processes and a probable mechanism of SO_2 influence was deduced based on XPS and TG results. The kinetic model of the Hg^0 adsorption process was examined as well.

2. Results and Discussion

2.1. Structure Characterization

Scanning electron microscopy (SEM) and transmission electron microscopy (TEM) were performed to investigate the morphologic and structural properties of αMnO_2-NR and αMnO_2-NR@TiO_2. Figure 1a,a' show SEM and TEM images of αMnO_2-NR. It can be seen that αMnO_2-NR has a uniform nanorod structure with an average diameter of about 100 nm. As shown in Figure 1b, for αMnO_2-NR@TiO_2, the uniform nanorod structure is well-retained after being coated with TiO_2 and the packing state of this sample is similar to αMnO_2-NR. The surface of αMnO_2-NR@TiO_2 are rougher when compared to αMnO_2-NR, and the average diameter increases to 150 nm due to the TiO_2 coating. The average length of the αMnO_2-NR@TiO_2 is about 2–3 μm (shown in Figure 1c). As shown in Figure 1b', an obvious dividing line can be detected between MnO_2 core and TiO_2 shell, and the shell with thickness of about 30 nm is well dispersed outside of the αMnO_2-NR.

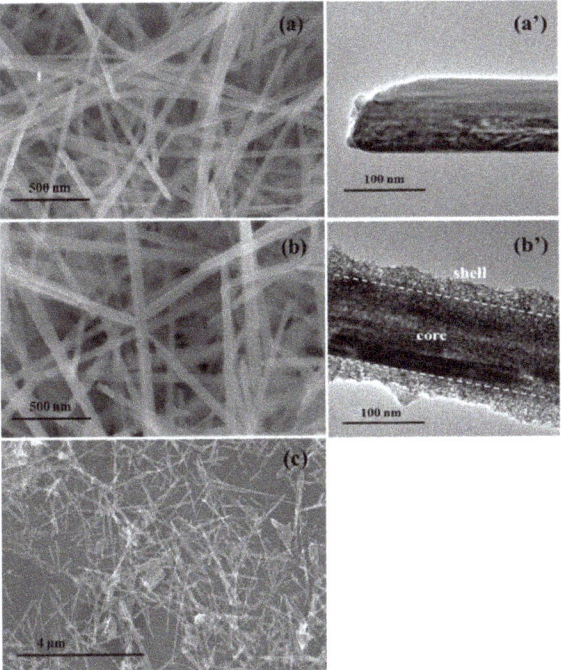

Figure 1. Scanning electron microscopy (SEM), and transmission electron microscopy (TEM) images of (**a,a'**) αMnO$_2$-NR; (**b,b'**), and (**c**) αMnO$_2$-NR@TiO$_2$.

N$_2$ sorption-desorption isotherms of the samples are shown in Figure 2. Both αMnO$_2$-NR and αMnO$_2$-NR@TiO$_2$ exhibit a type IV adsorption isotherm, according to the definition of IUPAC, which means that αMnO$_2$-NR and αMnO$_2$-NR@TiO$_2$ have a mesoporous structure. The surface areas, pore volumes, and average pore diameters of the sorbents are illustrated in Table 1. BET surface areas of the two sorbents are similar, suggesting that TiO$_2$ coating does not change the structure of αMnO$_2$-NR a lot. This result consists with SEM results.

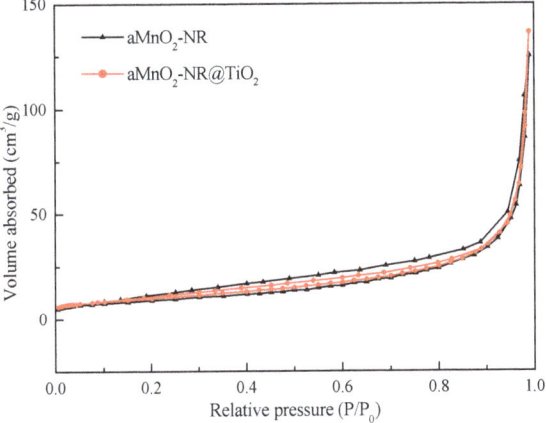

Figure 2. N$_2$ sorption-desorption isotherms for the sorbents.

Table 1. Pore structure analysis of the sorbents.

Samples	BET Surface Area (m²/g)	Pore Volume (cm³/g)	Average Pore Diameter (nm)
αMnO$_2$-NR	29.103	0.192	5.428
αMnO$_2$-NR@TiO$_2$	32.985	0.207	4.186

X-ray diffractometer (XRD) patterns of the two catalysts are shown in Figure 3. All the peaks in XRD pattern of αMnO$_2$-NR and αMnO$_2$-NR@TiO$_2$ were indexed to cryptomelane type α-MnO$_2$ (JCPDS 44-0141, tetragonal, I4/m, a = b = 0.978 nm, c = 0.286 nm). The intensity of diffraction peaks for the two samples is almost the same. It means that TiO$_2$ shell does not influence the dispersion of αMnO$_2$-NR, which is great agreement with BET and SEM results.

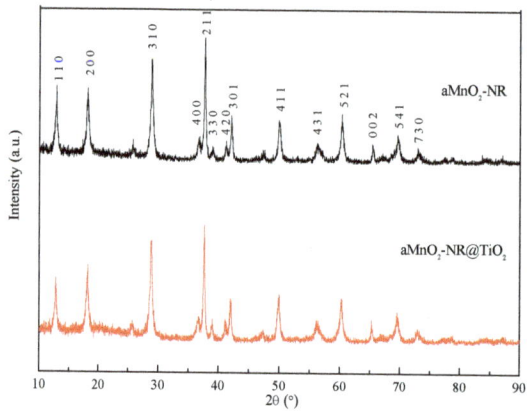

Figure 3. X-ray diffractometer (XRD) patterns of αMnO$_2$-NR and αMnO$_2$-NR@TiO$_2$.

2.2. Hg^0 Adsorption

2.2.1. Hg^0 Adsorption Performance

Breakthrough experiments were performed to investigate the Hg^0 adsorption performance of the two sorbents. A blank test was also performed and the results is shown in Figure S1. It can be seen that the outlet Hg^0 concentration is stable when no sorbent was loaded in the fixed-bed reactor. As shown in Figure 4, the Hg^0 removal efficiency of αMnO$_2$-NR is about 92% at the beginning of the test and it decreases to 41% after 130 min reaction. When it comes to αMnO$_2$-NR@TiO$_2$, the Hg^0 removal efficiency at the beginning of the test is about 81% which is lower than that of αMnO$_2$-NR. But it is about 43% at the end of the test suggesting a more stable removal activity. These results indicate that TiO$_2$ shell does not inhibit the Hg^0 diffusion from gas phase to the surface of αMnO$_2$-NR.

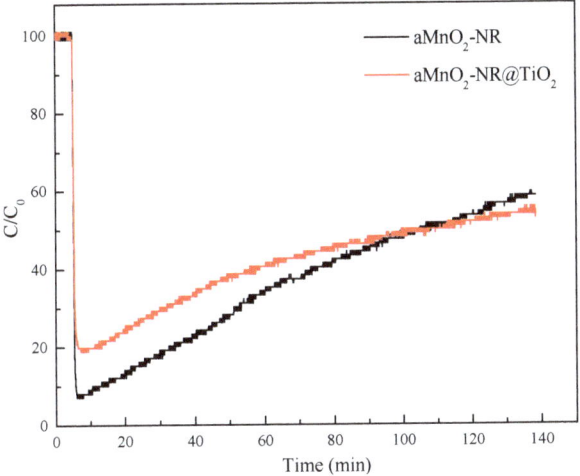

Figure 4. Hg^0 breakthrough curves of αMnO$_2$-NR and αMnO$_2$-NR@TiO$_2$ under pure N$_2$ atmosphere. Reaction condition: 150 °C, GHSV = 180,000 h^{-1}.

Figure 5 shows the effects of SO$_2$ on Hg^0 adsorption performance. For αMnO$_2$-NR, Hg^0 removal efficiency sharply declines from 55% to 14% during the 35 min reaction, when SO$_2$ is injected into flue gas. However, for αMnO$_2$-NR@TiO$_2$, the downward trend of Hg^0 removal efficiency is much slower and decreases from 76% to 43% in a 30 min test, and still has a Hg^0 removal efficiency of 25% after 80 min. These results confirm that TiO$_2$ shell can inhibit the direct interaction between SO$_2$ and MnO$_2$ surface, which will efficiently protect MnO$_2$ core from SO$_2$ poisoning.

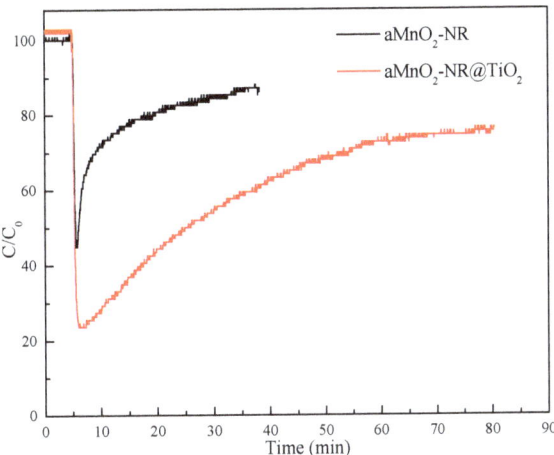

Figure 5. Hg^0 breakthrough curves of αMnO$_2$-NR and αMnO$_2$-NR@TiO$_2$ under 100 ppm SO$_2$, N$_2$ balanced. Reaction condition: 150 °C, GHSV = 180,000 h^{-1}.

αMnO$_2$-NR@TiO$_2$ was used to investigate reusability for Hg^0 removal. The results are shown in Figure 6. After 10 h Hg^0 adsorption test, αMnO$_2$-NR@TiO$_2$ reaches a Hg^0 adsorption equilibrium. And then, the sorbent was heated at 450 °C for 2 h to release the HgO on sorbent surface. It can be found that, after heated treatment, the Hg^0 adsorption efficiency and capacity of αMnO$_2$-NR@TiO$_2$ recovers to its original level. After two recycling, it still shows a good Hg^0 adsorption efficiency. Furthermore,

SEM results of the fresh and used αMnO$_2$-NR@TiO$_2$ (shown in Figure S2) show that recycle have no effect on the microstructure. These results suggest an outstanding reusability of αMnO$_2$-NR@TiO$_2$. The Hg^0 adsorption capacity of αMnO$_2$-NR@TiO$_2$ is 0.11 mg/g, it is good enough compared to other sorbents (shown in Table S1). The surface areas of the sorbents in the present work are relatively low thereby lowering the available surface active sites. αMnO$_2$-NR@TiO$_2$ with higher surface area will be studied in our following works, and may give a better Hg^0 adsorption capacity.

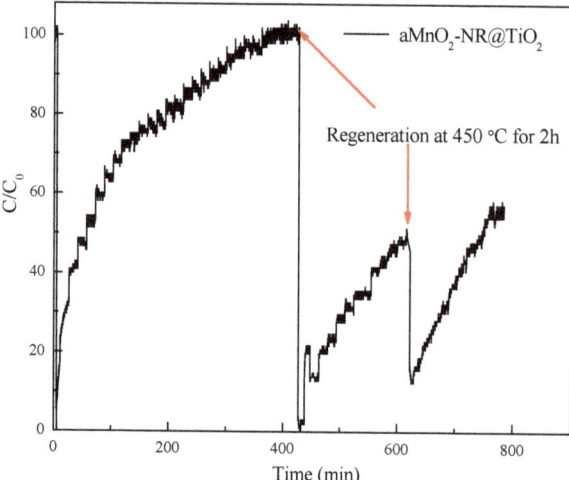

Figure 6. Hg^0 breakthrough curves of αMnO$_2$-NR@TiO$_2$ under pure N$_2$ atmosphere. Reaction condition: 150 °C, GHSV = 180,000 h^{-1}.

Hg^0 adsorption test of αMnO$_2$-NR@TiO$_2$ at different Hg^0 concentration was also investigated and the results are shown in Figure S3. With a doubled Hg^0 concentration, the breakthrough curve gets steep suggesting that αMnO$_2$-NR@TiO$_2$ will easily reach Hg^0 adsorption equilibrium at a higher Hg^0 concentration.

2.2.2. Structure-Activity Relationship

Fourier Transform Infrared Spectrometer (FTIR) was used to confirm the kind of surface active site for Hg^0 adsorption. As can be seen in Figure 7, the peaks at 429, 503, and 700 cm^{-1} correspond to Mn-O vibration [26], which becomes much weaker after reaction. It suggests that Mn-O group participates in Hg^0 adsorption process. According to previous work, the surface active oxygen species in Mn-O group should be the active sites for Hg^0 adsorption.

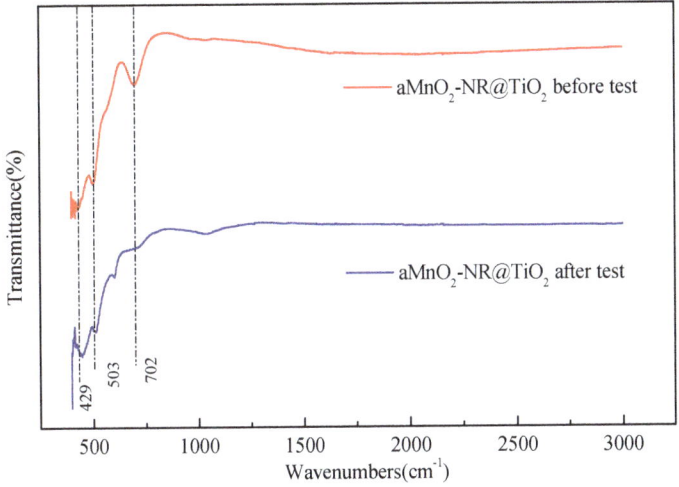

Figure 7. FTIR spectrum of αMnO$_2$-NR@TiO$_2$ before and after test.

2.3. Models of Adsorption Kinetics

In order to better illustrate the Hg^0 adsorption mechanisms of αMnO$_2$-NR and αMnO$_2$-NR@TiO$_2$, two popular models of pseudo-first order and pseudo-second order kinetic models, which have been widely used to investigate the adsorption process [27], were employed to fit the above experimental data. These two kinetic equations are displayed as follows [28]:

$$\lg(q_e - q_t) = \lg q_e - \frac{k_1}{2.303}t \quad \text{pseudo-first order} \tag{1}$$

$$\frac{t}{q_t} = \frac{1}{k_2 q_e^2} + \frac{1}{q_e} \quad \text{pseudo-second order kinetic} \tag{2}$$

where q_e and q_t are the adsorption capacity of Hg^0 on the sorbents at equilibrium, and at reaction time t (min), respectively. The parameters k_1 (min^{-1}) and k_2 (g/(μg·min)) are the rate constants of the pseudo-first order, and pseudo-second order models, respectively.

The fitting results are shown in Figure 8, and the obtained values of correlation coefficient (R^2) are summarized in Table 2. The values of R^2 of the pseudo-second order model for αMnO$_2$-NR and αMnO$_2$-NR@TiO$_2$ are 0.991, and 0.995, respectively, which are higher than those of pseudo-first order kinetic model (0.944 and 0.938 for αMnO$_2$-NR and αMnO$_2$-NR@TiO$_2$). It indicates that the pseudo-second order model can better fit the experimental data and Hg^0 removal process are dominantly controlled by chemisorption. After SO$_2$ introduction, the values of R^2 of the pseudo-second order model for αMnO$_2$-NR and αMnO$_2$-NR@TiO$_2$ are 0.997 and 0.992, which are still much higher than those of the pseudo-first order model. These results show that Hg^0 adsorption process in the presence of SO$_2$ atmosphere are also dominantly controlled by chemisorption.

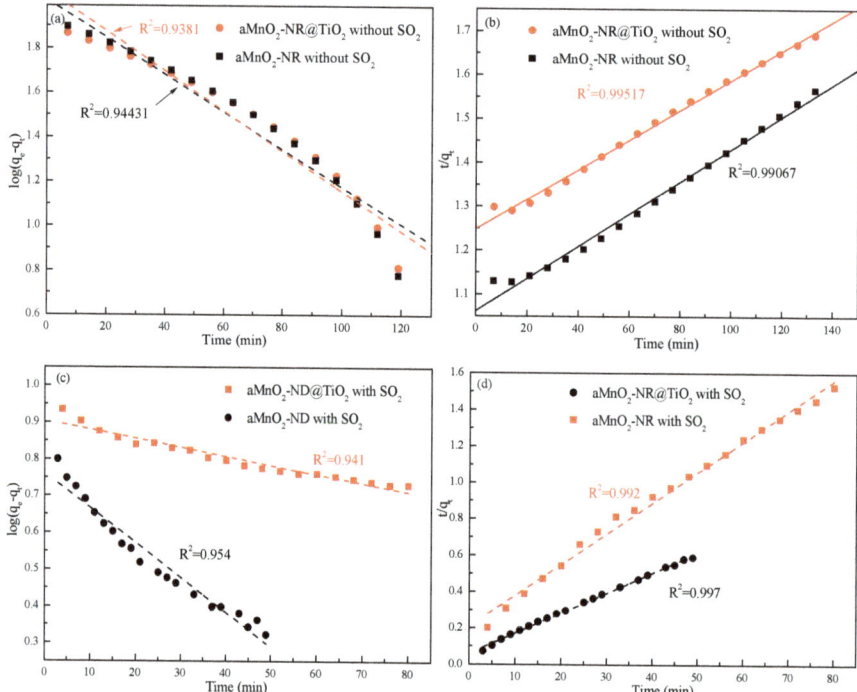

Figure 8. Kinetic analysis of Hg^0 adsorption on αMnO_2-NR and αMnO_2-NR@TiO_2 (**a**) pseudo-first order kinetic model without SO_2, (**b**) pseudo-second order kinetic model without SO_2, (**c**) pseudo-first order kinetic model with SO_2, (**d**) pseudo-second order kinetic model with SO_2.

Table 2. Kinetic parameters (R^2) of pseudo-first order and pseudo-second order models.

Kinetic Models	αMnO_2-NR without SO_2	αMnO_2-NR@TiO_2 without SO_2	αMnO_2-NR with SO_2	αMnO_2-NR@TiO_2 with SO_2
Pseudo-first (R^2)	0.944	0.938	0.954	0.941
Pseudo-second (R^2)	0.991	0.995	0.997	0.992

2.4. The Mechanism of SO_2 Effects on Hg^0 Adsorption

XPS analysis was employed to explore the relative proportion of elements on the sample surface. The XPS spectra of Mn 2p, O 1s and S 2p for the fresh and used samples are shown in Figure 9. The surface atomic concentrations and surface atomic ratios are summarized in Table 3.

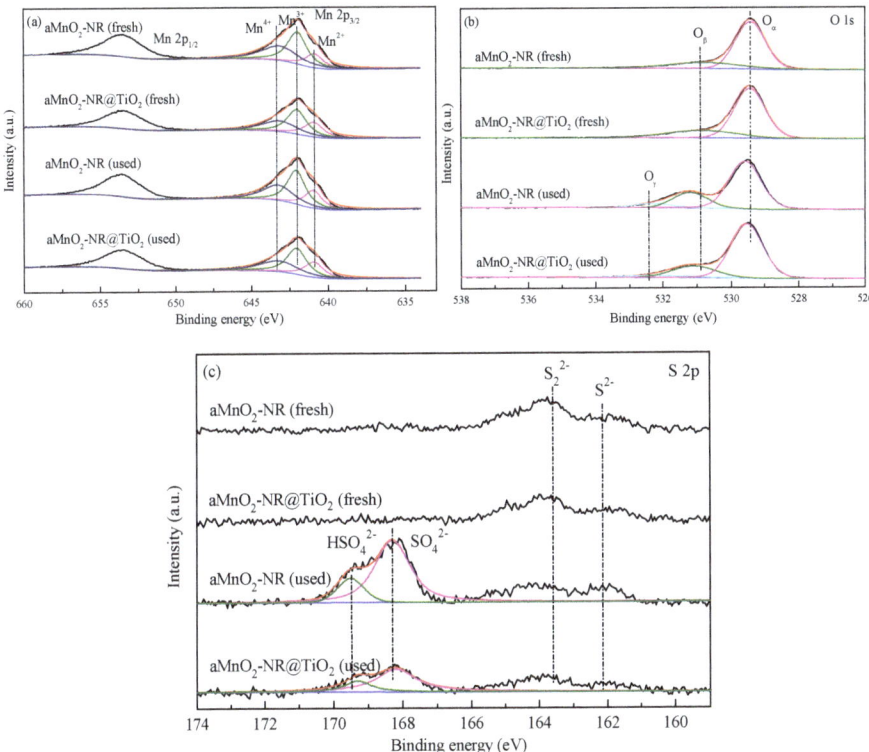

Figure 9. XPS spectra of (**a**) Mn 2p, (**b**) O 1s and (**c**) S 2p.

Table 3. The surface atomic concentrations and the relative concentration ratios of samples based on XPS.

Samples	S	Mn^{4+}/Mn	O_β/O
αMnO$_2$-NR (fresh)	3.17	37.8	26.0
αMnO$_2$-NR@TiO$_2$ (fresh)	2.27	33.4	24.7
αMnO$_2$-NR (used)	4.97	34.0	22.8
αMnO$_2$-NR@TiO$_2$ (used)	2.66	33.0	20.0

Figure 9a shows the XPS spectra of Mn 2p. A doublet due to spin orbital coupling can be detected which corresponds to Mn 2p$_{3/2}$ (around 641.24 eV) and Mn 2p$_{1/2}$ (around 652.82 eV). Due to the high intensity of Mn 2p$_{3/2}$, it was fitted to give detail information of valence state of Mn and it can be separated into three peaks at 640.2–641.2 eV, 641.2–642.1 eV, and 642.2–643.4 eV corresponding to Mn^{2+}, Mn^{3+}, and Mn^{4+}, respectively [29,30]. As shown in Table 3, the ratio of Mn^{4+}/Mn is about 37.8% for the fresh αMnO$_2$-NR and it decreases to 33.4% after the SO$_2$ resistance test. Compared to αMnO$_2$-NR, Mn^{4+} content is almost constant for αMnO$_2$-NR@TiO$_2$ before, and after, SO$_2$ resistance test. These results indicate that, for αMnO$_2$-NR, Mn^{4+} is easily reduced to Mn^{2+} during SO$_2$ resistance process via the reaction between SO$_2$ and MnO$_2$ [31]. For αMnO$_2$-NR@TiO$_2$, the interaction between SO$_2$ and MnO$_2$ is inhibited by the TiO$_2$ shell structure, which can efficiently protect active component Mn^{4+} in the core.

Figure 9b shows O 1s XPS spectra. For the fresh catalysts, O 1s bands can be split into two peaks, corresponding to lattice oxygen (peak at 529.5 eV, denoted as O_α) and chemisorbed oxygen (peak at 530.8 eV, denoted as O_β), respectively [32]. Whereas, a new peak appears around 532.3 eV after SO$_2$

treatment, which corresponds to SO_4^{2-} (denoted as O_γ) [33]. The intensity of the peak around 532.3 eV for αMnO_2-NR@TiO_2 is weaker than that for αMnO_2-NR suggesting a lower amount of SO_4^{2-} on the used αMnO_2-NR@TiO_2 surface. Furthermore, the peaks of O_α and O_β in αMnO_2-NR have an obvious slight shift to higher binding energy after SO_2 treatment. It might be due to the formation of sulfate salts during the sulfating process [34].

To determine the above deduction, S 2p bands was further investigated and the results are shown in Figure 9c. For the fresh αMnO_2-NR and αMnO_2-NR@TiO_2, two peaks around 162.2 eV and 163.2 eV attributed to S^{2-} and S_2^{2-} can be detected [35,36], which may come from $MnSO_4$ (the precursor of MnO_2). But for the used αMnO_2-NR and αMnO_2-NR@TiO_2, two new peaks at about 168.8 eV and 170.0 eV are observed, which may be assigned to SO_4^{2-}, and HSO_4^-, respectively [37,38]. The peak intensity of the used αMnO_2-NR is much higher than that of αMnO_2-NR@TiO_2. As shown in Table 3, for αMnO_2-NR, the surface atomic concentrations of S increases from 3.17% to 4.97% after SO_2 teatment while it increases from 2.27% to 2.66% for αMnO_2-NR@TiO_2. These results confirm that TiO_2 shell can inhibit the S accumulation on catalyst surface.

To obtain more information about the SO_2 poisoning mechanism, Thermo-gravimetric-differential thermos-gravimetry (TG-DTG) was performed to investigate the weight loss of αMnO_2-NR and αMnO_2-NR@TiO_2 after SO_2 treatment, and the results are presented in Figure 10. It can be seen that the used αMnO_2-NR has an obvious weight loss step in the temperature range of 680–780 °C with a weight loss of about 2.4%, which can be attributed to manganese sulfate decomposition [39–41]. There is no weight loss step between 680–780 °C with respect to αMnO_2-NR@TiO_2, but there is a new weak step around 780–850 °C can be detected, and it may be due to the decomposition of $Ti(SO_4)O$ [42]. This result demonstrates that SO_2 tends to react with titanium oxides instead of manganese oxides over αMnO_2-NR@TiO_2. Based on these results, TiO_2 shell can lead to the preferential adsorption of SO_2 on Ti surrounding forming $Ti(SO_4)O$ to protect Mn active component from being deactivated.

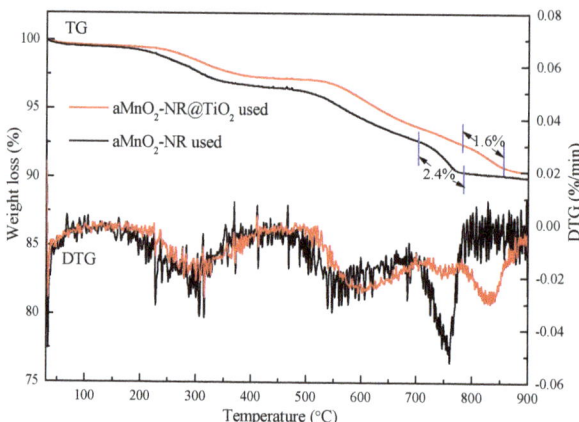

Figure 10. Thermo-gravimetric (TG) and differential thermos-gravimetry (DTG) of spectras of αMnO_2-NR and αMnO_2-NR@TiO_2 after SO_2 treatment.

3. Materials and Methods

3.1. Catalysts Preparation

The αMnO_2 nanorods were synthesized through a hydrothermal method [43]. $KMnO_4$ (2.5 g, AR) and $MnSO_4 \cdot H_2O$ (1.05 g, AR) were dissolved in 80 mL distilled water. The mixed solution was transferred into a Teflon-line stainless steel autoclave, sealed, and kept in an oven at 160 °C for 12 h. After cooling to room temperature, the precipitates were filtered off, washed several times using deionized water and dried at 110 °C overnight. Finally, the product was calcined at 400 °C in a muffle furnace for 4 h and the obtained sample is denoted as αMnO_2-NR.

$MnO_2@TiO_2$ core-shell nanorods were synthesized through a versatile kinetics-controlled coating method [44]. αMnO_2-NR (0.075 g) and aqueous ammonia (0.28 mL, 28 wt.%) were dispersed in 100 mL absolute ethanol under ultrasound for 30 min. Afterwards, titanium tetrabutoxide (TBOT) (0.75 mL) was added drop-wise into the mixture and then kept at 45 °C for 24 h. The mixed solution was filtered, washed and dried at 60 °C for 12 h. Finally, the solid was calcined under flow air at 500 °C for 2 h to obtain the sample (denoted as αMnO_2-NR@TiO_2).

ALL reagents are from Aladdin company, Shanghai, China.

3.2. Hg^0 Adsorption Experiments

The Hg^0 removal test has been described in detail in our previous work [45]. The experimental reactor contains a gas distribution system, a Hg^0 vapor generating device, a fixed-bed quartz reactor (ID = 8 mm), an online mercury analyzer and a tail gas treating unit. The mercury permeation tube was placed in a U-shape glass tube, which was immersed in a water bath at a constant-temperature (38 °C) to ensure a constant Hg^0 permeation rate. The total gas flow was 600 mL/min, and the sorbent volume was generally 0.2 mL, resulting in a GHSV of 1.8×10^5 h^{-1}. The concentrations of Hg^0 and SO_2 were monitored by a VM-3000 online mercury analyzer (Mercury Instruments, München, German), and flue gas analyzer (KM950, Kane International Ltd., London, United Kingdom), respectively.

During each test, the Hg^0 gas first bypassed the fixed-bed reactor, and then introduced into the reactor for 2 h to obtain a stable Hg^0 concentration. Hg^0 breakthrough ratio was quantified by the following formula,

$$Breakthrough\ ratio(\%) = \frac{C}{C_0} \times 100\% \qquad (3)$$

where C and C_0 represent the inlet and outlet Hg^0 concentrations ($\mu g/Nm^3$) in the fixed-bed reactor.

3.3. Characterization

The morphology and microstructure of the samples were observed using SEM (Nova NanoSEM 450, FEI) and TEM (Tecnai G2 F30 S-Twin, FEI). The surface areas and pore parameters of the samples were determined by Nitrogen adsorption/desorption method at liquid nitrogen temperature at −196 °C on an automated gas sorption analyzer (Autosorb-iQ-C, Quantachrome Instruments, Boynton Beach, FL, USA). The pore size and pore volume were derived from the desorption branches using the Barrette-Joynere-Halenda (BJH) model. The crystal structures of the samples were characterized by an XRD (XRD-7000S, SHIMADZU Corporation, Kyoto, Japan) operating at 40 kV and 100 mA using a Cu Kα radiation. The scanning range (2θ) was from 10° to 90° with a scan speed of 5°/min. The element (Mn, O, and Hg) valence state was analyzed by XPS (ESCALAB250 Thermo Fisher Scientific, Wilmington, DE, USA) with a monochromatic Al Kα source. The C 1s binding energy value of 284.8 eV was used to calibrate the observed spectra. TG was performed on TGA/DSC1 analyser (METTLER TOLEDO, Schwerzenbach, Switzerland), under a nitrogen flow of 20 mL/min, using a heating rate of 10 °C/min from room temperature to 900 °C (NETZSCH Corporation, Selb, Germany). DTG analysis was obtained based on residual weight of the sample with respect to time. FTIR spectra were obtained on a Nicolet Magana-IR 750 spectrometer to measure the surface groups of the samples (Thermo Nicolet Corporation, Madison, WI, USA).

4. Conclusions

αMnO_2-NR@TiO_2 was prepared by versatile kinetics-controlled coating method to compare with αMnO_2-NR in the Hg^0 removal process. SEM, BET, and XRD results showed that TiO_2 shell did not change the structure of αMnO_2-NR. Therefore, the two sorbents had similar Hg^0 removal performance in N_2 atmosphere. When SO_2 was introduced, αMnO_2-NR@TiO_2 had a much better performance than αMnO$_2$-NR. XPS and TG-DTG results showed that αMnO$_2$-NR@TiO$_2$ had lower surface S concentration after treatment of SO$_2$, and no manganese sulfate could be detected in αMnO$_2$-NR@TiO$_2$. It suggests that the TiO$_2$ shell can effectively protect MnO$_2$ from being deactivated by SO$_2$. Adsorption kinetic results showed that Hg^0 adsorption process over both the two sorbents obeys pseudo-second order model with, or without, SO$_2$.

Supplementary Materials: The following are available online at http://www.mdpi.com/2073-4344/10/1/72/s1, Figure S1: Outlet Hg^0 concentration without sorbent, Figure S2: The image of αMnO$_2$-NR@TiO$_2$ after adsorption, Figure S3: Breakthrough curve of αMnO$_2$-NR@TiO$_2$ with different Hg^0 feed concentration, Table S1: Comparison of the adsorption capacities of the sorbents.

Author Contributions: X.S. and H.Z. designed the experiments; X.H. and C.L. performed the experiments and analyzed the data; X.Z. wrote the paper; J.B., N.Z. and G.H. contributed reagents/materials/analysis tools. All authors have read and agreed to the published version of the manuscript.

Funding: This research was funded by [National Natural Science Foundation of China] grant number [51978124], [Liaoning Provincial Natural Science Foundation of China] grant number [20180510054].

Acknowledgments: We gratefully acknowledge the financial support of the National Natural Science Foundation of China (51978124), Liaoning Provincial Natural Science Foundation of China (20180510054), Shandong Provincial Natural Science Foundation PhD Programme (ZR2016EEB33), Foundation of China the Program for Changjiang Scholars (T2012049), Education Department of the Liaoning Province of China (LT2015007), the Fundamental Research Funds for the Central Universities (DUT18JC45).

Conflicts of Interest: The authors declare no conflict of interest.

References

1. Zhao, H.; Mu, X.; Yang, G.; George, M.; Cao, P.; Fanady, B.; Rong, S.; Gao, X.; Wu, T. Graphene-like MoS$_2$ containing adsorbents for Hg^0 capture at coal-fired power plants. *Appl. Energy* **2017**, *207*, 254–264. [CrossRef]
2. Zhang, J.; Duan, Y.; Zhou, Q.; Zhu, C.; She, M.; Ding, W. Adsorptive removal of gas-phase mercury by oxygen non-thermal plasma modified activated carbon. *Chem. Eng. J.* **2016**, *294*, 281–289. [CrossRef]
3. Zhang, B.; Zeng, X.; Xu, P.; Chen, J.; Xu, Y.; Luo, G.; Xu, M.; Yao, H. Using the Novel Method of Nonthermal Plasma To Add Cl Active Sites on Activated Carbon for Removal of Mercury from Flue Gas. *Environ. Sci. Technol.* **2016**, *50*, 11837–11843. [CrossRef] [PubMed]
4. Zhou, Q.; Duan, Y.F.; Hong, Y.G.; Zhu, C.; She, M.; Zhang, J.; Wei, H.Q. Experimental and kinetic studies of gas-phase mercury adsorption by raw and bromine modified activated carbon. *Fuel Process. Technol.* **2015**, *134*, 325–332. [CrossRef]
5. Liu, W.; Vidic, R.D.; Brown, T.D. Impact of Flue Gas Conditions on Mercury Uptake by Sulfur-Impregnated Activated Carbon. *Environ. Sci. Technol.* **2000**, *34*, 154–159. [CrossRef]
6. Bisson, T.M.; Xu, Z. Potential Hazards of Brominated Carbon Sorbents for Mercury Emission Control. *Environ. Sci. Technol.* **2015**, *49*, 2496–2502. [CrossRef]
7. Chen, C.; Jia, W.; Liu, S.; Cao, Y. The enhancement of CuO modified V$_2$O$_5$-WO$_3$/TiO$_2$ based SCR catalyst for Hg^0 oxidation in simulated flue gas. *Appl. Surf. Sci.* **2018**, *436*, 1022–1029. [CrossRef]
8. Li, H.; Zhang, W.; Wang, J.; Yang, Z.; Li, L.; Shih, K. Coexistence of enhanced Hg^0 oxidation and induced Hg^{2+} reduction on CuO/TiO$_2$ catalyst in the presence of NO and NH$_3$. *Chem. Eng. J.* **2017**, *330*, 1248–1254. [CrossRef]
9. Liu, T.; Xue, L.; Guo, X.; Liu, J.; Huang, Y.; Zheng, C. Mechanisms of Elemental Mercury Transformation on α-Fe$_2$O$_3$(001) Surface from Experimental and Theoretical Study: Influences of HCl, O$_2$, and SO$_2$. *Environ. Sci. Technol.* **2016**, *50*, 13585–13591. [CrossRef]
10. Zarei, S.; Niad, M.; Raanaei, H. The removal of mercury ion pollution by using Fe$_3$O$_4$-nanocellulose: Synthesis, characterizations and DFT studies. *J. Hazard. Mater.* **2018**, *344*, 258–273. [CrossRef]
11. He, C.; Shen, B.; Chi, G.; Li, F. Elemental mercury removal by CeO$_2$/TiO$_2$-PILCs under simulated coal-fired flue gas. *Chem. Eng. J.* **2016**, *300*, 1–8. [CrossRef]
12. Zhu, Y.; Han, X.; Huang, Z.; Hou, Y.; Guo, Y.; Wu, M. Superior activity of CeO$_2$ modified V$_2$O$_5$/AC catalyst for mercury removal at low temperature. *Chem. Eng. J.* **2018**, *337*, 741–749. [CrossRef]

13. Xu, H.; Jia, J.; Guo, Y.; Qu, Z.; Liao, Y.; Xie, J.; Shangguan, W.; Yan, N. Design of 3D MnO_2/Carbon sphere composite for the catalytic oxidation and adsorption of elemental mercury. *J. Hazard. Mater.* **2018**, *342*, 69–76. [CrossRef] [PubMed]
14. Cimino, S.; Scala, F. Removal of Elemental Mercury by MnO_x Catalysts Supported on TiO_2 or Al_2O_3. *Ind. Eng. Chem. Res.* **2016**, *55*, 5133–5138. [CrossRef]
15. Zhang, S.; Zhao, Y.; Yang, J.; Zhang, J.; Zheng, C. Fe-modified MnO_x/TiO_2 as the SCR catalyst for simultaneous removal of NO and mercury from coal combustion flue gas. *Chem. Eng. J.* **2018**, *348*, 618–629. [CrossRef]
16. Yao, T.; Duan, Y.; Zhu, C.; Zhou, Q.; Xu, J.; Liu, M.; Wei, H. Investigation of mercury adsorption and cyclic mercury retention over MnO_x/γ-Al_2O_3 sorbent. *Chemosphere* **2018**, *202*, 358–365. [CrossRef]
17. Xie, J.K.; Qu, Z.; Yan, N.Q.; Yang, S.J.; Chen, W.M.; Hu, L.G.; Huang, W.J.; Liu, P. Novel regenerable sorbent based on Zr–Mn binary metal oxides for flue gas mercury retention and recovery. *J. Hazard. Mater.* **2013**, *261*, 206–213. [CrossRef]
18. Li, H.; Wang, Y.; Wang, S.; Wang, X.; Hu, J. Removal of elemental mercury in flue gas at lower temperatures over Mn-Ce based materials prepared by co-precipitation. *Fuel* **2017**, *208*, 576–586. [CrossRef]
19. Yi, Y.; Li, C.; Zhao, L.; Du, X.; Gao, L.; Chen, J.; Zhai, Y.; Zeng, G. The synthetic evaluation of CuO-MnO_x-modified pinecone biochar for simultaneous removal formaldehyde and elemental mercury from simulated flue gas. *Environ. Sci. Pollut. Res.* **2017**, *25*, 4761–4775. [CrossRef]
20. Xu, H.; Qu, Z.; Zhao, S.; Mei, J.; Quan, F.; Yan, N. Different crystal-forms of one-dimensional MnO_2 nanomaterials for the catalytic oxidation and adsorption of elemental mercury. *J. Hazard. Mater.* **2015**, *299*, 86–93. [CrossRef]
21. Chalkidis, A.; Jampaiah, D.; Hartley, P.G.; Sabri, Y.M.; Bhargava, S.K. Regenerable α-MnO_2 nanotubes for elemental mercury removal from natural gas. *Fuel Process. Technol.* **2019**, *193*, 317–327. [CrossRef]
22. Zhang, X.; Li, Z.; Wang, J.; Tan, B.; Cui, Y.; He, G. Reaction mechanism for the influence of SO_2 on Hg^0 adsorption and oxidation with $Ce_{0.1}$-Zr-MnO_2. *Fuel* **2017**, *203*, 308–315. [CrossRef]
23. Yang, Z.; Li, H.; Liu, X.; Li, P.; Yang, J.; Lee, P.H.; Shih, K. Promotional effect of CuO loading on the catalytic activity and SO_2 resistance of MnO_x/TiO_2 catalyst for simultaneous NO reduction and Hg^0 oxidation. *Fuel* **2018**, *227*, 79–88. [CrossRef]
24. Li, H.; Wu, C.Y.; Li, Y.; Zhang, J. Superior activity of MnO_x-CeO_2/TiO_2 catalyst for catalytic oxidation of elemental mercury at low flue gas temperatures. *Appl. Catal. B* **2012**, *111*, 381–388. [CrossRef]
25. Mitsudome, T.; Yamamoto, M.; Maeno, Z.; Mizugaki, T.; Jitsukawa, K.; Kaneda, K. One-step Synthesis of Core-Gold/Shell-Ceria Nanomaterial and Its Catalysis for Highly Selective Semihydrogenation of Alkynes. *J. Am. Chem. Soc.* **2015**, *137*, 13452–13455. [CrossRef] [PubMed]
26. Wu, J.; Zhou, T.; Wang, Q.; Umar, A. Morphology and chemical composition dependent synthesis and electrochemical properties of MnO_2-based nanostructures for efficient hydrazine detection. *Sens. Actuators B* **2016**, *224*, 878–884. [CrossRef]
27. Tabish, T.A.; Memon, F.A.; Gomez, D.E.; Horsell, D.W.; Zhang, S. A facile synthesis of porous graphene for efficient water and wastewater treatment. *Sci. Rep.* **2018**, *8*, 1817. [CrossRef]
28. McKay, G.; Ho, Y.S. Pseudo-second order model for sorption processes. *Process Biochem.* **1999**, *34*, 451–465.
29. Zhang, D.; Hou, L.A.; Chen, G.; Zhang, A.; Wang, F.; Wang, R.; Li, C. Cr Doping MnO_x Adsorbent Significantly Improving Hg^0 Removal and SO_2 Resistance from Coal-Fired Flue Gas and the Mechanism Investigation. *Ind. Eng. Chem. Res.* **2018**, *57*, 17245–17258. [CrossRef]
30. Hu, H.; Cai, S.; Li, H.; Huang, L.; Shi, L.; Zhang, D. Mechanistic Aspects of $deNO_x$ Processing over TiO_2 Supported Co–Mn Oxide Catalysts: Structure–Activity Relationships and In Situ DRIFTs Analysis. *ACS Catal.* **2015**, *5*, 6069–6077. [CrossRef]
31. Xu, H.; Qu, Z.; Zong, C.; Quan, F.; Mei, J.; Yan, N. Catalytic oxidation and adsorption of Hg^0 over low-temperature NH_3-SCR $LaMnO_3$ perovskite oxide from flue gas. *Appl. Catal. B* **2016**, *186*, 30–40. [CrossRef]
32. Peng, Y.; Qu, R.; Zhang, X.; Li, J. The relationship between structure and activity of MoO_3-CeO_2 catalysts for NO removal: Influences of acidity and reducibility. *Chem. Commun.* **2013**, *49*, 6215–6217. [CrossRef] [PubMed]
33. Ma, Y.; Mu, B.; Zhang, X.; Yuan, D.; Ma, C.; Xu, H.; Qu, Z.; Fang, S. Graphene enhanced Mn-Ce binary metal oxides for catalytic oxidation and adsorption of elemental mercury from coal-fired flue gas. *Chem. Eng. J.* **2019**, *358*, 1499–1506. [CrossRef]

34. Reddy, B.M.; Sreekanth, P.M.; Yamada, Y.; Xu, Q.; Kobayashi, T. Surface characterization of sulfate, molybdate, and tungstate promoted TiO_2-ZrO_2 solid acid catalysts by XPS and other techniques. *Appl. Catal. A* **2002**, *228*, 269–278. [CrossRef]
35. Li, H.; Zhu, L.; Wang, J.; Li, L.; Shih, K. Development of Nano-Sulfide Sorbent for Efficient Removal of Elemental Mercury from Coal Combustion Fuel Gas. *Environ. Sci. Technol.* **2016**, *50*, 9551–9557. [CrossRef] [PubMed]
36. Liao, Y.; Chen, D.; Zou, S.; Xiong, S.; Xiao, X.; Dang, H.; Chen, T.; Yang, S. Recyclable Naturally Derived Magnetic Pyrrhotite for Elemental Mercury Recovery from Flue Gas. *Environ. Sci. Technol.* **2016**, *50*, 10562–10569. [CrossRef]
37. Yang, S.; Guo, Y.; Yan, N.; Wu, D.; He, H.; Xie, J.; Qu, Z.; Jia, J. Remarkable effect of the incorporation of titanium on the catalytic activity and SO_2 poisoning resistance of magnetic Mn–Fe spinel for elemental mercury capture. *Appl. Catal. B Environ.* **2011**, *101*, 698–708. [CrossRef]
38. Xu, W.; He, H.; Yu, Y. Deactivation of a Ce/TiO_2 Catalyst by SO_2 in the Selective Catalytic Reduction of NO by NH_3. *J. Phys. Chem. C* **2009**, *113*, 4426–4432. [CrossRef]
39. Yu, J.; Guo, F.; Wang, Y.; Zhu, J.; Liu, Y.; Su, F.; Gao, S.; Xu, G. Sulfur poisoning resistant mesoporous Mn-base catalyst for low-temperature SCR of NO with NH_3. *Appl. Catal. B Environ.* **2010**, *95*, 160–168. [CrossRef]
40. Zhang, A.; Zhang, Z.; Lu, H.; Liu, Z.; Xiang, J.; Zhou, C.; Xing, W.; Sun, L. Effect of Promotion with Ru Addition on the Activity and SO_2 Resistance of MnO_x–TiO_2 Adsorbent for Hg^0 Removal. *Ind. Eng. Chem. Res.* **2015**, *54*, 2930–2939. [CrossRef]
41. Wu, Z.; Jin, R.; Wang, H.; Liu, Y. Effect of ceria doping on SO_2 resistance of Mn/TiO_2 for selective catalytic reduction of NO with NH_3 at low temperature. *Catal. Commun.* **2009**, *10*, 935–939. [CrossRef]
42. Gardy, J.; Hassanpour, A.; Lai, X.; Ahmed, M.H. Synthesis of $Ti(SO_4)O$ solid acid nano-catalyst and its application for biodiesel production from used cooking oil. *Appl. Catal. A* **2016**, *527*, 81–95. [CrossRef]
43. Liang, S.; Teng, F.; Bulgan, G.; Zong, R.; Zhu, Y. Effect of phase structure of MnO_2 nanorod catalyst on the activity for CO oxidation. *J. Phys. Chem. C* **2008**, *112*, 5307–5315. [CrossRef]
44. Li, W.; Yang, J.; Wu, Z.; Wang, J.; Li, B.; Feng, S.; Deng, Y.; Zhang, F.; Zhao, D. A versatile kinetics-controlled coating method to construct uniform porous TiO_2 shells for multifunctional core–shell structures. *J. Am. Chem. Soc.* **2012**, *134*, 11864–11867. [CrossRef]
45. Zhang, X.; Cui, Y.; Wang, J.; Tan, B.; Li, C.; Zhang, H.; He, G. Simultaneous removal of Hg^0 and NO from flue gas by $Co_{0.3}$-$Ce_{0.35}$-$Zr_{0.35}O_2$ impregnated with MnO_x. *Chem. Eng. J.* **2017**, *326*, 1210–1222. [CrossRef]

© 2020 by the authors. Licensee MDPI, Basel, Switzerland. This article is an open access article distributed under the terms and conditions of the Creative Commons Attribution (CC BY) license (http://creativecommons.org/licenses/by/4.0/).

Review

Progress and Challenges of Mercury-Free Catalysis for Acetylene Hydrochlorination

Yanxia Liu [1,2,3], Lin Zhao [1], Yagang Zhang [1,2,3,*], Letao Zhang [3] and Xingjie Zan [3]

1. School of Materials and Energy, University of Electronic Science and Technology of China, Chengdu 611731, China; liuyanxia@ms.xjb.ac.cn (Y.L.); zhaolin316@uestc.edu.cn (L.Z.)
2. Department of Chemical and Environmental Engineering, Xinjiang Institute of Engineering, Urumqi 830026, China
3. Xinjiang Technical Institute of Physics and Chemistry, Chinese Academy of Sciences, Urumqi 830011, China; zhanglt@ms.xjb.ac.cn (L.Z.); zanxj@ms.xjb.ac.cn (X.Z.)
* Correspondence: ygzhang@uestc.edu.cn; Tel.: +86-28-61831516

Received: 9 September 2020; Accepted: 15 October 2020; Published: 20 October 2020

Abstract: Activated carbon-supported $HgCl_2$ catalyst has been used widely in acetylene hydrochlorination in the chlor-alkali chemical industry. However, $HgCl_2$ is an extremely toxic pollutant. It is not only harmful to human health but also pollutes the environment. Therefore, the design and synthesis of mercury-free and environmentally benign catalysts with high activity has become an urgent need for vinyl chloride monomer (VCM) production. This review summarizes research progress on the design and development of mercury-free catalysts for acetylene hydrochlorination. Three types of catalysts for acetylene hydrochlorination in the chlor-alkali chemical industry are discussed. These catalysts are a noble metal catalyst, non-noble metal catalyst, and non-metallic catalyst. This review serves as a guide in terms of the catalyst design, properties, and catalytic mechanism of mercury-free catalyst for the acetylene hydrochlorination of VCM. The key problems and issues are discussed, and future trends are envisioned.

Keywords: vinyl chloride monomer; acetylene hydrochlorination; mercury-free catalysts; noble metal catalyst; non-noble metal catalyst; non-metallic catalysts

---

1. Introduction

Polyvinyl chloride (PVC) is one of the most important general purpose plastics. It is widely used in daily necessities and industrial application. It has the advantages of cheap and affordable features while also having superior comprehensive performance [1]. PVC is used in building materials, floor leather, packing materials, wires and cables, commodities, and other aspects [2]. In 2016, the global demand for PVC exceeded 45 million tons, among which China's PVC production exceeded 23 million tons, making it the world's largest producer and consumer of PVC [1]. PVC is a polymer made from vinyl chloride monomer (VCM) through the free radical polymerization mechanism. At present, the synthesis of VCM is mainly divided into three kinds: Ethane oxychlorination, ethylene oxychlorination, and acetylene hydrochlorination, respectively. Ethane oxychlorination started with natural gas resources, which is a new clean production process [1]. The lack of catalyst with good performance, especially high selectivity, is one of the main limiting factors restricting the method [3]. Ethylene oxychlorination is an environmentally benign process that depends on petroleum fossil fuels. Acetylene hydrochlorination is an important alternative process for the production of VCM in the coal-rich areas and the country [4–6]. In recent years, the increasing price of crude oil has highlighted the economic advantages of acetylene hydrochlorination. In the reaction of acetylene hydrochlorination to VCM, the atomic utilization of reactants is 100% [1]. The reaction process is as follows:

$$CaO + 3C \rightarrow CaC_2 + CO \qquad (1)$$

$$CaC_2 + 2H_2O \rightarrow CH \equiv CH + Ca(OH)_2 \quad (2)$$

$$CH \equiv CH + HCl \rightarrow CH_2 = CHCl \quad (3)$$

This reaction is exothermic with high selectivity, and the optimum reaction temperature is 170–180 °C [6]. For more than 60 years, almost all the catalysts used in this process in industry are mercury-containing catalysts, with the dosage of $HgCl_2$ ranging from 5 wt % to 12 wt % [7]. However, Hg catalysts have several disadvantages. $HgCl_2$ is a linear triatomic aggregate, which is easily sublimed during the hydrochlorination reaction. The life cycle of the catalyst is correlated to the amount of $HgCl_2$ catalyst loaded on the support [8]. During the acetylene hydrochlorination reaction, high temperature (180–220 °C) and pressure induces the desorption and sublimation of the $HgCl_2$. This not only causes a loss of $HgCl_2$ from the support but also poses a serious pollution problem [7]. The high temperature generates hot spots that aggravate the situation, resulting in a significant amount of mercury loss [9]. As a cumulative toxin, $HgCl_2$ is extremely toxic and harmful. Therefore, the loss of $HgCl_2$ will not only lead to the deactivation of the catalysts but also cause serious environmental issues [10]. Although efforts have been taken to recycle the losing mercury, about 25% of it is released directly into the environment [11]. In order to solve this issue, worldwide, the Minamata Convention has agreed on controlling the mercury emission reduction act in 2013 by the United Nations Environment Protection Committee. The agreement came into effect on 16th August 2017.

In order to reduce the mercury compounds released from the hydrochlorination process, it is urgent and of paramount importance to develop environmentally friendly green mercury-free catalysts as substitutes. In this review, we discuss the recent research progress and challenges for mercury-free catalysts (Scheme 1). Perspectives and future trends for the development of greener and cleaner processes for VCM production with mercury-free catalysts are also discussed.

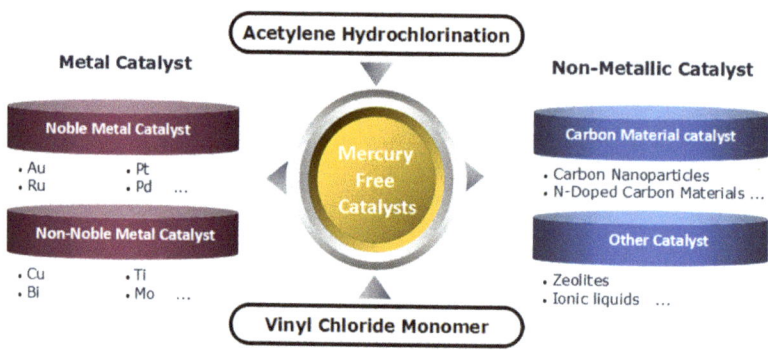

Scheme 1. Mercury-free catalysts for acetylene hydrochlorination.

2. Noble Metal Catalysts

The research of acetylene hydrochlorination catalysts supported by metal has been going on for 50 years. Smith et al. reported compound metal (Hg, Bi, Ni, Zn, Cd, Cu, Mn, and Ca) catalysts supported on silica for acetylene hydrochlorination in 1968 [12]. In 1975, Shinoda prepared metal chloride catalysts with activated carbon as the support. In their work, the results revealed that the catalytic activity of metal chloride was well correlated to the electron affinity of metal cations (Figure 1) [13]. One class of metal cations had its catalytic activity increase with the increase of electron affinity. Meanwhile, another class of metal cations demonstrated the opposite trend, and this was considered to be due to cations forming complexes with HCl of the type $H_m[MCl_{n+m}]$, which were Friedel–Crafts-type catalysts [14].

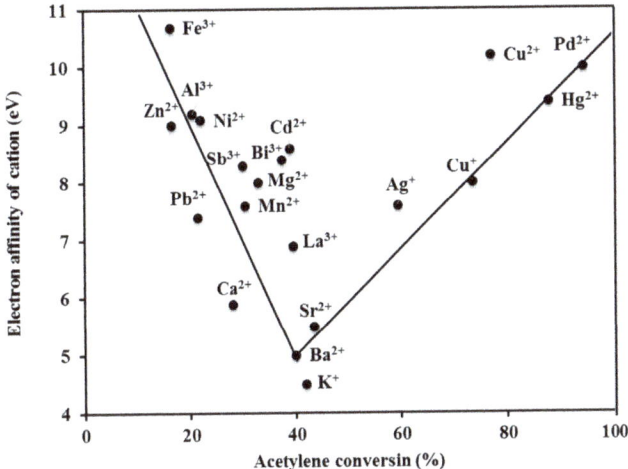

Figure 1. The relationship between acetylene hydrochlorination activity of metal chlorides and electron affinities of metal cations. Adopted from Ref [13].

In 1985, Hutchings evaluated a series of supported metal chloride catalysts analyzed by Shinoda and made different attribution associations. Hutchings reported that the interaction between acetylene and metal chloride may involve the transfer of more than one electron. Therefore, the standard reduction potential was considered to be a more appropriate parameter to correlate catalytic activity. Except for a few metals such as K, Ba, Mg, and La, most of the metals showed a positive correlation between its activity and the standard electrodes potential in acetylene hydrochlorination (Figure 2) [14]. The proposed correlation has certain guiding significance to the research of acetylene hydrochlorination. Based on these results, a large number of studies on mercury-free catalysts are conducted on noble metals.

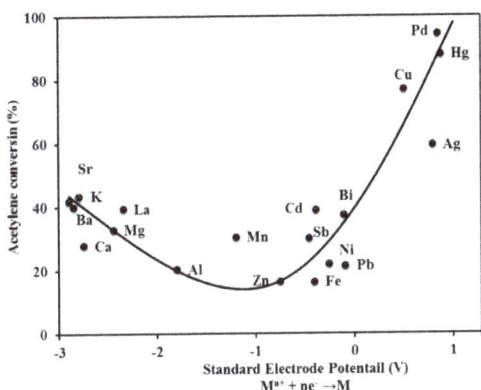

Figure 2. The relationship between acetylene hydrochlorination activity of metal chlorides and standard reduction potential [14].

2.1. Au Catalysts

In 1985, Hutchings predicted that Au might have good activity for acetylene hydrochlorination [14]. The prediction was verified in the article reported in 1988 [8]. Figure 3 further confirmed the correlation with the standard electrode potential. Since then, studies on supported Au catalysts have been increasing.

The earliest method for preparing Au catalysts was to dissolve the gold of high purity in aqua regia and on the carbon support, which could minimize the introduction of impurities and avoid the adverse effects of impurities on the catalysts [8,15,16].

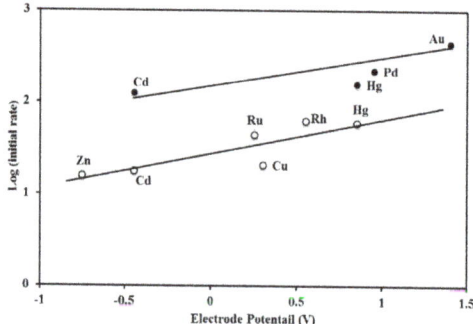

Figure 3. The correlation between catalytic activity and standard electrode potential of metal cations in acetylene hydrochlorination [16].

2.1.1. Au^{n+} Catalysts

Most studies on the catalytic mechanism of Au catalysts showed that the amount of Au^{3+} had a great correlation with the catalytic activity and was considered as the active species for acetylene hydrochlorination. For Au species, the order of catalytic activity was Au^{3+} > Au^{+} > Au^{0}. Although it has also been reported that AuCl can be used as a catalytic active species [17], due to the poor stability of AuCl in aqueous solution and reaction conditions, a disproportionation reaction occurs, resulting in the formation of AuCl$_3$ and elemental Au:

$$3AuCl \rightarrow AuCl_3 + 2Au \tag{4}$$

Since the equilibrium constant of the reaction is very large (10^{10} orders of magnitude), it can be considered that there is almost no free Au^{+} in aqueous solution, and it is difficult to prepare supported AuCl catalysts using conventional preparation methods. Therefore, Au^{3+} catalyst is the most studied and applied catalyst.

Conte et al. demonstrated that exposure to HCl before reaction led to enhanced catalyst activity, whereas exposure to acetylene resulted in a decrease of catalyst activity [5]. It was proposed that the catalytic activity of Au^{3+} could be explained by two possible mechanisms: (a) a nucleophilic electrophilic interaction between the Au^{3+} center and a triple bond of acetylene through π-coordination; and (b) the acidic protons of the terminal alkyne facilitated σ-coordination [5,18,19]. The reaction mechanism of HAuCl$_4$ (Au^{3+}) as the active species was proposed in Figure 4 [5]. This mechanism involved the presence of both acetylene and HCl for a six-member ring. In the six-member ring, the alkyne axially coordinated to the Au^{3+} center. Density functional theory (DFT) calculations implied that it was unlikely that the simultaneous coordination of acetylene and HCl to the Au^{3+} center would occur. The transition state for HCl addition to the π-complex of acetylene with AuCl$_3$ by calculation predicted that the stereochemistry of Cl addition was controlled by a hydrogen bond between HCl and a Cl ligand of Au. The anti-addition of HCl observed experimentally was a consequence of a sequential addition of Cl and H to the acetylene [5,6,20–23].

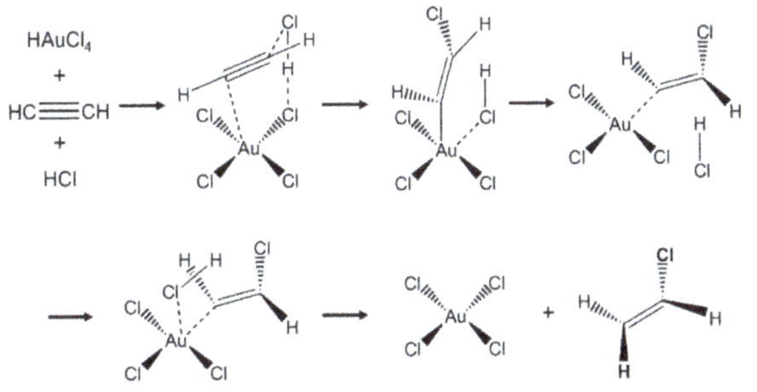

Figure 4. Acetylene hydrochlorination model of Au^{3+} catalyst [5].

The preparation method of catalysts has an important influence on its performance and activity. Conte et al. showed that Au could be highly dispersed on support with acid as the solvent [24]. They found that aqua regia was the most effective solvent for preparing the catalyst, while catalysts prepared using HCl or HNO_3 individually were less active. The superior activity of the catalyst prepared in aqua regia was proposed to be a combination of the oxidizing effect of HNO_3 and the nucleating effect of HCl. Both of these are helpful to promote the high dispersion of Au [5,7,25]. Although aqua regia is an effective solvent, it is extremely corrosive and dangerous, causing negative impacts on the environment and threatening the safety of the process. In addition, there are many difficulties in the treatment, recovery, and disposal of the used aqua regia, so it is not suitable for industrial application [26,27]. Therefore, researchers began to develop a similar activation protocol that could be realized by using other environmentally friendly solvents.

The general mechanism of catalyst deactivation is the reduction of Au^{3+} to Au^0. It is urgent to improve the stability of active Au^{3+} species and extend the service life of the catalyst. Zhao et al. reported that organic aqua regia (OAR, 1:10 $SOCl_2$: DMF) could be used as a substituent for conventional aqua regia to activate Au/AC catalyst with Au^{3+} as the active center. OAR could promote Au oxidation and help achieve high disparity. The content of Au^{3+} species in $Au(H_2O)/AC(OAR)$ was greater than that in the Au(aqua regia)/AC sample, indicating that OAR treatment could partially promote the oxidation of Au^0 into Au^{3+}. Residual S and N stabilized the Au^{3+} species and generated a more thermally stable catalyst by forming Au–S complexes, increasing the electron density of the Au center through electron transfer, enhancing the reduction temperature of Au^{3+}, and promoting the effective adsorption of HCl on the Au catalyst (Figure 5) [28].

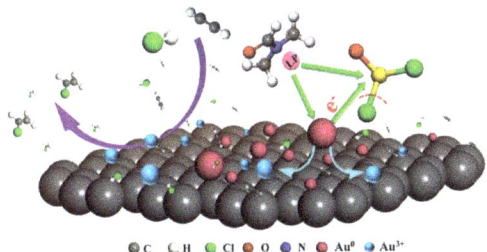

Figure 5. Au catalysts prepared by organic aqua regia [28].

Huang et al. studied the effect of the 1,10-phenanthroline ligand on Au catalyst performance. With an appropriate ligand, the $[AuCl_2(phen)]Cl$ catalyst achieved excellent acetylene conversion (90%) after 40 h of operation. The enhanced catalytic stability was attributed to the presence of the phen ligand; electron transfer from the phen species to the Au^{3+} center increased the Au^{3+} electron density, which inhibited the active Au^{3+} component deactivation [29].

Chao et al. prepared Au catalysts using activated carbon pretreated at different temperatures. It was shown that thermal treatments can change the surface functional groups on activated carbon. Ketone, lactone, and carbonyl tended to anchor Au on the surface with a good dispersion state. The phenolic and alcohol groups can easily reduce Au^{3+} to Au^0, which is not active for the reaction. Surface functional groups may be a constituent part of the active sites adsorbing and activating acetylene. This indicated that both Au^{3+} and the surface functional groups work in synergy as the active sites [30].

2.1.2. Au^0 Catalysts

There are some studies that have reported that Au^0 is the active center of Au catalysts. The catalytic efficiency of Au^0 catalysts is closely related with the dispersion degree of Au nanoparticles (Au NPs). Some researchers have demonstrated that the various edges or defects of multiple-twinned or polycrystalline particles formed on the surface of the Au NPs catalyst have high surface stress and are easily broken to reduce their surface energy, thus providing active sites for acetylene hydrochlorination.

Zhang et al. revealed that Au^0 catalyst activity was strongly associated with the properties of the used solvent. It was found that the catalyst's activity increased with the decreased polarity of solvents. Compared to polar solvents such as water and aqua regia, less polar solvent alcohols would influence the formation of crystallization of Au NPs. The mean particle size in Au-x/AC (x = alcohol) catalysts was determined to be 3.9–4.7 nm and showed peculiar characteristic of spherical Au NPs, demonstrating good dispersity on the catalysts. Au NPs exhibited good catalytic activity. Polycrystalline particles formed with various edges or defects could act as active sites. Weakly polar alcohols facilitated the interfacial interaction between the support and Au^0 species, made the active species highly dispersed, and anchored and inhibited the agglomeration and loss during the reaction. Moreover, the interaction also enhanced catalytic activity by increasing the adsorption capacity on the catalyst's surface [31].

Different preparation methods including ultrasonic, microwave, and incipient wetness impregnation were also studied and compared for acetylene hydrochlorination. Wittanadecha et al. investigated the catalytic performance of Au^0 catalysts using different preparation routes. It was found that the ultrasonic-assisted method improved the dispersion of the active component and increased the pore volume and aperture of the support. Noticeably and interestingly, only the initial activity of the catalyst was affected with different pretreatment processes, while the overall catalyst activity was almost unaffected. They suspected that the active Au species is Au^0. The Cl-containing substance is chemically adsorbed and reacts with acetylene to form a VCM [32].

Tian et al. prepared an Au catalyst (MIV-1Au/C1) with significantly increased activity. The catalysts were synthesized using mixed solvents and vacuum drying instead of the traditional impregnation method. Au^0 with a mean size of 5.2 nm was in the face-centered cubic form and was the only active species of MIV-1Au/C1 at the initial/highest point of the testing. The active species of Au^0 could be oxidized to Au^{3+} and became deactivated during the acetylene hydrochlorination reaction. The particle size of Au NPs prepared by this method was much smaller than that by the immersion method [33].

2.1.3. Modification of Au Catalysts

Efforts were devoted to improving the efficiency and enhancing the stability of active sites of Au catalysts. Adding synergistic metals, forming ligand coordination, and support modification have been attempted to solve the problems of catalysts deactivation, coke deposits, and Au particles aggregation (Scheme 2).

Scheme 2. Modification of Au catalysts.

Addition Synergistic Metal

Au^{3+} catalysts suffer from poor stability both during preparation and application. One way to solve this problem is to dope Au with other elements so that the electronic state of the active sites can be tuned. In 2008, Conte et al. studied the effect of different metal additives on the catalytic performance of Au^{3+} catalysts. The metals used include Pd, Pt, Ir, Rh, and Ru. The results showed that dispersed pure Au was actually the most active one for acetylene hydrochlorination in a long run. The addition of other metals did not improve the catalytic reactivity [34]. However, in recent years, studies have shown the opposite observations and conclusions when Au was doped with other elements, such as La, Cu, Bi, Sn, Ce, Sr, etc. to modified Au catalysts, which may have higher catalytic stability due to synergy effects [35]. On the one hand, the loading capacity of Au can be reduced, which led to reducing the cost of catalyst by adding metal additives. On the other hand, these metal additives provide electrons to Au^{3+}, absorb more oxidizing HCl, and slow down the reduction of Au^{3+}. Table 1 lists some AuM catalysts prepared with $HAuCl_4$ in collaboration with other synergistic metal precursors. The results show that the addition of metal additives can improve the dispersion of Au^{3+}, stabilize the chemical valence of the active center, enhance the electron cloud density of the Au^{3+} active center, inhibit carbon deposition, and finally improve the catalytic efficiency for acetylene hydrochlorination.

Table 1. Performance of AuM catalysts for acetylene hydrochlorination.

Year	Catalyst	Synergistic Metal	Au (wt %)	GHSV (h^{-1})	Temp (°C)	Acetylene Conv. (%)	Running Time (h)	Ref.
2019	AuCe	CeO_2	0.1	60	180	99.9	70	[4]
2016	AuCe	CeO_2	1.0	852	180	98.4	20	[36]
2017	AuY	YCl_3	1.0	800	180	87.8	10	[37]
2016	AuSn	$SnCl_2$	0.9	720	170	95	48	[38]
2017	AuCu	$CuCl_2$	0.1	740	180	98.5	500	[39]
2016	AuCu	$CuCl_2$	0.1	120	150	97	4	[40]
2015	AuCu	$CuCl_2$	0.25	120	150	97	2	[41]
2017	AuCuK	$CuCl_2/KCl$	0.2	40	165	89	1600	[42]
2014	AuCoCu	$Co(NH_3)_6Cl_3/CuCl_2$	1.0	720	150	99	5	[43]
2018	AuSr	$SrCl_2$	1.0	1806	180	99.7	180	[44]
2016	AuSr	$SrCl_2$	1.0	762	180	87.7	20	[45]
2015	AuBa	$BaCl_2$	1.0	360	200	98.4	50	[46]
2015	AuCs	CsCl	1.0	740	180	94	50	[47]
2015	AuCs	CsCl	1.0	1480	180	90.1	50	[48]
2015	AuInCs	$CsCl/InCl_3$	1.0	1480	180	92.8	50	[49]
2014	AuBi	$BiCl_3$	1.0	600	180	85	10	[17]
2014	AuBi	$BiCl_3$	1.0	120	150	96	10	[50]
2014	AuTi	TiO_2	1.0	870	180	92	10	[51]
2014	AuNi	$NiCl_2$	1.5	900	170	95.4	46	[52]
2013	AuCo	$Co(NH_3)_6Cl_3$	1.0	360	150	99.9	36	[53]
2012	AuLa	$LaCl_3$	1.0	360	150	98	50	[54]

Ligand Coordination

It is a common modification method to introduce suitable ligands to stabilize the high valence Au^{3+} in Au catalysts. Results showed that the activity and stability of the Au^{3+} catalyst could be significantly improved with the addition of chelating donors and N-containing ligands. The donor groups include thiosulfate, thiocyanate, thiourea, and cyanides [55,56]. The catalyst complexes can be prepared by synthesizing complexes and then loading them on the support; or they can be formed by directly adding the precursor of the complexes to the $HAuCl_4$ during the preparation process and then forming the Au complexes with catalytic activity in situ. Zhou et al. reported an Au catalyst with thiocyanate (–SCN) donors with 0.25 wt % Au loading. The donor significantly decreased the electrode potential of Au^{3+} from 0.926 to 0.662 V. Complexion prevented Au catalyst deactivation by increasing the reaction energy barrier [57]. Table 2 shows some Au catalysts prepared with different ligand precursors; Au^{3+} has high catalytic activity for acetylene hydrochlorination.

The catalytic activities of Au^0 catalysts (Au NPs) could be influenced by many factors. These factors include the Au cluster size and shape [58,59], the cluster charge [60,61], the structure of the support material [62], and the ligand [63]. Recently, it was found that doped Au clusters can form distinctive structures and achieve enhanced properties. Zhao et al. proposed a mechanism for pristine Au_7 and Au_8 clusters catalysts and, on the Si-doped Au clusters, Au_6Si and Au_7Si based on DFT. The results showed that the reaction process of acetylene hydrochlorination catalyzed by Au^0 involved two steps. In the first step, HCl disassociated and Cl was added, which was followed by H addition through the transfer of protons from the Au cluster to acetylene chloride. Doping can not only change the size and shape of Au clusters but also improve the efficiency for acetylene hydrochlorination [64].

Table 2. Au catalysts prepared with different ligand precursors.

Year	Catalyst	Ligand	Au (wt %)	GHSV (h^{-1})	Temp (°C)	Acetylene Conv. (%)	Running Time (h)	Ref.
2016	Au/TCCA	Trichloroisocyanuric acid	0.2	90	180	98	24	[65]
2015	$HAu(C_3Cl_3-N_3O_3)_3Cl$	Trichloroisocyanuric acid	1.0	500	130	52	24	[8]
2015	$Au(CS(NH_2)_2$	Thiourea	0.1	500	130	95	24	[8]
2015	Au/SCN	KSCN	0.25	1200	180	99	10	[57]
2014	$[AuCl_2(phen)]Cl$	1,10–phenanthroline	0.49	603	180	90	40	[29]
2013	$AuCl_3$/PPy-MWCNT	Pyrrole	1.47	120	150	90	10	[66]

Support Modification

Support is an important component of catalysts and has a great influence on the performance of Au catalysts. Choosing an appropriate support material is crucial under rigorous reaction conditions. The catalysts need to be robust and stable enough at high temperature. Support is also critical in view of stabilizing the highly dispersed metal chloride and preventing the agglomeration of active components [7]. For supported metal catalysts, support modification is common to increase the guest–host interaction. One example is to dope carbon support with heteroatoms such as N, B, and P.

The catalytic performance of supported metal catalysts is greatly dependent on the properties of the support. Carbon materials are recognized as ideal supports because of their large surface area and good electrical conductivity. Chen et al. observed that mesoporous carbon supports could increase the activity of Au^{3+} catalysts prepared with an $HAuCl_4$ precursor. The pore size of supports was important because large pore sizes allowed rapid diffusion and suppressed coke formation [67]. Jia et al. reported that the addition of a B species was favorable for the stabilization of active Au^{3+} species and the inhibition of a transition from Au^{3+} to Au^0. In addition, the B species also inhibited carbon deposition and catalyst sintering during the reaction [68].

Zhang et al. reported several Au–bimetallic catalysts for acetylene hydrochlorination. The catalysts were prepared with $HAuCl_4$ and $LaCl_3$ precursors on different supports. The results indicated that compared with SiO_2 and TiO_2, coconut shell activated carbon and pitch-based spherical activated carbon were desirable. These supports showed more developed pore structures and had larger specificity. Pitch-based spherical activated carbon was found to be more desirable for acetylene hydrochlorination than coconut shell activated carbon because it had a higher N content. Au particles' aggregation and the loss of Au^{3+} would decrease the catalyst activity. Furthermore, the valence change of Au^{3+} and carbon deposition were also responsible for catalyst deactivation. The addition of La to Au can help stabilize Au^{3+} by inhibiting the valence change and carbon deposition [54].

N-doped carbon is one of the hot topics in support modification. N-doped carbon material as a non-metal catalyst has shown promising properties for acetylene hydrochlorination [69,70]. The introduction of N species changed the electronic structure of the adjacent carbon atoms so to adsorb HCl or acetylene easily, thereby enhancing the catalytic activity of carbon-supported catalysts. Dai et al. reported $AuCl_3$/PPy–multiwall carbon nanotubes (MWCNTs) with enhanced catalytic performance compared with $AuCl_3$/MWCNTs. This was attributed to electron transfer from polypyrrole to the Au^{3+} center, which facilitated the adsorption of HCl (Figure 6) [66]. Zhao et al. incorporated N-containing functional groups into carbon support by the post-modification of activated carbon with urea. The results showed that electron transfer from N atoms to the Au^{3+} center accelerated the adsorption of HCl by increasing the electron density of Au^{3+} [71].

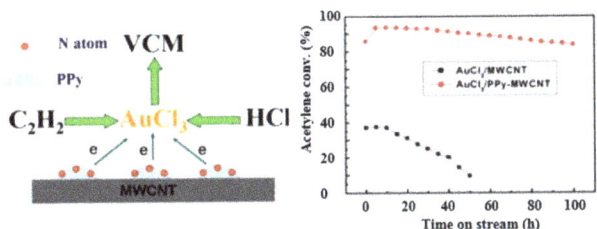

Figure 6. $AuCl_3$/PPy–multiwall carbon nanotube (MWCNT) catalyst for acetylene hydrochlorination [66].

ILs (Ionic liquids) are also used to stabilize Au^{3+} catalysts. Zhao et al. obtained Au^{3+} complexes, Au ILs (1-propyl-3-methylimidazolium tetrachloroaurate ([Prmim]$AuCl_4$)). The Au^{3+}-IL/C catalyst showed higher activity and stability than IL-free Au/C [72]. Zhao et al. reported that the reduced Au^0 could be regenerated in situ to Au^{3+} by $CuCl_2$ and further stabilized by the electron transfer from Cu^{2+} to these active species. The Au^{3+}-Cu^{2+}-IL/C catalyst demonstrated excellent activity. With this catalyst, over 99.8% selectivity was achieved for the VCM product (Figure 7) [39].

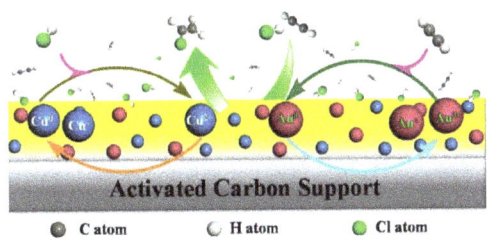

Figure 7. Au^{3+}-Cu^{2+}–IL/C catalyst for acetylene hydrochlorination [39]. IL: Ionic liquid.

Modification of the support can also improve the activity of the Au^0 catalyst. Tian et al. found that amorphous silica could be dispersed on the carbon surface uniformly as spherical particles. Silica deposition achieved a better distribution of Au NPs, which improved the catalytic activity of the Au^0 catalyst, although the surface area was decreased [73]. Dai et al. found that a mesoporous carbon nitride material can control the Au NPs sizes and was active for acetylene hydrochlorination [74]. Kang et al. reported that graphene substrate can improve the adsorption of HCl and C_2H_2, and N-doping significantly weakened the interaction with the Au_3 cluster for acetylene hydrochlorination [75]. Gong et al. reported that adding the heteroatom N to Au embedded in graphene (AuG) could reduce the reaction activation energy based on DFT calculations (Figure 8). With N-doping in AuG-SAC, the adsorption ability was increased, and the interaction between the starting materials HCl and acetylene was significantly enhanced. The energy band gap (ΔE_g) between the Highest Occupied Molecular Orbital (HOMO) of AuG-N_n-acetylene and the Lowest Unoccupied Molecular Orbital (LUMO) of the HCl, N-doped AuG-SACs decreased. Overall, this could reduce the activation energy of acetylene hydrochlorination [76].

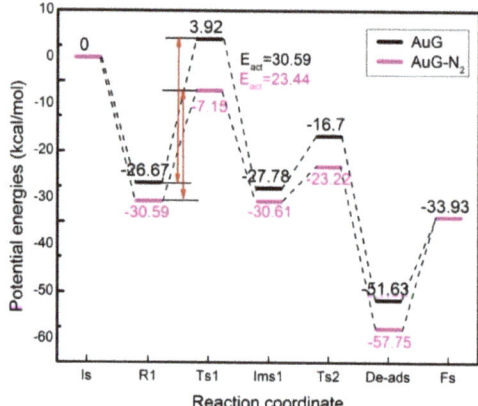

Figure 8. Energy diagrams of acetylene hydrochlorination on AuG and AuG–N_4 spheric active carbons (SACs) [76].

2.1.4. Deactivation and Regeneration of Au Catalysts

In 1988, Hutchings et al. investigated the reactivation by treatment of used catalysts in situ in the reactor with HCl. This treatment was applied to Au^{3+} and Hg^{2+} catalysts (Figure 9). The results indicated that the Hg catalyst cannot be reactivated, and 25% of the mercury had been lost from the catalyst. However, the activity of Au catalysts treated with HCl can be restored to the initial activity and the Au content of the deactivated catalyst did not decrease [16].

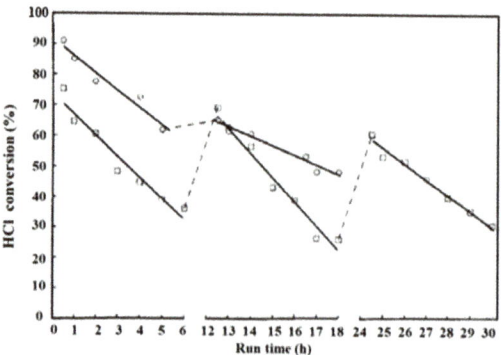

Figure 9. Effect of reactivation of catalyst with HCl treatment; ○ $HgCl_2$/C, 9% Hg; □ $HAuCl_4$/C, 1.9% Au; (——) HCl treatment [16].

The deactivation of a supported Au catalyst with Au^{3+} as the active species can be attributed to the following factors: the reduction, aggregation of the active particles of the Au catalyst, and the carbon deposition on the catalyst surface area, which may occur simultaneously [9,77]. Nkosi et al. reported that at low temperature (60–100 °C), carbon deposition led to the deactivation of the Au catalyst. At higher temperatures (120–180 °C), the active component Au^{3+}/Au^+ was reduced to Au^0, resulting in the deactivation of the Au catalyst (Figure 10) [78]. It was also observed that a minimum deactivation rate occurred at 100 °C. However, at this temperature, the activity of the Au catalyst was very low; it was not the optimum reaction temperature [26,31,79]. The deactivation of Au catalysts caused by a reduction of active components can be explained from the reaction mechanism of Au catalysts. Conte et al. reported that $AuCl_3$ could absorb acetylene and HCl, but the adsorption on acetylene was stronger. Acetylene has higher reducibility, which was one of the reasons for catalyst deactivation [5].

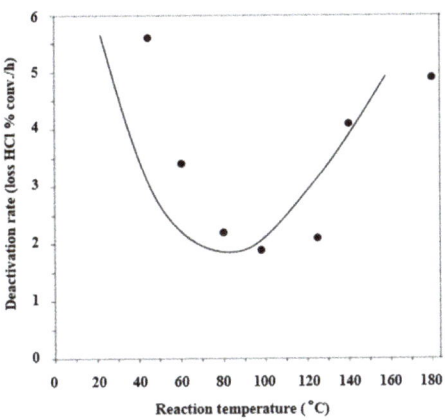

Figure 10. Effect of reaction temperature on rate of Au^{3+} catalyst deactivation [78].

For catalysts with Au^{3+} as the active center, the reduction of Au^{3+} is observed to be detrimental to catalyst activity [80]. One approach to improve the stability of Au^{3+} catalysts is to reactivate them. A large number of studies have shown that the deactivated Au catalyst can be regenerated by re-oxidation. Conte et al. demonstrated that a used catalyst can be regenerated by boiling aqua regia and a deactivation of Au catalyst was due to a loss of Au^{3+}, which was restored by the aqua regia treatment [81]. Cl_2, NO, and N_2O have been found to be effective in re-oxidation. These oxidants could alleviate the deactivation rate of catalyst [80]. Although Au^{3+} catalyst can be regenerated, the actual industrial application is still limited for various reasons.

The main reason for the deactivation of Au NPs with Au^0 as the active species is the aggregation of Au NPs during acetylene hydrochlorination [82]. Malta et al. reported that single-site Au was the active site, but the formation of metallic Au particles by single-site Au^0 was one reason for catalyst deactivation. Inelastic neutron scattering studies of Au^0 catalysts exposed to acetylene showed that an oligomeric acetylene species formed on the catalyst surface. These oligomers could be precursors for carbon formation and become a deactivation process, thus significantly slowing down the reaction process (Figure 11) [83].

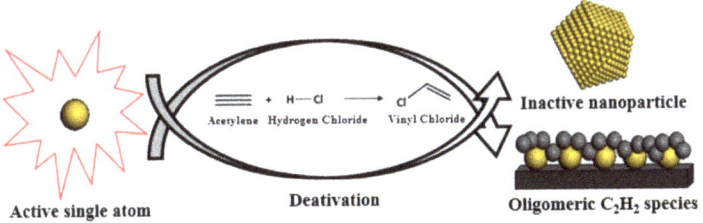

Figure 11. Deactivation of a single-site Au^0 catalyst [83].

2.2. Ru Catalysts

The research on Ru catalysts has drawn considerable attention. In 2013, Zhu et al. calculated the activity parameters of Au, Hg, and Ru for acetylene hydrochlorination by DFT. The results showed that the calculated activation barrier was 16.3, 11.9, and 9.1 kcal mol^{-1} for $HgCl_2$, $AuCl_3$, and $RuCl_3$, respectively. These results indicated that $RuCl_3$ was a good candidate for the hydrochlorination of acetylene [84]. Table 3 summarizes and shows the performance of some Ru catalysts for acetylene hydrochlorination.

Table 3. Performance of Ru catalysts for acetylene hydrochlorination.

Year	Catalyst	Material	Ru (wt %)	GHSV (h^{-1})	Temp (°C)	Acetylene Conv. (%)	Running Time (h)	Ref.
2018	Ru(III)-ChCl/AC	$RuCl_3$ + ChCl	0.2	900	170	99.3	25	[85]
2018	Φ-P-Ru/AC-HNO_3	$RuCl_3$ + tris-(triphenylphosphine) ruthenium dichloride	1	180	180	99.2	48	[86]
2018	$RuCl_3$-A/AC	$RuCl_3$ + $NH_3 \cdot H_2O$	2	100	180	95.8	8	[87]
2017	$(NH_4)_2RuCl_6$/AC	$(NH_4)_2RuCl_6$	1	180	170	90.5	12	[88]
2017	TPAP/AC-HCl	$C_{12}H_{28}NO_4Ru$	1	180	180	97	48	[89]
2016	Ru/N-AC	$RuCl_3$	1	57	250	95.2	180	[90]
2013	Ru_1Co_3/SAC	$RuCl_3$ + $CoCl_2$	1	180	170	95	48	[91]
2014	Ru/SAC-C300	$RuCl_3$	1	180	170	96.5	48	[92]
2016	Ru-Co(III)-Cu(II)/SAC	$RuCl_3$ + $CuCl_2$ + $Co(NH_3)_6Cl_3$	0.1	180	170	99	48	[93]
2017	Ru10%[BMIM]BF_4/AC	$RuCl_3$ + 1-Butyl-3-methylimidazolium tetrafluoroborate	1	180	170	98.9	24	[94]

Zhang et al. reported a series of Ru catalysts for acetylene hydrochlorination. Monometallic Ru, bimetallic Ru-Cu, and Ru-Co were prepared on spheric active carbon (SAC) support. It was proposed that the addition of Co could enhance the catalytic activity and Ru species (RuO_2, Ru^0, RuO_x, and $RuCl_3$) of the Ru catalyst [91]. It was found that the active ingredient of RuO_2 was critical for improving the catalytic performance of Ru catalysts [92]. A series of trimetallic Ru-Co^{3+}-Cu^{2+}/SAC catalysts were synthesized and evaluated for acetylene hydrochlorination. Compared with the monometallic and bimetallic catalysts, the trimetallic Ru-Co^{3+}-Cu^{2+}/SAC catalyst with 0.1 wt % Ru loading showed a superior catalytic performance with the acetylene conversion of 99.0%. It was found that the addition of Co^{3+} and Cu^{2+} can help form highly dispersed Ru on the support, and Cu^{2+} species can not only inhibit the reduction of $RuCl_3$ precursors but also facilitate the formation of Co^{3+} from Co^{2+} or Co^0 (Figure 12) [93].

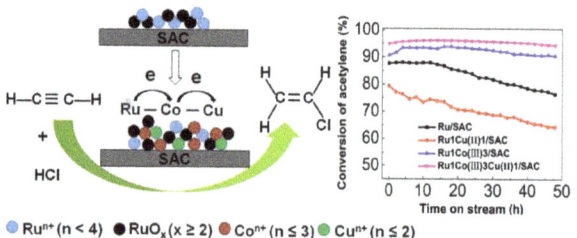

Figure 12. Ru-Co^{3+}-Cu^{2+}/SAC catalyst for acetylene hydrochlorination [93].

In 2017, imidazolium-based ILs (IBILs) were reported. The optimal Ru10% [BMIM]BF_4/C catalyst achieved the acetylene conversion of 98.9% and the selectivity to VCM of 99.8%. It was proposed that IBILs additives could improve the dispersion of Ru species and prevent coke deposition due to the interactions between Ru species and [BMIM]BF_4. Moreover, the O-containing functional groups on the carbon support were involved in the interactions between Ru species and [BMIM]BF_4 [94].

2.3. Pt Catalysts

There are a few reports on the catalytic hydrochlorination of acetylene by Pt catalysts, and the research mainly focuses on the catalytic mechanism. Mitchenko et al. conducted a series of studies on Pt catalysts. In their study, K_2PtCl_6 salt was treated under acetylene, ethylene, or propylene atmosphere to form a heterogeneous catalyst. It was shown that with this approach, the formation of Pt^{2+} complexes and Pt complexes with vacancies were observed. It was hypothesized that the active centers were defected in the form of impure Pt^{2+} ions in the K_2PtCl_6 matrix [95]. In 2004, Mitchenko et al. found that trans-d-VCM was formed in the DCl atmosphere. According to the established catalytic reaction mechanism on the active sites, a K_2PtCl_6 lattice that defected in the form of topologically bound coupled $PtCl_4^{2-}$–$PtCl_5^-$ was formed by mechanical pre-activation under acetylene. Acetylene chloroplatination started first by coordination with an unsaturated Pt^{4+} complex producing an intermediate β-chlorovinyl Pt^{4+} derivative. The reduction of this intermediate yielded a Pt^{2+} organometallic derivative. The catalytic mechanism of acetylene hydrochlorination over the Pt catalyst is shown in Figure 13 [96].

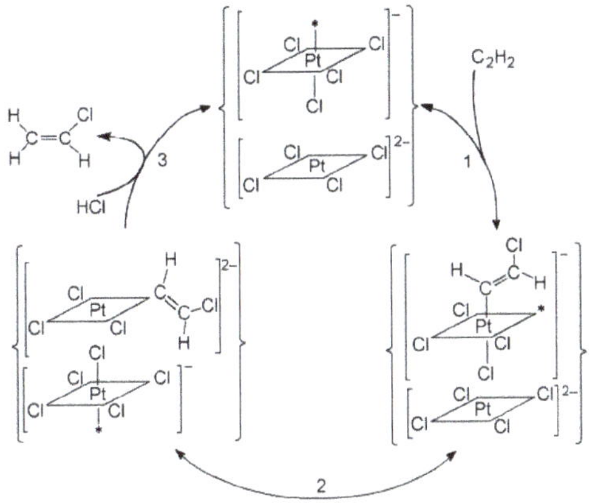

Figure 13. Catalysis mechanism of acetylene hydrochlorination over the Pt catalyst [96].

Mitchenko et al. also found that the catalytic reaction of acetylene chloroplatination proceeds in the form of π-acetylene complexes. The limiting step of the reaction was acetylene chloroplatination. It involved a π-acetylene complex and HCl, producing a new π-acetylene Pt^{2+} complex and an intermediate β-chlorovinyl Pt^{2+} derivative. The protonolysis of the intermediates resulted in VCM formation [23]. In 2014, Mitchenko et al. reported that HCl was involved in two steps of acetylene hydrochlorination. With K_2MCl_4 (M = Pt or Pd) and K_2PtCl_6 catalysts, the intermediate β-chlorovinyl derivative of Pt^{2+} was observed in protodemetallation over the Pt-containing catalysts [97].

2.4. Pd Catalysts

In 1985, Hutchings made Pd catalysts supported on carbon with relatively high initial activity for acetylene hydrochlorination but also found that Pd catalysts would be rapidly deactivated due to Pd^{2+} being reduced into Pd^0 and carbon deposition [14]. Nkosi et al. reported that Pd catalysts were found to be as active as Hg catalysts on a mole basis. However, loadings of $PdCl_2$ catalysts with durable stability were very low (<1 wt %). At higher loadings, $PdCl_2$ was rapidly lost from the catalyst presumably as a result of the formation of a volatile Pd acetylene complex. The loss of the average rate of conversion was 6.8% conv. h^{-1} during the first 3 h [16]. Strebelle et al. reported that 57.3% yield and a 96% selectivity for VCM was achieved with $PdCl_2$ catalyst without a detailed description of catalyst stability [98].

Table 4 shows the performance of some Pd catalysts for acetylene hydrochlorination, in which the Pd^{2+} species is the active species.

Table 4. Performance of Pd catalysts for acetylene hydrochlorination.

Year	Catalyst	Material	Pd (wt %)	GHSV (h^{-1})	Temp (°C)	Acetylene Conv. (%)	Running Time	Ref.
2016	Pd/PANI-HY	H_2PdCl_4 + polyaniline	0.9	110	160	95	300 h	[99]
2016	Pd-K/NFY	$PdCl_2$ + KCl + NH_4F	0.9	110	160	99	50 h	[100]
2015	Pd/NH4F-HY	$PdCl_2$ + NH_4F	0.9	110	160	99.92	400 min	[101]
2013	Pd/HY	H_2PdCl_4	0.5	110	160	95	160 min	[102]
2010	$PdCl_2$-KCl-$LaCl_3$/C	$PdCl_2$ + KCl + $LaCl_3$	0.9	120	160	99	3 h	[103]

Wang et al. reported that the presence of NH_4F in Pd/HY catalysts can significantly improve the catalyst performance (acetylene conversion >95% and VCM selectivity >90%). The hypothesis was that NH_4F could partially enhance the surface acidity, prevent carbon deposition, and slow down the loss of Pd (Figure 14) [101]. Wang et al. prepared a Pd catalyst (Pd/HY) with Y zeolite (HY) support. The catalyst exhibited excellent catalytic performance with >95% acetylene conversion and >90% VCM selectivity. The results showed that with the aid of ultrasonic-assisted impregnation, Pd was uniformly dispersed on the Y zeolite surface [102].

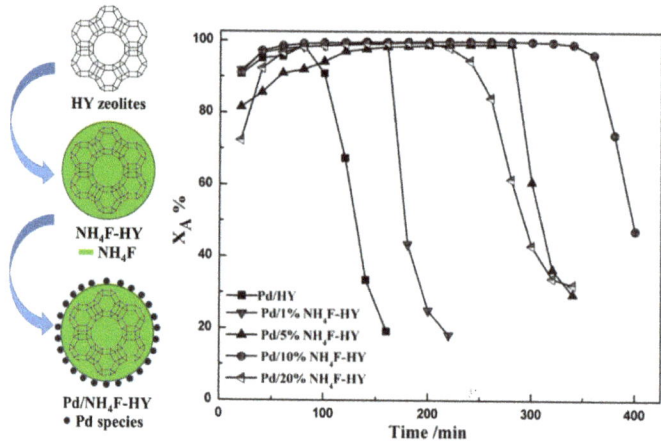

Figure 14. Pd/HY catalysts modified with NH_4F for acetylene hydrochlorination. Adopted from Ref [101].

Song et al. reported that the deactivation of Pd catalysts were mainly from the loss of an active component, surface carbon deposition, and decrease in specific surface area. The results showed that the addition of KCl and $LaCl_3$ can reduce the loss of $PdCl_2$, increase the conversion of acetylene, and improve the selectivity of VCM. When the used $PdCl_2$-KCl-$LaCl_3$/C catalyst was treated with HNO_3 and HCl, it can be re-activated to some extent by oxidation, but it is not as good as the original one [103].

N-doped carbon materials and ILs were used to improve the catalytic performance of Pd catalysts. Li et al. demonstrated that N doping can prevent carbon deposition and improve the catalyst stability. The doped N atoms can act as a ligand, stabilizing $PdCl_2$ and improving the surface activity [104]. Yang et al. reported IL-stabilized metal NPs (NPs@IL)–cosolvent liquid–liquid biphasic Pd NPs catalysts with improved performance for acetylene hydrochlorination. The NPs@IL droplets served as microreactors. The low viscosity organic phase enabled the rapid mass transfer. The acetylene conversion of 98% and VCM selectivity of 99.5% were achieved with this type of Pd catalyst [105]. Zhao et al. chose 1-butyl-3-methylimidazolium chloride ([Bmim]Cl) IL and activated carbon support to immobilize volatile catalytically active Pd complexes. The 0.5Pd-10IL/C catalyst showed superior performance with 98.6% conversion of acetylene and above 99.8% selectivity to VCM. The influence of IL on the catalyst properties such as activity and stability was mainly due to the stabilization and dispersion of active Pd species in the IL layer, the possible coordination of the IL to the Pd atoms, and the confinement of Pd complexes within the pore's supports (Figure 15) [99].

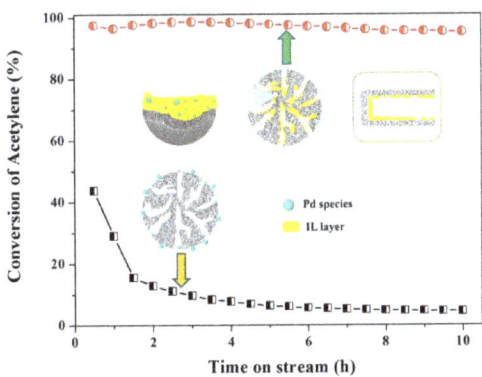

Figure 15. Supported IL-Pd catalyst for acetylene hydrochlorination [106].

2.5. Challenge of Noble Metal Catalysts

For the noble metal catalysts, the main problems of their industrialization are the high cost and short service life. There are mainly two problems that need to be solved: On the one hand, reducing the load of noble metals, which needs to be optimized from the catalyst formulation, and on the other hand, improving the stability of the noble metal catalyst. Therefore, the industrial application of noble metal catalysts should consider the reduction of noble metal catalyst loading capacity; at the same time, it should include a comprehensive consideration of support optimization, activated carbon pretreatment, additives, and catalyst preparation methods.

3. Non-Noble Metal Catalysts

Supported noble metal catalysts have the characteristics of high price and poor stability. Even the relatively good performance of Au catalyst is not able to meet the industrialization requirements for the economy and effect. Non-noble metal catalysts have become one of the catalysts with great potential for industrialization due to their advantages of being cheap and easy to obtain. At present, researches on non-noble metals are mainly focused on supported catalysts with Cu, Bi, and Sn as active centers.

3.1. Cu Catalysts

Compared with noble metals, Cu has the advantage of low price and easy availability with good thermal stability. Its price is relatively low, even in non-noble metals. In Hutchings' study, Cu^{2+} has a high potential of catalyzing acetylene hydrochlorination due to its high electrode potential. Cu is often used as an auxiliary agent in the modification of other metal active center catalysts [14]. In fact, acetylene hydrochlorination with Cu as the main active center has been reported as early as in 1933. US patent 1934324 first introduced CuCl catalyzed acetylene hydrochlorination as an active component [107]. Table 5 shows the performance of some Cu catalysts for acetylene hydrochlorination.

Table 5. Performance of Cu catalysts for acetylene hydrochlorination.

Year	Catalyst	Material	Cu (wt %)	GHSV (h^{-1})	Temp (°C)	Acetylene Conv. (%)	Running Time	Ref.
2018	Cu-Cs/AC	CuCl$_2$ + CsCl	1	50	200	92	200	[108]
2016	Cu-g-C$_3$N$_4$/AC	CuCl$_2$ + dicyan-diamide	4.15	72	180	79	450 min	[109]
2015	Cu$_2$P$_2$O$_7$/SAC	Cu$_2$P$_2$O$_7$	15	180	140	40.5	500 min	[110]
2015	Cu400Ru/MWCNTs	CuCl$_2$ + RuCl$_3$	4.24	180	180	51.6	6 h	[111]
2014	Cu-NCNT	CuCl$_2$ + N-doped carbon nanotubes	5	180	180	45.8	4 h	[112]

Zhao et al. found that Cu-g-C$_3$N$_4$/AC significantly improved catalytic performance compared to the catalyst without N doping. The authors hypothesized that after doping, pyrrolic N promoted the adsorption of HCl and acetylene, and it inhibited the coke deposition on the catalyst surface [109]. Li et al. reported that the P-doped Cu demonstrated good stability and catalytic activity. The amount of P used was critical and significantly influenced the catalyst activity. It was assumed that P doping enhanced the interaction between metal and support, improved the dispersity of Cu, retarded the reduction of the active components (Cu^{2+}/Cu$^+$) to Cu^0, and effectively prevented the aggregation of Cu particles [110]. Zhou et al. reported that N-doped carbon nanotubes (N-CNTs) improved the electron conductivity ability of the carbon sheets as well as the interaction between Cu and N-CNTs, and the improvement of the adsorption capacity of acetylene led to high catalytic activity and excellent VCM selectivity. The active components were proposed to be a mixture of Cu$^+$ and Cu^{2+}. It was speculated that acetylene hydrochlorination preferred to take place at binding sites of Cu and N (Figure 16) [112].

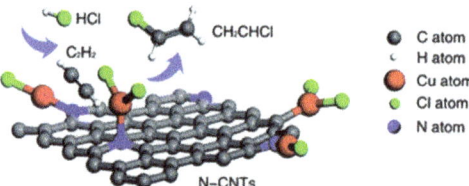

Figure 16. Cu-N-CNT catalyst for acetylene hydrochlorination [112]. N-CNT: N-doped carbon nanotubes.

In the preparation of Cu catalysts, the addition of promoters such as metal chlorides and rare earth oxides can improve the catalytic performance. Shi et al. reported that the catalyst La$_{1.7}$K$_{0.3}$NiMnO$_6$-CuCl$_2$/γ-Al$_2$O$_3$ doped with K had a large amount of Mn^{4+} and surface-adsorbed oxygen. These species accelerated the oxidation of Cu$^+$ and released Cl$_2$ [3]. Xu et al. reported a catalyst consisting of 400 ppm Ru and 4.24 wt % Cu. Carbon nanotubes were chosen as the support. The synergistic effect between Ru and Cu led to a high activity (TOF = 1.81 min^{-1}) [111]. Zhai et al. synthesized a perovskite-like catalyst for acetylene hydrochlorination. The CsCuCl$_3$ NPs were supported on activated carbon with 1 wt % Cu content, which exhibited superior performance compared with a pure Cu catalyst without Cs. The formation of the perovskite-like CsCuCl$_3$ complex structure stabilized the active Cu^{2+} [108].

In order to improve the performance of Cu catalysts, efforts were attempted by modifying the support and adding promoters for improving the dispersion of Cu. The modification did not change the structure of the Cu active center. The preparation of a highly dispersed Cu catalyst is complex and is limited for large-scale industrial production. The addition of expensive noble metals can increase the catalytic activity of Cu catalysts; however, the advantage of cheap and easy availability of non-noble metal catalysts has been lost.

3.2. Other Non-Noble Metal Catalysts

Zhou et al. synthesized a Bi/Cu/H$_3$PO$_4$ catalyst supported by silica gel. The catalyst exhibited acceptable initial activity (30% activity of Hg catalyst) and stability at 200 °C and GHSV 360 h^{-1}. The catalyst could be recovered by a two-step regeneration method [11]. Hu et al. found that the doping of S increased the Brunauer-Emmett-Teller (BET) surface areas and decreased the active species particle size of the Bi catalyst, which led to more accessible active sites and consequently boosted the catalytic activity in hydrochlorination. The main active components of the catalyst is BiOCl crystal particle, and the coke deposition was responsible for the deactivation of the Bi catalyst [113].

Dai et al. developed a series of transitional metal catalysts. In their work, Mo and Ti were chosen as the active ingredients in a catalyst for acetylene hydrochlorination. With this catalyst, 89% acetylene conversion and >98.5% VCM selectivity can be obtained. The doping of Mo and Ti reduced the adsorption capacity for acetylene while increasing the adsorption of HCl [114]. The authors reported three transition metal nitride VN/C, Mo$_2$N/C, and W$_2$N/C catalysts on activated carbon. With these catalysts, 98% VCM selectivity was achieved [115]. For W$_2$N, two catalysts were prepared. The first was prepared by depositing W onto activated carbon, whereas the second was prepared in the same procedure except for extra plasma treatment. The author claimed that plasma treatment increased the interaction between active ingredients and support material. The results showed that plasma treatment increased the number of W–N bonds and restricted the deposition of coke, thus enhancing the catalytic performance [116]. They found that the Co-N-AC catalyst effectively improved the adsorption of HCl. The presence of Co-N$_x$ played a major role in acetylene hydrochlorination (Figure 17) [117].

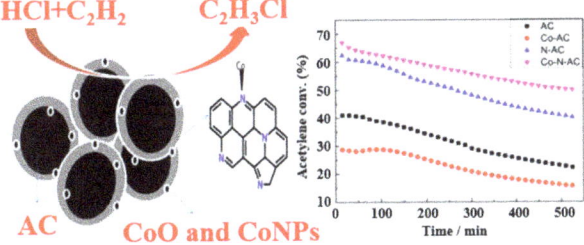

Figure 17. Co-N-AC catalyst for acetylene hydrochlorination [117].

3.3. Challenge of Non-Noble Metal Catalysts

At present, the research of non-noble metals catalysts is largely confined to the laboratory stage. Non-noble metal catalysts are constantly modified and improved to increase their activity and stability by adding additives or using bimetallic or trimetallic catalysts. However, there are still challenges and limitations, such as low monomer conversion rate, poor selectivity of VCM, severe carbon accumulation, loss of catalytic activity over time, unsatisfactory stability, and short lifetime.

4. Non-Metallic Catalysts

An alternative approach to replace Hg catalysts is to develop the non-metallic catalysts. Non-metallic catalysts are newly emerging green catalytic materials that have attracted much attention in recent years due to their advantages of environmental friendliness and low cost.

4.1. Carbon Material Catalysts

The research of carbon catalysts is mainly about the structural defect sites, the regulation of functional groups on the carbon materials surface, and the doping of heteroatoms. Non-metallic nanocarbon materials for acetylene hydrochlorination have been intensively investigated by many researchers (Table 6).

Table 6. Performance of carbon material catalysts for acetylene hydrochlorination.

Year	Catalyst	Material	GHSV (h^{-1})	Temp (°C)	Acetylene Conv. (%)	Running Time (h)	Ref.
2019	ND@G	Nanodiamond + graphene	300	220	50	10	[118]
2018	ZIF-8/SAC	ZIF-8 + spherical activated carbon	30	220	81	2	[119]
2018	NS-C-NH$_3$	S + N-doped carbon + NH$_3$	35	220	80	9	[120]
2017	p-BN	H$_3$BO$_3$ + melamine + NH$_3$	-	280	99	50	[121]
2016	Z4M1	ZIF-8 + melamine	50	180	60	20	[122]
2015	B,N-G	Graphene oxide + H$_3$BO$_3$ + NH$_3$	360	150	94.87	4	[123]
2014	N-OMC-700	N-doped ordered mesoporous carbon	-	200	77	100	[124]
2014	N-CNTs	C$_2$H$_4$ + NH$_3$	180	180	7.2	3	[125]
2014	g-C$_3$N$_4$	Activated carbon + cyanamide	50	180	76.52	7	[69]
2014	PSAC-N	Pitch-based spherical activated carbon + melamine	120	250	68	-	[126]

Efforts have been devoted to improving the activity and stability of catalysts by modifying carbon-based catalysts. The most common materials studied are N-doped carbon materials. N-containing carbon materials are often prepared by heating different N-containing precursors under N$_2$ atmospheres or direct carbonization [7,70]. It was hypothesized that the incorporation of N atoms could tune the basicity of carbon materials surface. The N atoms could also provide available lone pairs, which can facilitate a hydrochlorination reaction.

Li et al. prepared a series of zeolitic imidazolate framework (ZIF)-derived N-doped carbon using melamine as N sources. Additions of melamine not only facilitated forming a special morphological structure and pore structure but also adjusted the relative content ratio of three N species in ZIF-derived N-doped carbon materials [122]. Li et al. reported that N-doped ordered mesoporous carbon (N-OMC) with high surface areas processed high activity and stability. A 77% acetylene conversion and 98% VCM selectivity was obtained over the 100 h test period. The N-OMC activity increased with the increasing of the catalyst surface area and N content [124]. Wang et al. prepared an N-doped pitch-based spherical activated carbon catalyst (PSAC-N) with melamine. The active site of PSAC-N had a special N-6v structure in which quaternary N bonded between two 6-membered rings. Compared to controls, the adsorption capacity of HCl on PSAC-N was the highest, which was beneficial for acetylene hydrochlorination. The reaction energy of N-6v(7) was calculated as 236.2 kJ mol^{-1} [126]. Li et al. reported that the g-C$_3$N$_4$/C catalyst prepared with cyanamide as a precursor displayed considerable catalytic performance. C$_3$N$_4$ was able to adsorb and activate acetylene and HCl simultaneously, and the strong C(1)-C(2) interaction led to the formation of a stable CHCl=CH-C$_3$N$_4$H unit that provided a rate-controlling step (E_{act} = 77.94 kcal mol^{-1} [69]). Qian et al. synthesized a g-C$_3$N$_4$ framework with rich N defects using melamine formaldehyde as a precursor. The fragmentary g-C$_3$N$_4$ showed high catalytic activity with 94.5% acetylene conversion, which was 30 times higher than that of pure g-C$_3$N$_4$. The improvements were attributed to the etching of the g-C$_3$N$_4$ framework by O species in the melamine formaldehyde resin. Moreover, DFT calculations showed that the N defects in the g-C$_3$N$_4$ framework greatly improved the adsorption of HCl and acetylene and brought down the energy barrier from 62.0 to 38.1 kcal mol^{-1} [127]. Zhang et al. reported that the N-doped carbon catalyst with polyaniline had significantly improved catalytic activity and stability (Figure 18). The catalytic performance of this catalyst was closely related to pyrrolic N (pyrrolic N > graphitic N > pyridinic N) [128].

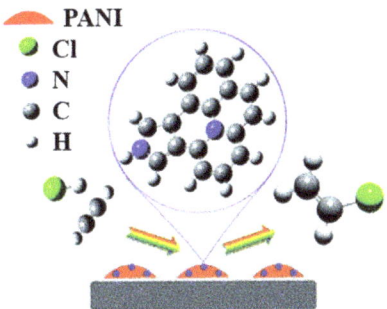

Figure 18. N-doped active carbon catalyst for acetylene hydrochlorination [128].

Lan et al., reported nanodiamond–raphene composed of a nanodiamond core and graphitic shell with defects. The catalyst showed pretty high catalytic activity, which was comparable with metal catalysts for acetylene hydrochlorination. Figure 19 shows that the defects on the carbon surface and edge were assumed to be desirable for acetylene hydrochlorination [118]. Zhou et al. found that N-CNTs featured good catalytic activity (TOF = 2.3×10^{-3} s^{-1}) and high VCM selectivity (>98%). There was an observable correlation between the quaternary N content and acetylene conversion. DFT calculation revealed that N doping enhanced the interaction between acetylene and N-CNTs compared with the CNTs without N [125]. Cyanamide and its derivatives are common raw materials for preparing N-doped carbon catalysts. Li et al. reported ZIF-8-derived NPs featuring more active sites, which were easily accessible for reactants HCl and acetylene. Bamboo-shaped CNTs promoted the efficiency of active site N, changed the porosity of the catalyst, and inhibited the formation of coke deposition [119]. Zhao et al. proposed that the defects in graphene could be useful to improve the catalytic performance for acetylene hydrochlorination based on the DFT calculated [129].

Figure 19. Defective nanodiamond–graphene catalyst for acetylene hydrochlorination [118].

Dual heteroatom doping was also attempted for catalyst design. Dong et al. reported that an S and N co-doped carbon catalyst exhibited a higher acetylene conversion than the N-doped carbon. The C atoms adjacent to pyridinic N were found to be active sites, and S improved the pyridinic N content in N-doped carbon materials (Figure 20) [120]. Li et al. reported that porous boron nitride (p-BN) was active for acetylene hydrochlorination. The activity of p-BN was originated from the defects and edge sites. Particularly, the armchair edges of p-BN can polarize and activate acetylene, which react with HCl to form VCM [121]. Dai et al. synthesized B and N heteroatoms dual doped on a graphene oxide (B,N-G) catalyst. With this catalyst, acetylene conversion is near 95%, which is significantly higher than that of singly B- or N-doped graphene. The authors proposed that the synergistic effect of B and N doping promoted HCl adsorption (Figure 21) [123]. Li et al. reported an

N-doped carbon nanocomposite derived from silicon carbide. It was found that the catalyst activated acetylene directly in which C atoms bonded with pyrrolic N were the active sites [130]. Zhao et al. found that the doping of C atoms on boron nitride (BN) cages can significantly promote the adsorption ability of the acetylene molecule compared with the BN cages without doping. Acetylene was adsorbed onto $B_{12-n}N_{11+n}C$ ($n = 0, 1$) clusters prior to HCl and then formed three adsorption states: two *trans* configuration and one *cis* configuration. The energy barrier of the minimum energy pathway based on $B_{11}N_{12}C$ catalyst was as low as 36.08 kcal mol^{-1} [131].

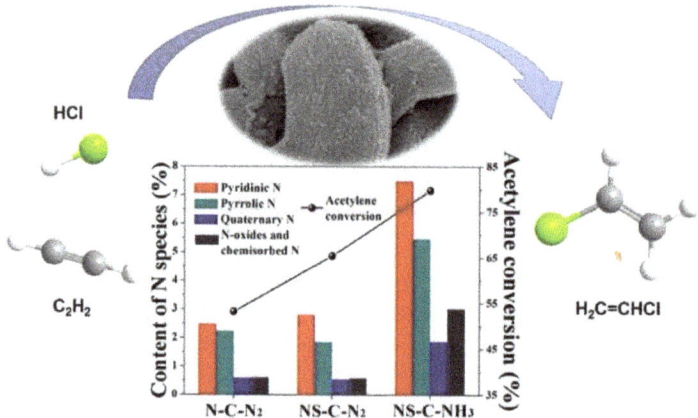

Figure 20. S and N co-doped carbon catalyst for acetylene hydrochlorination [120].

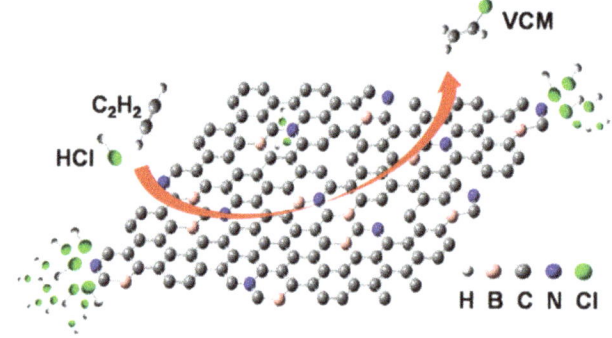

Figure 21. B,N-graphene catalyst for acetylene hydrochlorination [123].

The deactivation mechanism of N-doped carbon catalysts has been studied. Li et al. reported that a PDA/SiC composite with high activity and stability can be prepared by coating silicon carbide with polydopamine (PDA). The catalyst deactivation was caused presumably by the carbon deposition, which can be regenerated with NH$_3$ treatment at high temperature [132]. Chao et al. reported that C atoms adjacent to the pyridinic N were the active sites of N-doped carbon catalyst derived from ZIF-8. Coke deposition over pyridinic N was responsible for catalyst deactivation [133].

4.2. Other Non-Metallic Catalysts

Zeolites not only have a large specific surface area and unique pore structure, but also have abundant active sites and acid centers. Therefore, zeolites are widely adopted as catalysts in the chemical industry. Song et al. investigated the activity of faujasite (FAU) zeolite 13X for acetylene hydrochlorination without loading any metal. It was found that the FAU zeolite 13X demonstrated observable activity presumably from its abundant micropores and unique supercage; acetylene conversion was up to 98.5%. Na cations in the zeolite framework could facilitate acetylene hydrochlorination. The deactivation of 13X was probably due to the collapse of its crystal structure and severe carbon deposit. The spent catalyst could be partly regenerated by calcination [134].

ILs are well-known green solvents. They have unique physicochemical properties involving high solubility, high thermal stability, and negligible volatility, and they have been applied in many fields including catalysis, adsorption, and organic synthesis. Li et al. found that tetraphenylphosphonium bromide (TPPB) ILs exhibited strong adsorption for HCl but weak adsorption for acetylene. At the catalytic active site of TPPB, the activation energy of acetylene hydrochlorination was 21.15 kcal mol^{-1}, which is much lower than that without a catalyst (44.29 kcal mol^{-1}). During the reaction, the H-Cl bond was preferentially activated through the electrons transfer from the anion of TPPB; then, acetylene was activated to undergo the addition of H and Cl. Such unique preferential activation toward the H–Cl bond significantly promoted the catalytic activity and the stability of TPPB catalysts. The catalyst has good catalytic activity and stability, and the highest acetylene conversion rate was 97.1% [135]. Wang et al. investigated the catalytic performance of imidazolium-based ILs 1-ethyl-3-methylimidazolium tetrafluoroborate ([Emim][BF$_4$]) (IL/CaX) for acetylene hydrochlorination with a nearly 90% acetylene conversion. The reaction proceeded via a two-site mechanism. HCl adsorbed on the Ca^{2+} in zeolite reacted with acetylene adsorbed on IL cations to form VCM. The catalytic reaction occurred at the IL/CaX interface, while the IL/CaX in the upper interphase had no activity. Catalyst deactivation was mostly caused by by-products dissolved in the IL and can be reactivated by a simple vacuum process [136]. Moreover, Zhou et al. reported that 1-alkyl-3-methyimidazolium-based ILs as metal-free catalysts have to consist of the chloride anion due to its participation in acetylene hydrochlorination with the formation of [HCl$_2$]$^-$ by the activation of HCl molecules. The structural modification of conjugated cation in ILs was the key to promote the catalytic activity (Figure 22) [137].

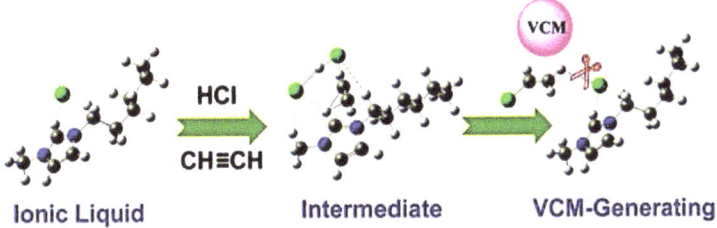

Figure 22. ILs catalyst for acetylene hydrochlorination [137].

4.3. Challenge of Non-Metallic Catalysts

The design and development of non-metallic catalysts for acetylene hydrochlorination has been widely studied in recent years. It is worth looking forward to the application of non-metallic catalysts in industry as an economical and environmentally friendly approach. However, most of the non-metallic catalytic materials reported at present have some intrinsic disadvantages, which limit the practical industrial applications. Researchers are working vigorously to meet the needs and requirements of viable catalysts in the production of the VCM industry. Although non-metallic doped catalysts, especially N-doped carbon catalysts, have been used in many reactions, there are still some problems and hurdles for acetylene hydrochlorination. The first is that the reaction mechanism of different N species (pyridine N, pyrrole N, and graphite N) has not yet been determined. In addition, the highest conversion rate of acetylene for N-doped carbon catalysts is about 95%, which reasonably matches the industrial space velocity. Compared with the acetylene conversion rate of Hg or even Au catalyst approaching 100% at extreme space velocity, a significant amount of work needs to be done to improve both acetylene conversion and VCM selectivity.

Compared with solid-phase catalysts, ILs have some advantages in activity and stability. However, it is difficult to separate the reaction products from the catalysts, and there are also issues in the recovery and reuse of deactivated catalysts.

5. Future Prospects

Acetylene hydrochlorination is an important reaction in the traditional coal (chlor-alkali) chemical industry. The main problems of mercury-free catalyst with noble metals are the high application cost and short catalyst life.

High cost has become the biggest issue to the industrial application of noble metal catalysts. Reducing the content of noble metal alongside improving the stability and life of the catalysts are very critical in the research and development of noble metal catalysts. The following aspects of research need to be strengthened as well. The first is to develop or find better support materials. The commonly used activated carbon has well-developed pores. The distribution of pores is very extensive, and the existence of functional groups is very rich. At present, research on the microstructure and internal properties of activated carbon are relatively few. It is necessary and meaningful to continue to develop high-performance carbon supports that are suitable for acetylene hydrochlorination with a large specific surface area, appropriate pore size, and suitable internal chemical properties. At the same time, it is important to find ways to improve the reaction efficiency of noble metal catalysts by actively developing new supports other than carbon materials. The second is to look for non-noble metal additives that could be substituents of noble metal active components. By adding non-noble metal active components or additives, one would expect to see coordination and collective effects in improving catalyst performance. Meanwhile, the use of noble metals can also be reduced to lower the cost of the catalyst. The third is to find a better way to support the noble metal components so as to have dispersity at the nanometer level while maintaining durable stability and robustness. It is easy for the traditional simple impregnation method to cause the aggregation and agglomeration of the active components, which quickly bring down the catalytic activity of noble metal species and limit their applications. A reasonable loading mode can provide more active centers for noble metal components per unit mass, so as to improve the catalytic efficiency of noble metal particles and reduce noble metal content. Finally, the regeneration and recycling of noble metal catalysts are also very important. If the noble metal can be reused well, the problem of high price and high cost of catalyst may be solved.

The problems of non-noble metal and non-metallic catalysts are that although they are cheap, they suffer from poor stability and selectivity, low activity, and a low conversion rate. An in-depth study of the reaction mechanisms of the catalytic process would be very helpful. The interaction and collaboration between active components and supports, in other words, the multi-component composite catalysts may be prepared to further improve the catalysts' performance. Optimizing the formula and loading process of active components with a well-established computational approach and artificial intelligence is challenging and exciting. The traditional production process of VCM adopts the fixed bed reaction technology, which is well established and mature; however, the catalysts in this process suffer large loss, easily get pulverized, and suffer poor heat transfer, which limits the catalytic performance of the catalysts.

With continuing efforts and deeper understanding of the catalytic mechanism, the design and development of non-mercury catalysts with satisfactory performance is viable. The industrial application of mercury-free acetylene hydrochlorination catalysts is expected to be realized in the near future.

Funding: This work was financially supported by UESTC Talent Startup Funds (A1098 5310 2360 1208), National Natural Science Foundation of China (21464015, 21472235), and Xinjiang Tianshan Talents Program (2018xgytsyc 2–3).

Conflicts of Interest: The authors declare no conflict of interest.

References

1. Xu, H.; Luo, G.H. Green production of PVC from laboratory to industrialization: State–of–the–art review of heterogeneous non–mercury catalysts for acetylene hydrochlorination. *J. Ind. Eng. Chem.* **2018**, *65*, 13–25. [CrossRef]
2. He, W.; Zhu, G.Q.; Gao, Y.; Wu, H.; Fang, Z.; Guo, K. Green plasticizers derived from epoxidized soybean oil for poly (vinyl chloride): Continuous synthesis and evaluation in PVC films. *Chem. Eng. J.* **2020**, *380*, 122532. [CrossRef]
3. Shi, D.Z.; Hu, R.S.; Zhou, Q.H.; Yang, L.R. Catalytic activities of supported perovskite promoter catalysts La_2NiMnO_6–$CuCl_2$/γ–Al_2O_3 and $La_{1.7}K_{0.3}NiMnO_6$–$CuCl_2$/γ–Al_2O_3 for ethane oxychlorination. *Chem. Eng. J.* **2016**, *288*, 588–595. [CrossRef]
4. Ye, L.; Duan, X.P.; Wu, S.; Wu, T.S.; Zhao, Y.X.; Robertson, A.W.; Chou, H.L.; Zheng, J.W.; Ayvali, T.; Day, S.; et al. Self–regeneration of Au/CeO_2 based catalysts with enhanced activity and ultra–stability for acetylene hydrochlorination. *Nat. Commun.* **2019**, *10*, 914. [CrossRef] [PubMed]
5. Conte, M.; Carley, A.F.; Heirene, C.; Willock, D.J.; Johnston, P.; Herzing, A.A.; Kiely, C.J.; Hutchings, G.J. Hydrochlorination of acetylene using a supported gold catalyst: A study of the reaction mechanism. *J. Catal.* **2007**, *250*, 231–239. [CrossRef]
6. Zhang, J.L.; Liu, N.; Li, W.; Dai, B. Progress on cleaner production of vinyl chloride monomers over non–mercury catalysts. *Front. Chem. Sci. Eng.* **2011**, *5*, 514–520. [CrossRef]
7. Malta, G.; Freakley, S.J.; Kondrat, S.A.; Hutchings, G.J. Acetylene hydrochlorination using Au/carbon: A journey towards single site catalysis. *Chem. Commun.* **2017**, *53*, 11733–11746. [CrossRef]
8. Johnston, P.; Carthey, N.; Hutchings, G.J. Discovery, development, and commercialization of gold catalysts for acetylene hydrochlorination. *J. Am. Chem. Soc.* **2015**, *137*, 14548–14557. [CrossRef]
9. Ren, W.; Duan, L.; Zhu, Z.W.; Du, W.; An, Z.Y.; Xu, L.J.; Zhang, C.; Zhuo, Y.Q.; Chen, C.H. Mercury transformation and distribution across a polyvinyl chloride (PVC) production line in China. *Environ. Sci. Technol.* **2014**, *48*, 2321–2327. [CrossRef]
10. Hutchings, G.J. Gold catalysis in chemical processing. *Catal. Today* **2002**, *72*, 11–17. [CrossRef]
11. Zhou, K.; Jia, J.C.; Li, X.G.; Pang, X.D.; Li, C.H.; Zhou, J.; Luo, G.H.; Wei, F. Continuous vinyl chloride monomer production by acetylene hydrochlorination on Hg–free bismuth catalyst: From lab–scale catalyst characterization, catalytic evaluation to a pilot–scale trial by circulating regeneration in coupled fluidized beds. *Fuel Process. Technol.* **2013**, *108*, 12–18. [CrossRef]
12. Smith, D.M.; Walsh, P.M.; Slager, T.L. Studies of silica–supported metal chloride catalysts for the vapor–phase hydrochlorination of acetylene. *J. Catal.* **1968**, *11*, 113–130. [CrossRef]
13. Shinoda, K. The vapor–phase hidrochlorination of acetylene over metal chlorides supported on activated carbon. *Chem. Lett.* **1975**, *4*, 219–220. [CrossRef]
14. Hutchings, G.J. Vapor phase hydrochlorination of acetylene: Correlation of catalytic activity of supported metal chloride catalysts. *J. Catal.* **1985**, *96*, 292–295. [CrossRef]
15. Nkosi, B.; Coville, N.J.; Hutchings, G.J. Reactivation of a supported gold catalyst for acetylene hydrochlorination. *J. Chem. Soc. Chem. Commun.* **1988**, 71–72. [CrossRef]
16. Nkosi, B.; Coville, N.J.; Hutchings, G.J. Vapour phase hydrochlorination of acetylene with group VIII and IB metal chloride catalysts. *Appl. Catal.* **1988**, *43*, 33–39. [CrossRef]
17. Zhou, K.; Wang, W.; Zhao, Z.; Luo, G.; Miller, J.T.; Wong, M.S.; Wei, F. Synergistic gold–bismuth catalysis for non–mercury hydrochlorination of acetylene to vinyl chloride monomer. *ACS Catal.* **2014**, *4*, 3112–3116. [CrossRef]
18. Hashmi, A.S.K. Homogeneous gold catalysts and alkynes: A successful liaison. *Gold Bull.* **2003**, *36*, 3–9. [CrossRef]
19. Schmidbaur, H. *Gold: Progress in Chemistry, Biochemistry, and Technology*; John Wiley & Sons Inc: New York, NY, USA, 1999; p. 894.
20. Krasnyakova, T.V.; Zhikharev, I.V.; Mitchenko, R.S.; Burkhovetski, V.I.; Korduban, A.M.; Kryshchuk, T.V.; Mitchenko, S.A. Acetylene catalytic hydrochlorination over mechanically pre–activated K_2PdCl_4 salt: A study of the reaction mechanism. *J. Catal.* **2012**, *288*, 33–43. [CrossRef]

21. Mitchenko, S.A.; Ananikov, V.P.; Beletskaya, I.P. Mechanoactivation of acetylene hydrochlorination in the presence of K$_2$PtCl$_6$. *Zhurnal. Org. Khimii.* **1998**, *34*, 1859–1860. [CrossRef]
22. Mitchenko, S.A. Acetylene hydrochlorination by gaseous hydrogen chloride on the surface of mechanically activated K$_2$PtCl$_6$ salt. *Kinet. Catal.* **1998**, *39*, 859–862. [CrossRef]
23. Mitchenko, S.A.; Krasnyakova, T.V.; Mitchenko, R.S.; Korduban, A.N. Acetylene catalytic hydrochlorination over powder catalyst prepared by pre–milling of K$_2$PtCl$_4$ salt. *J. Mol. Catal. A–Chem.* **2007**, *275*, 101–108. [CrossRef]
24. Conte, M.; Davies, C.J.; Morgan, D.J.; Carley, A.F.; Johnston, P.; Hutchings, G.J. Characterization of Au^{3+} species in Au/C catalysts for the hydrochlorination reaction of acetylene. *Catal. Lett.* **2014**, *144*, 1–8. [CrossRef]
25. Liu, X.; Conte, M.; Elias, D.; Lu, L.; Morgan, D.J.; Freakley, S.J.; Johnston, P.; Kiely, C.J.; Hutchings, G.J. Investigation of the active species in the carbon–supported gold catalyst for acetylene hydrochlorination. *Catal. Sci. Technol.* **2016**, *6*, 5144–5153. [CrossRef]
26. Nkosi, B.; Adams, M.D.; Coville, N.J.; Hutchings, G.J. Hydrochlorination of acetylene using carbon–supported gold catalysts: A study of catalyst reactivation. *J. Catal.* **1991**, *128*, 378–386. [CrossRef]
27. Conte, M.; Davies, C.J.; Morgan, D.J.; Davies, T.E.; Elias, D.J.; Carley, A.F.; Johnston, P.; Hutchings, G.J. Aqua regia activated Au/C catalysts for the hydrochlorination of acetylene. *J. Catal.* **2013**, *297*, 128–136. [CrossRef]
28. Zhao, J.; Wang, B.L.; Xu, X.L.; Yu, Y.; Di, S.X.; Xu, H.; Zhai, Y.Y.; He, H.H.; Guo, L.L.; Pan, Z.Y.; et al. Alternative solvent to aqua regia to activate Au/AC catalysts for the hydrochlorination of acetylene. *J. Catal.* **2017**, *350*, 149–158. [CrossRef]
29. Huang, C.F.; Zhu, M.Y.; Kang, L.H.; Dai, B. A novel high–stability Au(III)/schiff–based catalyst for acetylene hydrochlorination reaction. *Catal. Commun.* **2014**, *54*, 61–65. [CrossRef]
30. Chao, S.L.; Guan, Q.X.; Li, W. Study of the active site for acetylene hydrochlorination in AuCl$_3$/C catalysts. *J. Catal.* **2015**, *330*, 273–279. [CrossRef]
31. Zhang, C.M.; Zhang, H.Y.; Man, B.C.; Li, X.; Dai, H.; Zhang, J.L. Hydrochlorination of acetylene catalyzed by activated carbon supported highly dispersed gold nanoparticles. *Appl. Catal. A Gen.* **2018**, *566*, 15–24. [CrossRef]
32. Wittanadecha, W.; Laosiripojana, N.; Ketcong, A.; Ningnuek, N.; Praserthdam, P.; Monnier, J.R.; Assabumrungrat, S. Preparation of Au/C catalysts using microwave–assisted and ultrasonic–assisted methods for acetylene hydrochlorination. *Appl. Catal. A Gen.* **2014**, *475*, 292–296. [CrossRef]
33. Tian, X.H.; Hong, G.T.; Jiang, B.B.; Lu, F.P.; Liao, Z.W.; Wang, J.D.; Yang, Y.R. Efficient Au^0/C catalyst synthesized by a new method for acetylene hydrochlorination. *RSC Adv.* **2015**, *5*, 46366–46371. [CrossRef]
34. Conte, M.; Carley, A.F.; Attard, G.; Herzing, A.A.; Kiely, C.J.; Hutchings, G.J. Hydrochlorination of acetylene using supported bimetallic Au–based catalysts. *J. Catal.* **2008**, *257*, 190–198. [CrossRef]
35. Zhu, M.Y.; Wang, Q.Q.; Chen, K.; Wang, Y.; Huang, C.F.; Dai, H.; Yu, F.; Kang, L.H.; Dai, B. Development of a heterogeneous non–mercury catalyst for acetylene hydrochlorination. *ACS Catal.* **2015**, *5*, 5306–5316. [CrossRef]
36. Li, G.B.; Li, W.; Zhang, J.L. Non–mercury catalytic acetylene hydrochlorination over activated carbon–supported Au catalysts promoted by CeO$_2$. *Catal. Sci. Technol.* **2016**, *6*, 1821–1828. [CrossRef]
37. Ke, J.H.; Zhao, Y.X.; Yin, Y.; Chen, K.; Duan, X.P.; Ye, L.M.; Yuan, Y.Z. Yttrium chloride–modified Au/AC catalysts for acetylene hydrochlorination with improved activity and stability. *J. Rare Earths* **2017**, *35*, 1083–1091. [CrossRef]
38. Dong, Y.Z.; Zhang, H.Y.; Li, W.; Sun, M.X.; Guo, C.L.; Zhang, J.L. Bimetallic Au–Sn/AC catalysts for acetylene hydrochlorination. *J. Ind. Eng. Chem.* **2016**, *35*, 177–184. [CrossRef]
39. Zhao, J.; Yu, Y.; Xu, X.L.; Di, S.X.; Wang, B.L.; Xu, H.; Ni, J.; Guo, L.L.; Pan, Z.Y.; Li, X.N. Stabilizing Au(III) in supported–ionic–liquid–phase (SILP) catalyst using CuCl$_2$ via a redox mechanism. *Appl. Catal. B Environ.* **2017**, *206*, 175–183. [CrossRef]
40. Du, Y.F.; Hu, R.S.; Jia, Y.; Zhou, Q.H.; Meng, W.W.; Yang, J. CuCl$_2$ promoted low–gold–content Au/C catalyst for acetylene hydrochlorination prepared by ultrasonic–assisted impregnation. *J. Ind. Eng. Chem.* **2016**, *37*, 32–41. [CrossRef]
41. Zhao, J.G.; Zeng, J.J.; Cheng, X.G.; Wang, L.; Yang, H.H.; Shen, B.X. An Au–Cu bimetal catalyst for acetylene hydrochlorination with renewable γ–Al$_2$O$_3$ as the support. *RSC Adv.* **2015**, *5*, 16727–16734. [CrossRef]
42. Wang, L.; Shen, B.X.; Zhao, J.G.; Bi, X.T. Trimetallic Au–Cu–K/AC for acetylene hydrochlorination. *Can. J. Chem. Eng.* **2017**, *95*, 1069–1075. [CrossRef]

43. Zhang, H.Y.; Dai, B.; Li, W.; Wang, X.G.; Zhang, J.L.; Zhu, M.Y.; Gu, J.J. Non–mercury catalytic acetylene hydrochlorination over spherical activated–carbon–supported Au–Co(III)–Cu(II) catalysts. *J. Catal.* **2014**, *316*, 141–148. [CrossRef]
44. Li, G.B.; Li, W.; Zhang, J.L.; Zhang, W.; Zhou, H.; Si, C.L. The effect of N–doping in activated carbon–supported Au–Sr catalysts for acetylene hydrochlorination to vinyl chloride. *ChemistrySelect* **2018**, *3*, 3561–3569. [CrossRef]
45. Li, G.B.; Li, W.; Zhang, J.L. Strontium promoted activated carbon–supported gold catalysts for non–mercury catalytic acetylene hydrochlorination. *Catal. Sci. Technol.* **2016**, *6*, 3230–3237. [CrossRef]
46. Zhang, H.Y.; Li, W.; Li, X.Q.; Zhao, W.; Gu, J.J.; Qi, X.Y.; Dong, Y.Z.; Dai, B.; Zhang, J.L. Non–mercury catalytic acetylene hydrochlorination over bimetallic Au–Ba(ii)/AC catalysts. *Catal. Sci. Technol.* **2015**, *5*, 1870–1877. [CrossRef]
47. Zhao, J.; Xu, J.T.; Xu, J.H.; Ni, J.; Zhang, T.T.; Xu, X.L.; Li, X.N. Activated–carbon–supported gold–cesium(I) as highly effective catalysts for hydrochlorination of acetylene to vinyl chloride. *ChemPlusChem* **2015**, *80*, 196–201. [CrossRef]
48. Zhao, J.; Zhang, T.T.; Di, X.X.; Xu, J.T.; Xu, J.; Feng, F.; Ni, J.; Li, X.N. Nitrogen–modified activated carbon supported bimetallic gold–cesium(i) as highly active and stable catalyst for the hydrochlorination of acetylene. *RSC Adv.* **2015**, *5*, 6925–6931. [CrossRef]
49. Zhao, J.; Zhang, T.T.; Di, X.X.; Xu, J.T.; Gu, S.C.; Zhang, Q.F.; Ni, J.; Li, X.N. Activated carbon supported ternary gold–cesium(i)–indium(iii) catalyst for the hydrochlorination of acetylene. *Catal. Sci. Technol.* **2015**, *5*, 4973–4984. [CrossRef]
50. Zhao, J.; Cheng, X.; Wang, L.; Ren, R.; Zeng, J.; Yang, H.; Shen, B. Free–mercury catalytic acetylene hydrochlorination over bimetallic Au–Bi/γ–Al$_2$O$_3$:A low gold content catalyst. *Catal. Lett.* **2014**, *144*, 2191–2197. [CrossRef]
51. Huang, C.F.; Zhu, M.Y.; Kang, L.H.; Li, X.Y.; Dai, B. Active carbon supported TiO$_2$–AuCl$_3$/AC catalyst with excellent stability for acetylene hydrochlorination reaction. *Chem. Eng. J.* **2014**, *242*, 69–75. [CrossRef]
52. Pu, Y.F.; Zhang, J.L.; Wang, X.; Zhang, H.Y.; Yu, L.; Dong, Y.Z.; Li, W. Bimetallic Au–Ni/CSs catalysts for acetylene hydrochlorination. *Catal. Sci. Technol.* **2014**, *4*, 4426–4432. [CrossRef]
53. Zhang, H.Y.; Dai, B.; Wang, X.G.; Li, W.; Han, Y.; Gu, J.J.; Zhang, J.L. Non–mercury catalytic acetylene hydrochlorination over bimetallic Au–Co(iii)/SAC catalysts for vinyl chloride monomer production. *Green Chem.* **2013**, *15*, 829–836. [CrossRef]
54. Zhang, H.Y.; Dai, B.; Wang, X.G.; Xu, L.L.; Zhu, M.Y. Hydrochlorination of acetylene to vinyl chloride monomer over bimetallic Au–La/SAC catalysts. *J. Ind. Eng. Chem.* **2012**, *18*, 49–54. [CrossRef]
55. Goguet, A.; Hardacre, C.; Harvey, I.; Narasimharao, K.; Saih, Y.; Sa, J. Increased dispersion of supported gold during methanol carbonylation conditions. *J. Am. Chem. Soc.* **2009**, *131*, 6973–6975. [CrossRef]
56. Zhong, J.W.; Xu, Y.P.; Liu, Z.M. Heterogeneous non–mercury catalysts for acetylene hydrochlorination: Progress, challenges, and opportunities. *Green Chem.* **2018**, *20*, 2412–2427. [CrossRef]
57. Zhou, K.; Jia, J.C.; Li, C.H.; Xu, H.; Zhou, J.; Luo, G.H.; Wei, F. A low content Au–based catalyst for hydrochlorination of C$_2$H$_2$ and its industrial scale–up for future PVC processes. *Green Chem.* **2015**, *17*, 356–364. [CrossRef]
58. Li, X.B.; Wang, H.Y.; Yang, X.D.; Zhu, Z.H.; Tang, Y.J. Size dependence of the structures and energetic and electronic properties of gold clusters. *J. Chem. Phys.* **2007**, *126*, 084505. [CrossRef]
59. Li, Y.F.; Mao, A.J.; Li, Y.; Kuang, X.Y. Density functional study on size–dependent structures, stabilities, electronic and magnetic properties of Au$_n$M (M = Al and Si, n = 1–9) clusters: Comparison with pure gold clusters. *J. Mol. Model.* **2012**, *18*, 3061–3072. [CrossRef]
60. Bürgel, C.; Reilly, N.M.; Johnson, G.E.; Mitrić, R.; Kimble, M.L.; Castleman, A.W.; Bonačić–Koutecký, V. Influence of charge state on the mechanism of co oxidation on gold clusters. *J. Am. Chem. Soc.* **2008**, *130*, 1694–1698. [CrossRef]
61. Socaciu, L.D.; Hagen, J.; Bernhardt, T.M.; Wöste, L.; Heiz, U.; Häkkinen, H.; Landman, U. Catalytic co oxidation by free Au$_2^-$: Experiment and theory. *J. Am. Chem. Soc.* **2003**, *125*, 10437–10445. [CrossRef]
62. Gao, M.; Lyalin, A.; Taketsugu, T. Role of the support effects on the catalytic activity of gold clusters: A density functional theory study. *Catalysts* **2011**, *1*, 18–39. [CrossRef]
63. Spivey, K.; Williams, J.I.; Wang, L. Structures of undecagold clusters: Ligand effect. *Chem. Phys. Lett.* **2006**, *432*, 163–166. [CrossRef]

64. Zhao, Y.; Zhao, F.; Kang, L.H. Catalysis of the acetylene hydrochlorination reaction by Si–doped Au clusters: A DFT study. *J. Mol. Model.* **2018**, *24*, 61. [CrossRef] [PubMed]
65. Xu, H.; Zhou, K.; Si, J.; Li, C.; Luo, G. A ligand coordination approach for high reaction stability of an Au–Cu bimetallic carbon–based catalyst in the acetylene hydrochlorination process. *Catal. Sci. Technol.* **2016**, *6*, 1357–1366. [CrossRef]
66. Li, X.; Zhu, M.; Dai, B. AuCl$_3$ on polypyrrole–modified carbon nanotubes as acetylene hydrochlorination catalysts. *Appl. Catal. B Environ.* **2013**, *142*, 234–240. [CrossRef]
67. Chen, K.; Kang, L.H.; Zhu, M.Y.; Dai, B. Mesoporous carbon with controllable pore sizes as a support of the AuCl3 catalyst for acetylene hydrochlorination. *Catal. Sci. Technol.* **2015**, *5*, 1035–1040. [CrossRef]
68. Jia, Y.; Hu, R.S.; Zhou, Q.H.; Wang, H.Y.; Gao, X.; Zhang, J. Boron–modified activated carbon supporting low–content Au–based catalysts for acetylene hydrochlorination. *J. Catal.* **2017**, *348*, 223–232. [CrossRef]
69. Li, X.Y.; Wang, Y.; Kang, L.H.; Zhu, M.Y.; Dai, B. A novel, non–metallic graphitic carbon nitride catalyst for acetylene hydrochlorination. *J. Catal.* **2014**, *311*, 288–294. [CrossRef]
70. Yang, Y.; Lan, G.J.; Wang, X.L.; Li, Y. Direct synthesis of nitrogen–doped mesoporous carbons for acetylene hydrochlorination. *Chin. J. Catal.* **2016**, *37*, 1242–1248. [CrossRef]
71. Zhao, J.; Xu, J.; Xu, J.; Zhang, T.; Di, X.; Ni, J.; Li, X. Enhancement of Au/AC acetylene hydrochlorination catalyst activity and stability via nitrogen–modified activated carbon support. *Chem. Eng. J.* **2015**, *262*, 1152–1160. [CrossRef]
72. Zhao, J.; Gu, S.C.; Xu, X.L.; Zhang, T.T.; Yu, Y.; Di, X.X.; Ni, J.; Pan, Z.Y.; Li, X.N. Supported ionic–liquid–phase–stabilized Au(iii) catalyst for acetylene hydrochlorination. *Catal. Sci. Technol.* **2016**, *6*, 3263–3270. [CrossRef]
73. Tian, X.H.; Hong, G.T.; Liu, Y.; Jiang, B.B.; Yang, Y.R. Catalytic performance of AuIII supported on SiO$_2$ modified activated carbon. *RSC Adv.* **2014**, *4*, 36316–36324. [CrossRef]
74. Dai, B.; Li, X.; Zhang, J.; Yu, F.; Zhu, M. Application of mesoporous carbon nitride as a support for an Au catalyst for acetylene hydrochlorination. *Chem. Eng. Sci.* **2015**, *135*, 472–478. [CrossRef]
75. Wang, Y.; Zhu, M.; Kang, L.; Dai, B. Neutral Aun (n = 3–10) clusters catalyze acetylene hydrochlorination: A density functional theory study. *RSC Adv.* **2014**, *4*, 38466–38473. [CrossRef]
76. Gong, W.Q.; Zhao, F.; Kang, L.H. Novel nitrogen–doped Au–embedded graphene single–atom catalysts for acetylene hydrochlorination: A density functional theory study. *Comput. Theor. Chem.* **2018**, *1130*, 83–89. [CrossRef]
77. Zhang, J.L.; He, Z.H.; Li, W.; Han, Y. Deactivation mechanism of AuCl$_3$ catalyst in acetylene hydrochlorination reaction: A DFT study. *RSC Adv.* **2012**, *2*, 4814–4821. [CrossRef]
78. Nkosi, B.; Coville, N.J.; Hutchings, G.J.; Adams, M.D.; Friedl, J.; Wagner, F.E. Hydrochlorination of acetylene using gold catalysts: A study of catalyst deactivation. *J. Catal.* **1991**, *128*, 366–377. [CrossRef]
79. Conte, M.; Davies, C.J.; Morgan, D.J.; Davies, T.E.; Carley, A.F.; Johnston, P.; Hutchings, G.J. Modifications of the metal and support during the deactivation and regeneration of Au/C catalysts for the hydrochlorination of acetylene. *Catal. Sci. Technol.* **2013**, *3*, 128–134. [CrossRef]
80. Hashmi, A.S.K.; Blanco, M.C.; Fischer, D.; Bats, J.W. Gold catalysis: Evidence for the in–situ reduction of gold(iii) during the cyclization of allenyl carbinols. *Eur. J. Org. Chem.* **2006**, *2006*, 1387–1389. [CrossRef]
81. Conte, M.; Carley, A.F.; Hutchings, G.J. Reactivation of a carbon–supported gold catalyst for the hydrochlorination of acetylene. *Catal. Lett.* **2008**, *124*, 165–167. [CrossRef]
82. Dai, B.; Wang, Q.Q.; Yu, F.; Zhu, M.Y. Effect of Au nano–particle aggregation on the deactivation of the AuCl$_3$/AC catalyst for acetylene hydrochlorination. *Sci Rep* **2015**, *5*, 10553. [CrossRef]
83. Malta, G.; Kondrat, S.A.; Freakley, S.J.; Davies, C.J.; Dawson, S.; Liu, X.; Lu, L.; Dymkowski, K.; Fernandez–Alonso, F.; Mukhopadhyay, S.; et al. Deactivation of a single–site gold–on–carbon acetylene hydrochlorination catalyst: An X–ray absorption and inelastic neutron scattering study. *ACS Catal.* **2018**, *8*, 8493–8505. [CrossRef]
84. Zhu, M.Y.; Kang, L.H.; Su, Y.; Zhang, S.Z.; Dai, B. MCl$_x$ (M = Hg, Au, Ru; x = 2, 3) catalyzed hydrochlorination of acetylene—A density functional theory study. *Can. J. Chem.* **2013**, *91*, 120–125. [CrossRef]
85. Li, H.; Wu, B.T.; Wang, F.M.; Zhang, X.B. Efficient and stable Ru(III)/choline chloride catalyst system with low Ru content for non–mercury acetylene hydrochlorination. *Chin. J. Catal.* **2018**, *39*, 1770–1781. [CrossRef]

86. Li, X.; Zhang, H.; Man, B.; Zhang, C.; Dai, H.; Dai, B.; Zhang, J. Synthesis of vinyl chloride monomer over carbon–supported tris–(triphenylphosphine) ruthenium dichloride catalysts. *Catalysts* **2018**, *8*, 276. [CrossRef]
87. Wang, X.; Lan, G.; Liu, H.; Zhu, Y.; Li, Y. Effect of acidity and ruthenium species on catalytic performance of ruthenium catalysts for acetylene hydrochlorination. *Catal. Sci. Technol.* **2018**, *8*, 6143–6149. [CrossRef]
88. Gu, J.; Gao, Y.; Zhang, J.; Li, W.; Dong, Y.; Han, Y. Hydrochlorination of acetylene catalyzed by an activated carbon–supported ammonium hexachlororuthenate complex. *Catalysts* **2017**, *7*, 17. [CrossRef]
89. Li, X.; Zhang, H.; Man, B.; Hou, L.; Zhang, C.; Dai, H.; Zhu, M.; Dai, B.; Dong, Y.; Zhang, J. Activated carbon–supported tetrapropylammonium perruthenate catalysts for acetylene hydrochlorination. *Catalysts* **2017**, *7*, 311. [CrossRef]
90. Zhao, W.; Li, W.; Zhang, J. Ru/N–AC catalyst to produce vinyl chloride from acetylene and 1,2–dichloroethane. *Catal. Sci. Technol.* **2016**, *6*, 1402–1409. [CrossRef]
91. Zhang, J.L.; Sheng, W.; Guo, C.L.; Li, W. Acetylene hydrochlorination over bimetallic Ru–based catalysts. *RSC Adv.* **2013**, *3*, 21062–21068. [CrossRef]
92. Pu, Y.F.; Zhang, J.L.; Yu, L.; Jin, Y.H.; Li, W. Active ruthenium species in acetylene hydrochlorination. *Appl. Catal. A Gen.* **2014**, *488*, 28–36. [CrossRef]
93. Zhang, H.Y.; Li, W.; Jin, Y.H.; Sheng, W.; Hu, M.C.; Wang, X.Q.; Zhang, J.L. Ru–Co(III)–Cu(II)/SAC catalyst for acetylene hydrochlorination. *Appl. Catal. B–Environ.* **2016**, *189*, 56–64. [CrossRef]
94. Li, Y.; Dong, Y.Z.; Li, W.; Han, Y.; Zhang, J.L. Improvement of imidazolium–based ionic liquids on the activity of ruthenium catalyst for acetylene hydrochlorination. *Mol. Catal.* **2017**, *443*, 220–227. [CrossRef]
95. Mitchenko, S.A.; Khomutov, E.V.; Shubin, A.A.; Shul'ga, Y.M. Mechanochemical activation of K_2PtCl_6: Heterogeneous catalyst for gas–phase hydrochlorination of acetylene. *Theor. Exp. Chem.* **2003**, *39*, 255–258. [CrossRef]
96. Mitchenko, S.A.; Khomutov, E.V.; Shubin, A.A.; Shul'ga, Y.M. Catalytic hydrochlorination of acetylene by gaseous HCl on the surface of mechanically pre–activated K_2PtCl_6 salt. *J. Mol. Catal. A–Chem.* **2004**, *212*, 345–352. [CrossRef]
97. Mitchenko, S.A.; Krasnyakova, T.V. Acetylene hydrochlorination over mechanically activated K_2MCl_4 (M = Pt, Pd) and K_2PtCl_6 catalysts: The HCl/DCl kinetic isotope effect and reaction mechanisms. *Kinet. Catal.* **2014**, *55*, 722–728. [CrossRef]
98. Strebelle, M.; Devos, A. Catalytic Hydrochlorination System and Process for the Manufacture of Vinyl Chloride from Acetylene and Hydrogen Chloride in the Presence of this Catalytic System. U.S. Patent NO. 5,254,777, 17 October 1993.
99. Wang, L.; Wang, F.; Wang, J. Enhanced stability of hydrochlorination of acetylene using polyaniline–modified Pd/HY catalysts. *Catal. Commun.* **2016**, *74*, 55–59. [CrossRef]
100. Wang, L.; Wang, F.; Wang, J. Effect of K promoter on the stability of Pd/NFY catalysts for acetylene hydrochlorination. *Catal. Commun.* **2016**, *83*, 9–13. [CrossRef]
101. Wang, L.; Wang, F.; Wang, J.D. Catalytic properties of Pd/HY catalysts modified with NH_4F for acetylene hydrochlorination. *Catal. Commun.* **2015**, *65*, 41–45. [CrossRef]
102. Wang, L.; Wang, F.; Wang, J.D.; Tang, X.L.; Zhao, Y.L.; Yang, D.; Jia, F.M.; Hao, T. Hydrochlorination of acetylene to vinyl chloride over Pd supported on zeolite Y. *React. Kinet. Mech. Catal.* **2013**, *110*, 187–194. [CrossRef]
103. Song, Q.L.; Wang, S.J.; Shen, B.X.; Zhao, J.G. Palladium–based catalysts for the hydrochlorination of acetylene: Reasons for deactivation and its regeneration. *Pet. Sci. Technol.* **2010**, *28*, 1825–1833. [CrossRef]
104. Li, P.; Ding, M.Z.; He, L.M.; Tie, K.; Ma, H.; Pan, X.L.; Bao, X.H. The activity and stability of $PdCl_2$/C–N catalyst for acetylene hydrochlorination. *Sci. China Chem.* **2018**, *61*, 444–448. [CrossRef]
105. Yang, L.F.; Yang, Q.W.; Hu, J.Y.; Bao, Z.B.; Su, B.G.; Zhang, Z.G.; Ren, Q.L.; Xing, H.B. Metal nanoparticles in ionic liquid–cosolvent biphasic systems as active catalysts for acetylene hydrochlorination. *Aiche. J.* **2018**, *64*, 2536–2544. [CrossRef]
106. Zhao, J.; Yue, Y.X.; Sheng, G.F.; Wang, B.L.; Lai, H.X.; Di, S.X.; Zhai, Y.Y.; Guo, L.L.; Li, X.N. Supported ionic liquid–palladium catalyst for the highly effective hydrochlorination of acetylene. *Chem. Eng. J.* **2019**, *360*, 38–46. [CrossRef]
107. Perkins, G.A. Preparation of Vinyl Chloride. U.S. Patent No. 1,934,324, 7 November 1933.

108. Zhai, Y.Y.; Zhao, J.; Di, X.X.; Di, S.X.; Wang, B.L.; Yue, Y.X.; Sheng, G.F.; Lai, H.X.; Guo, L.L.; Wang, H.; et al. Carbon–supported perovskite–like CsCuCl$_3$ nanoparticles: A highly active and cost–effective heterogeneous catalyst for the hydrochlorination of acetylene to vinyl chloride. *Catal. Sci. Technol.* **2018**, *8*, 2901–2908. [CrossRef]
109. Zhao, W.L.; Zhu, M.Y.; Dai, B. The preparation of Cu–g–C$_3$N$_4$/AC catalyst for acetylene hydrochlorination. *Catalysts* **2016**, *6*, 193. [CrossRef]
110. Li, H.; Wang, F.M.; Cai, W.F.; Zhang, J.L.; Zhang, X.B. Hydrochlorination of acetylene using supported phosphorus–doped Cu–based catalysts. *Catal. Sci. Technol.* **2015**, *5*, 5174–5184. [CrossRef]
111. Xu, J.T.; Zhao, J.; Zhang, T.T.; Di, X.X.; Gu, S.C.; Ni, J.; Li, X.N. Ultra–low Ru–promoted CuCl$_2$ as highly active catalyst for the hydrochlorination of acetylene. *RSC Adv.* **2015**, *5*, 38159–38163. [CrossRef]
112. Zhou, K.; Si, J.K.; Jia, J.C.; Huang, J.Q.; Zhou, J.; Luo, G.H.; Wei, F. Reactivity enhancement of N–CNTs in green catalysis of C$_2$H$_2$ hydrochlorination by a Cu catalyst. *RSC Adv.* **2014**, *4*, 7766–7769. [CrossRef]
113. Hu, D.; Wang, L.; Wang, F.; Wang, J.D. Active carbon supported S–promoted Bi catalysts for acetylene hydrochlorination reaction. *Chin. Chem. Lett.* **2018**, *29*, 1413–1416. [CrossRef]
114. Dai, H.; Zhu, M.Y.; Zhang, H.Y.; Yu, F.; Wang, C.; Dai, B. Activated carbon supported Mo–Ti–N binary transition metal nitride as catalyst for acetylene hydrochlorination. *Catalysts* **2017**, *7*, 200. [CrossRef]
115. Dai, H.; Zhu, M.Y.; Zhang, H.Y.; Yu, F.; Wang, C.; Dai, B. Activated carbon supported VN, Mo$_2$N, and W$_2$N as catalysts for acetylene hydrochlorination. *J. Ind. Eng. Chem.* **2017**, *50*, 72–78. [CrossRef]
116. Dai, H.; Zhu, M.Y.; Zhao, D.; Yu, F.; Dai, B. Effective catalytic performance of plasma–enhanced W$_2$N/AC as catalysts for acetylene hydrochlorination. *Top. Catal.* **2017**, *60*, 1016–1023. [CrossRef]
117. Zhao, W.L.; Zhu, M.Y.; Dai, B. Cobalt–nitrogen–activated carbon as catalyst in acetylene hydrochlorination. *Catal. Commun.* **2017**, *98*, 22–25. [CrossRef]
118. Lan, G.; Qiu, Y.; Fan, J.; Wang, X.; Tang, H.; Han, W.; Liu, H.; Liu, H.; Song, S.; Li, Y. Defective graphene@diamond hybrid nanocarbon material as an effective and stable metal–free catalyst for acetylene hydrochlorination. *Catal. Commun.* **2019**, *55*, 1430–1433. [CrossRef]
119. Li, X.; Zhang, J.; Han, Y.; Zhu, M.; Shang, S.; Li, W. MOF–derived various morphologies of N–doped carbon composites for acetylene hydrochlorination. *J. Mater. Sci.* **2018**, *53*, 4913–4926. [CrossRef]
120. Dong, X.B.; Chao, S.L.; Wan, F.F.; Guan, Q.X.; Wang, G.C.; Li, W. Sulfur and nitrogen co–doped mesoporous carbon with enhanced performance for acetylene hydrochlorination. *J. Catal.* **2018**, *359*, 161–170. [CrossRef]
121. Li, P.; Li, H.B.; Pan, X.L.; Tie, K.; Cui, T.T.; Ding, M.Z.; Bao, X.H. Catalytically active boron nitride in acetylene hydrochlorination. *ACS Catal.* **2017**, *7*, 8572–8577. [CrossRef]
122. Li, X.Y.; Zhang, J.L.; Li, W. MOF–derived nitrogen–doped porous carbon as metal–free catalysts for acetylene hydrochlorination. *J. Ind. Eng. Chem.* **2016**, *44*, 146–154. [CrossRef]
123. Dai, B.; Chen, K.; Wang, Y.; Kang, L.H.; Zhu, M.Y. Boron and nitrogen doping in graphene for the catalysis of acetylene hydrochlorination. *ACS Catal.* **2015**, *5*, 2541–2547. [CrossRef]
124. Li, X.Y.; Pan, X.L.; Bao, X.H. Nitrogen doped carbon catalyzing acetylene conversion to vinyl chloride. *J. Energy Chem.* **2014**, *23*, 131–135. [CrossRef]
125. Zhou, K.; Li, B.; Zhang, Q.; Huang, J.Q.; Tian, G.L.; Jia, J.C.; Zhao, M.Q.; Luo, G.H.; Su, D.S.; Wei, F. The catalytic pathways of hydrohalogenation over metal–free nitrogen–doped carbon nanotubes. *ChemSusChem* **2014**, *7*, 723–728. [CrossRef]
126. Wang, X.G.; Dai, B.; Wang, Y.; Yu, F. Nitrogen–doped pitch–based spherical active carbon as a nonmetal catalyst for acetylene hydrochlorination. *ChemCatChem* **2014**, *6*, 2339–2344. [CrossRef]
127. Qiao, X.L.; Zhou, Z.Q.; Liu, X.Y.; Zhao, C.Y.; Guan, Q.X.; Li, W. Constructing of fragmentary g–C$_3$N$_4$ framework with rich nitrogen defects as highly efficient metal–free catalyst for acetylene hydrochlorination. *Catal. Sci. Technol.* **2019**, *9*, 3753–3762. [CrossRef]
128. Zhang, C.L.; Kang, L.H.; Zhu, M.Y.; Dai, B. Nitrogen–doped active carbon as a metal–free catalyst for acetylene hydrochlorination. *RSC Adv.* **2015**, *5*, 7461–7468. [CrossRef]
129. Zhao, F.; Kang, L.H. The neglected significant role for graphene–based acetylene hydrochlorination catalysts — intrinsic graphene defects. *Chem. Sel.* **2017**, *2*, 6016–6022. [CrossRef]
130. Li, X.Y.; Pan, X.L.; Yu, L.; Ren, P.J.; Wu, X.; Sun, L.T.; Jiao, F.; Bao, X.H. Silicon carbide–derived carbon nanocomposite as a substitute for mercury in the catalytic hydrochlorination of acetylene. *Nat. Commun.* **2014**, *5*, 7. [CrossRef] [PubMed]

131. Zhao, F.; Wang, Y.; Zhu, M.Y.; Kang, L.H. C–doped boron nitride fullerene as a novel catalyst for acetylene hydrochlorination: A DFT study. *RSC Adv.* **2015**, *5*, 56348–56355. [CrossRef]
132. Li, X.Y.; Li, P.; Pan, X.L.; Ma, H.; Bao, X.H. Deactivation mechanism and regeneration of carbon nanocomposite catalyst for acetylene hydrochlorination. *Appl. Catal. B–Environ.* **2017**, *210*, 116–120. [CrossRef]
133. Chao, S.L.; Zou, F.; Wan, F.F.; Dong, X.B.; Wang, Y.L.; Wang, Y.X.; Guan, Q.X.; Wang, G.C.; Li, W. Nitrogen–doped carbon derived from ZIF–8 as a high–performance metal–free catalyst for acetylene hydrochlorination. *Sci. Rep.* **2017**, *7*, 39789. [CrossRef]
134. Song, Z.J.; Liu, G.Y.; He, D.W.; Pang, X.D.; Tong, Y.S.; Wu, Y.Q.; Yuan, D.H.; Liu, Z.M.; Xu, Y.P. Acetylene hydrochlorination over 13X zeolite catalysts at high temperature. *Green Chem.* **2016**, *18*, 5994–5998. [CrossRef]
135. Li, X.Y.; Nian, Y.; Shang, S.S.; Zhang, H.Y.; Zhang, J.L.; Han, Y.; Li, W. Novel nonmetal catalyst of supported tetraphenylphosphonium bromide for acetylene hydrochlorination. *Catal. Sci. Technol.* **2019**, *9*, 188–198. [CrossRef]
136. Wang, B.L.; Lai, H.X.; Yue, Y.X.; Sheng, G.F.; Deng, Y.Q.; He, H.H.; Guo, L.L.; Zhao, J.; Li, X.N. Zeolite supported ionic liquid catalysts for the hydrochlorination of acetylene. *Catalysts* **2018**, *8*, 351. [CrossRef]
137. Zhou, X.F.; Xu, S.G.; Liu, Y.L.; Cao, S.K. Mechanistic study on metal–free acetylene hydrochlorination catalyzed by imidazolium–based ionic liquids. *Mol. Catal.* **2018**, *461*, 73–79. [CrossRef]

Publisher's Note: MDPI stays neutral with regard to jurisdictional claims in published maps and institutional affiliations.

© 2020 by the authors. Licensee MDPI, Basel, Switzerland. This article is an open access article distributed under the terms and conditions of the Creative Commons Attribution (CC BY) license (http://creativecommons.org/licenses/by/4.0/).

Article

The Role of Iodine Catalyst in the Synthesis of 22-Carbon Tricarboxylic Acid and Its Ester: A Case Study

Yanxia Liu [1,2,3], Yagang Zhang [1,2,3,4,*], Lulu Wang [1,2,3], Xingjie Zan [1,*] and Letao Zhang [1]

1. Xinjiang Technical Institute of Physics and Chemistry, Chinese Academy of Sciences, Urumqi 830011, China; liuyanxia@ms.xjb.ac.cn (Y.L.); wanglulu@ms.xjb.ac.cn (L.W.); zhanglt@ms.xjb.ac.cn (L.Z.)
2. University of Chinese Academy of Sciences, Beijing 100049, China
3. Department of chemical and environmental engineering, Xinjiang Institute of Engineering, Urumqi 830026, China
4. School of Materials and Energy, University of Electronic Science and Technology of China, Chengdu 611731, China
* Correspondence: ygzhang@ms.xjb.ac.cn (Y.Z.); zanxj@ms.xjb.ac.cn (X.Z.); Tel.: +86-18129307169 (Y.Z.); +86-18100160595 (X.Z.)

Received: 1 November 2019; Accepted: 18 November 2019; Published: 20 November 2019

Abstract: Here, 22-carbon tricarboxylic acid (C22TA) and its ester (C22TAE) were prepared via the Diels–Alder reaction of polyunsaturated fatty acids (PUFAs) and their esters (PUFAEs) as dienes with fumaric acid (FA) and dimethyl fumarate (DF) as dienophiles, respectively. The role of an iodine catalyst for the synthesis of C22TA and C22TAE in the Diels–Alder type reaction was investigated using a spectroscopic approach. The chemical structures of the products were characterized using proton nuclear magnetic resonance (^{1}H-NMR) and electrospray ionization mass spectrometry (ESI-MS) analysis. Results showed that nonconjugated dienes can react with dienophiles through a Diels–Alder reaction with an iodine catalyst, and that iodine transformed the nonconjugated double bonds of dienes into conjugated double bonds via a radical process. DF was more favorable for the Diels–Alder reaction than FA. This was mainly because the dienophile DF contained an electron-withdrawing substituent, which reduced the highest and lowest occupied molecular orbital (HOMO–LUMO) energy gap and accelerated the Diels–Alder reaction. By transforming nonconjugated double bonds into conjugated double bonds, iodine as a Lewis acid increased the electron-withdrawing effect of the carbonyl group on the carbon–carbon double bond and reduced the energy difference between the HOMO of diene and the LUMO of dienophile, thus facilitating the Diels–Alder reaction.

Keywords: Diels–Alder reaction; 22-carbon tricarboxylic acid; iodine; ^{1}H-NMR; ESI-MS

1. Introduction

With the dramatic decrease in global fossil fuels and the increasing environmental problems and concerns, the use of renewable resources received great attention. Promoting sustainable environmental, social, and economic development is now a global consensus [1]. Sustainable bio-based resources are particularly desirable. One approach is the development and utilization of vegetable oil-based lubricants to partially or even comprehensively substitute petrochemical resources. Mineral oil is refined from natural crude oil, and its biodegradability is poor. Mineral oil cannot be used as the base oil of biodegradable lubricating oil because it causes serious pollution when discharged into the environment. It stays in water, soil, forest, and other ecosystems for a long time. It would be highly desirable to develop biodegradable, environmentally benign lubricating oil and its derivatives from plant oil and vegetable oil as raw materials [2].

In 2018, the total output of China's eight major plant oils and vegetable oils, including rapeseed, peanut, soybean, cottonseed, sunflower seed, sesame, flaxseed, and *Camellia oleracea*, was about 60 million tons. With an abundant source of plant oil and vegetable oil, as well as low-cost and well-established refining and processing technology, it could be an excellent renewable resource for preparing biodegradable lubricating oil [3]. Vegetable oil is rich in polyunsaturated fatty acids (PUFAs); PUFAs are hydrocarbon chains with a carboxyl group at one end and a terminal methyl group at the other with two or more double bonds present [4,5]. PUFAs include linoleic acid (LA) with two double bonds, linolenic acid, and eleostearic acid with three double bonds. Polyunsaturated fatty acid esters (PUFAEs) are the products of esterification of PUFAs with alcohols. In particular, 22-carbon tricarboxylic acid (C22TA) and its ester (C22TAE) have excellent lubricating properties which can be prepared via a Diels–Alder reaction of PUFA with PUFAE as diene with dienophile, respectively.

C22TA and C22TAE are new types of bio-based chemicals with superior properties. C22TA and C22TAE have low toxicity, they are environmentally benign with excellent biodegradability, and they are obtained from renewable raw materials, allowing them to be used as substitutes for fossil fuel-based petroleum products [6]. They can be widely used in lubricating oil, resin additive, printing ink, cutting fluid, green surfactant, corrosion inhibitor, and many other chemical industrial products [7–9]. The preparation of C22TA and C22TAE not only reduces the dependence on fossil fuels such as petroleum, but also reduces the pollution to the environment. In addition, preparing value-added products such as C22TA using bio-resources based on vegetable oil constitutes an important field in green chemistry and sustainable engineering. The synthesis of C22TA and C22TAE can be carried out via two main routes. One is a Diels–Alder reaction with vegetable oils, such as linoleic acid, linolenic acid, and their esters which do not contain conjugated double bonds. This reaction is carried out under the catalysis of iodine with a dienophile. Another type of Diels–Alder reaction uses starting materials containing conjugated double bonds such as conjugated linoleic acid (CLA), eleostearic acid, and their esters. In this route, compounds containing conjugated double bonds directly react with a dienophile at high temperature. Both processes are safe and easy to operate, and they have great application potential [10].

Huang et al. prepared C22TA via a Diels–Alder reaction using tung oil fatty acids with fumaric acid (FA), and subsequently converted it to the corresponding C22TAE which was then used for epoxy resins [7]. Vijayalakshmi et al. prepared C22TA via a Diels–Alder reaction of dehydrated castor oil fatty acids (containing 48% conjugated and 42% nonconjugated dienes) with FA. The reaction temperature, time, catalyst concentration, and molar ratio of reactants were varied to get maximum yield of the C22TA. The conditions found for maximum yield of C22TA (82.2%) were as follows: temperature of 200 °C, reaction time of 2 h, molar ratio of the reactants of 1:1.1, and catalyst iodine concentration of 0.3% [8]. A Diels–Alder reaction was applied to fatty acids containing both CLA and LA, wherein the nonconjugated portion of LA was also converted to C22TA in the presence of small amounts of iodine in United States (US) Patent. No. 4,081,462. In this invention, the fatty acids were naturally occurring fatty acids high in LA content (above 50%), such as distilled tall oil fatty acid, safflower fatty acid, and sunflower fatty acid. This invention provided a process for making C22TA from available LA in a fatty acid mixture [11]. Previous researchers prepared C22TA and C22TAE with excellent lubricating properties via Diels–Alder reactions. However, no one conducted a systematic and in-depth study on the reaction mechanism. How do PUFAs and their derivatives, dienophiles, and catalysts affect the Diels–Alder reaction? What is the role of the iodine catalyst? Does it only act as a catalyst to convert nonconjugated double bonds to conjugated double bonds? On the other hand, does it also promote and facilitate the Diels–Alder reaction? In this study, C22TA and C22TAE were synthesized and compared using the two routes discussed above. The effects of raw materials and catalysts on the Diels–Alder reaction were compared and studied to further understand the reaction mechanism and to provide theoretical guidance for the industrial production of C22TA and C22TAE.

Photoisomerization of linoleic acid alkyl esters or vegetable oils was suggested as a way of introducing conjugation [12]. Jain et al. used iodine as a catalyst, whereby 22% conjugated linoleic acid was obtained via irradiation of undiluted oil with ultraviolet (UV) light at 48 °C for 12 h [13]. Andrew et al. used a similar approach, whereby soy oil was converted to 25% total CLA [14]. There were other reports on the preparation of CLA by photoisomerization [15–17], but the reported methods had strict requirements on equipment, with a long reaction time and low conversion rate. Under the catalysis of iodine and without the use of a light source, it remains to be studied whether nonconjugated double bonds can be converted into conjugated double bonds by raising the reaction temperature, so as to promote the preparation efficiency of C22TA and C22TAE. Therefore, in this study, conjugated dienes were also used as raw materials to investigate the yield of C22TA and C22TAE with and without the iodine catalyst, and the role of iodine in the Diels–Alder reaction was discussed based on the data.

Along these lines, in the work reported here, we focused on answering the following questions: What is the role of iodine in the Diels–Alder reaction? What is the effect of substituents of reactants on the reaction efficiency of dienophiles? What is the difference between conjugated double bonds and nonconjugated double bonds when participating in the Diels–Alder reaction? We took a spectroscopy approach by combining the results of ^{1}H-NMR, ^{13}C-NMR, and electrospray ionization mass spectrometry (ESI-MS), and we systematically investigated the reaction processes of different PUFAs, PUFAEs, and dienophiles under different catalytic conditions. To the best of our knowledge, our work is the first attempt at answering these important questions, which have important theoretical and practical significance for the value-added product development of vegetable oil.

2. Results and Discussion

2.1. Synthesis and Characterization

C22TA and C22TAE were prepared via the Diels–Alder reaction of PUFAs and PUFAEs as dienes with fumaric acid (FA) and dimethyl fumarate (DF) as dienophiles, respectively, under various experimental conditions (Table 1). LA and its ester have two double bonds located on carbons 9 and 12, while conjugated double bonds of CLA and its ester were in the 9,11 or 10,12 positions. Each of the two double bonds can be in either *cis* or *trans* configuration [12]. In theory, the products of C22TA and C22TAE have different configurations. However, the purpose of this study was to unveil the mechanism of the Diels–Alder reaction; therefore, no attempt was made to separate the isomers. The ^{1}H-NMR spectra of dienes, as well as C22TA and C22TAE samples, are shown in Figures 1–4. The isomeric structures are displayed in the figures for dienes, C22TA, and C22TAE, respectively. The attributions of chemical shifts of the protons are labeled in the figure in detail.

Figure 1 shows the ^{1}H-NMR spectra of LA and the reaction compounds with dienophiles under different catalyst loading. Compound **1** was the reaction product of LA and FA with iodine catalysis. Compound **2** was the reaction product of LA and FA without iodine catalysis. Compound **3** was the reaction product of LA and DF with iodine catalysis. Compound **4** was the reaction product of LA and DF without iodine catalysis.

Table 1. Preparation and yield of 22-carbon tricarboxylic acid (C22TA) and its ester (C22TAE) via Diels–Alder reaction. FA—fumaric acid; DF—dimethyl fumarate; LA—linoleic acid; ML—methyl linoleate; CLA—conjugated linoleic acid; CLAEE—conjugated linoleic acid ethyl ester.

Entry	Reaction Conditions			Compounds	
	Dienes (1 mol)	Dienophiles (1.1 mol)	Catalyst (0.3 wt.%) [1]	Category	Yield (%)
1	LA	FA	Iodine	22-Carbon tricarboxylic acid (C22TA)	71.9
2	LA	FA	/	/	/
3	LA	DF	Iodine	22-Carbon tricarboxylic acid dimethyl ester (C22TADME)	81.3
4	LA	DF	/	/	/
5	ML	FA	Iodine	22-Carbon tricarboxylic acid monomethyl ester (C22TAMME)	65.4
6	ML	FA	/	/	/
7	ML	DF	Iodine	22-Carbon tricarboxylic acid trimethyl ester (C22TATME)	76.3
8	ML	DF	/	/	/
9	CLA	DF	Iodine	22-Carbon tricarboxylic acid dimethyl ester (C22TADME)	82.0
10	CLA	DF	/		79.4
11	CLAEE	FA	Iodine	22-Carbon tricarboxylic acid monoethyl ester (C22TAMEE)	74.4
12	CLAEE	FA	/		71.4
13	CLAEE	DF	Iodine	22-Carbon tricarboxylic acid dimethyl monoethyl ester (C22TADMMEE)	83.3
14	CLAEE	DF	/		77.5

[1] The amount of iodine was the mass percentage of diene.

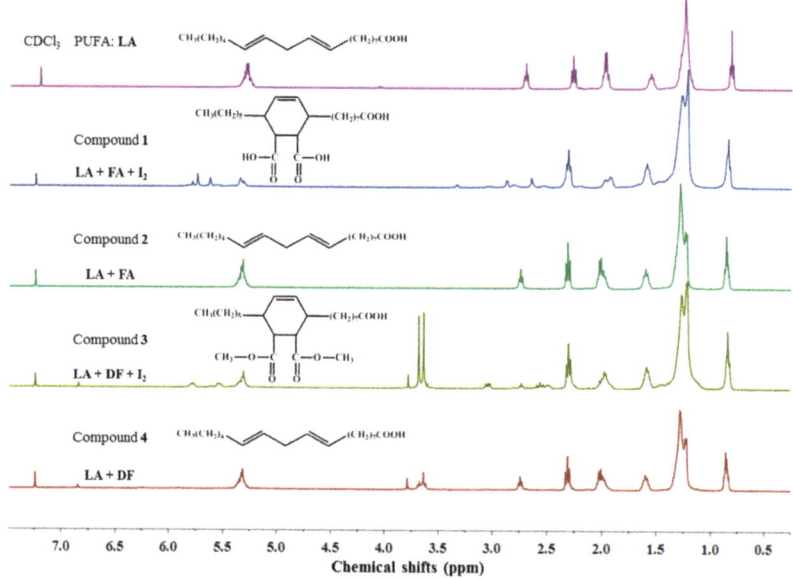

Figure 1. ^{1}H-NMR spectra of linoleic acid (LA) and reaction compounds with dienophiles in chloroform-d under different catalyst conditions: compound 1, reaction product of LA and fumaric acid (FA) with iodine catalysis; compound 2, reaction product of LA and FA without iodine catalysis; compound 3, reaction product of LA and dimethyl fumarate (DF) with iodine catalysis; compound 4, reaction product of LA and DF without iodine catalysis.

Compound **1** was obtained as C22TA. Its molecular formula $C_{22}H_{36}O_6$ was established from ESI-MS, m/z 395.24396 $[M - H]^-$. The proton signals at δ H 5.85–5.58 (m) revealed the presence of a double bond of cyclohexene, and δ H 2.95–2.57 (m) revealed the presence of a methine of cyclohexene. The chemical shifts at 129.97, 44.16, and 39.29 ppm were correspondingly attributed to the carbons in the cyclohexene (Figure S21, Supplementary Materials). These results implied that a Diels–Alder reaction occurred between LA and FA in the presence of iodine catalysis. Compound **3** was obtained as 22-carbon tricarboxylic acid dimethyl ester (C22TADME). Its molecular formula $C_{24}H_{40}O_6$ was established from ESI-MS, m/z 425.28860 $[M + H]^+$, 447.27042 $[M + Na]^+$, 463.24426 $[M + K]^+$. The proton signals at δ H 5.85–5.50 (m) revealed the presence of a double bond of cyclohexene. Signals at δ H 3.04 (dd) and 2.63–2.46 (m) revealed the presence of a methine of cyclohexene. Signals at δ H 3.82–3.53 (m) revealed the presence of a methoxy of dimethyl ester attached to cyclohexene. The chemical shifts at 129.96, 46.92, and 39.68 ppm were correspondingly attributed to the carbons in the cyclohexene (Figure S22, Supplementary Materials). These results implied that a Diels–Alder reaction occurred between LA and DF with iodine catalysis. However, the ^{1}H-NMR spectra of compound **2** and compound **4** were similar to that of LA, and no new peak was generated. These results indicated that no Diels–Alder reaction took place between LA and dienophiles (FA or DF) in the absence of iodine catalyst under the same heating conditions. The ^{1}H-NMR chemical shifts of C22TA and C22TADME in this study were in agreement with the data reported in the literature [18].

Figure 2 shows the ^{1}H-NMR spectra of methyl linoleate (ML) and the reaction compounds with dienophiles under different catalyst conditions. Compound **5** was the reaction product of ML and FA with iodine catalysis; compound **6** was the reaction product of ML and FA without iodine catalysis; compound **7** was the reaction product of ML and DF with iodine catalysis; compound **8** was the reaction product of ML and DF without iodine catalysis.

Compound **5** was obtained as 22-carbon tricarboxylic acid monomethyl ester (C22TAMME). Its molecular formula $C_{23}H_{38}O_6$ was established from ESI-MS, m/z 409.25961 $[M - H]^-$. The proton signal at δ H 5.85–5.50 (m) revealed the presence of a double bond of cyclohexene, and that at δ H 3.08–2.48 (m) revealed the presence of a methine of cyclohexene. The chemical shifts at 129.94, 44.15, and 39.20 ppm were correspondingly attributed to the carbons in the cyclohexene (Figure S23, Supplementary Materials). These results implied that a Diels–Alder reaction occurred between ML and FA with iodine catalysis. Compound **7** was obtained as 22-carbon tricarboxylic acid trimethyl ester (C22TATME). Its molecular formula $C_{25}H_{42}O_6$ was established from ESI-MS, m/z 439.30475 $[M + H]^+$, 461.28638 $[M + Na]^+$, 477.26047 $[M + K]^+$. The proton signal at δ H 5.99–5.50 (m) revealed the presence of a double bond of cyclohexene, and that at δ H 3.04(dd), 2.67–2.43 (m) revealed the presence of a methine of cyclohexene. The chemical shifts at 129.97, 46.91, and 39.67 ppm were correspondingly attributed to the carbons in the cyclohexene (Figure S24, Supplementary Materials). These results implied that a Diels–Alder reaction occurred between ML and DF with iodine catalysis. However, the ^{1}H-NMR spectra of compound **6** and compound **8** were similar to that of starting material ML, and no new peak was observed. These results indicated that no Diels–Alder reaction took place between ML and dienophiles (FA or DF) in the absence of iodine catalyst and under the same heating conditions.

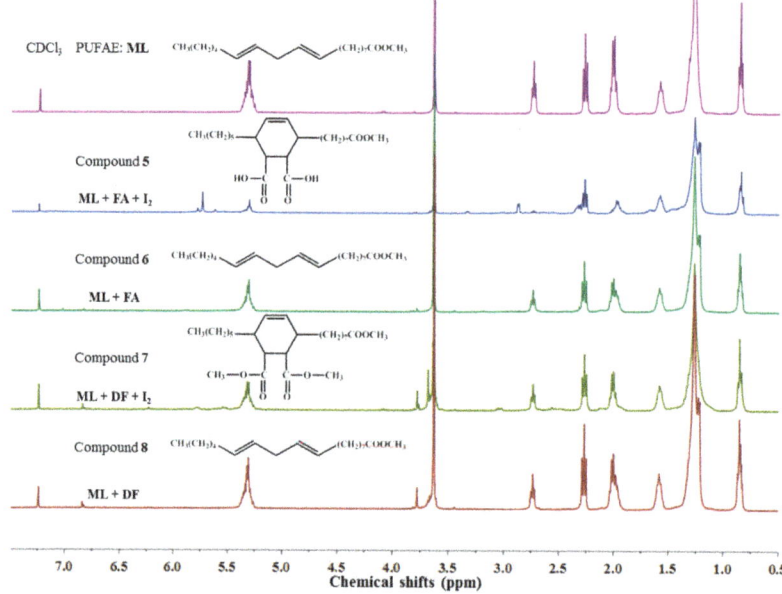

Figure 2. ^{1}H-NMR spectra of methyl linoleate (ML) and the reaction compounds with dienophiles in chloroform-d under different catalyst conditions: compound 5, reaction products of ML and FA with iodine catalysis; compound 6, reaction products of ML and FA without iodine catalysis; compound 7, reaction products of ML and DF with iodine catalysis; compound 8, reaction products of ML and DF without iodine catalysis.

The results in Figures 1 and 2 showed that dienes without conjugated double bonds must have iodine as a catalyst to facilitate the Diels–Alder reaction. Under heating conditions, without a catalyst, the nonconjugated double bonds of dienes cannot be converted efficiently to conjugated double bonds. In order to further investigate the role of iodine in the Diels–Alder reaction, more studies were carried out by investigating the reaction of PUFAs containing conjugated double bonds as dienes with dienophiles under different conditions.

Figure 3 shows the ^{1}H-NMR spectra of CLA (conjugated linoleic acid) with conjugated double bonds and the reaction compounds with dienophiles under different catalyst conditions. Compound 9 was the reaction product of CLA and DF with iodine catalysis; compound 10 was the reaction product of CLA and FA without iodine catalysis.

Both compound 9 and compound 10 were obtained as C22TADME, the same as Compound 3. Its molecular formula $C_{24}H_{40}O_6$ was established from ESI-MS: compound 9, m/z 425.28876 [M + H]$^+$, 447.27066 [M + Na]$^+$, 463.24426 [M + K]$^+$; compound 10, m/z 425.28879 [M + H]$^+$, 447.27072 [M + Na]$^+$, 463.24438 [M + K]$^+$. The proton signal of compound 9 at δ H 3.13–2.98 (m) and the proton signal of compound 10 at δ H 2.63–2.42 (m) revealed the presence of a methine of cyclohexene, while the proton signals of compound 9 and compound 10 at δ H 3.72–3.61 (m) revealed the presence of a methoxy of dimethyl ester attached to cyclohexene. The chemical shifts of compound 9 and compound 10 at 129.95, 46.91, and 39.67 ppm were correspondingly attributed to the carbons in the cyclohexene (Figures S25 and S26, Supplementary Materials). These results implied that a Diels–Alder reaction occurred between CLA and DF with and without iodine catalysis. Dienes containing conjugated double bonds such as CLA can react with dienes under heating conditions.

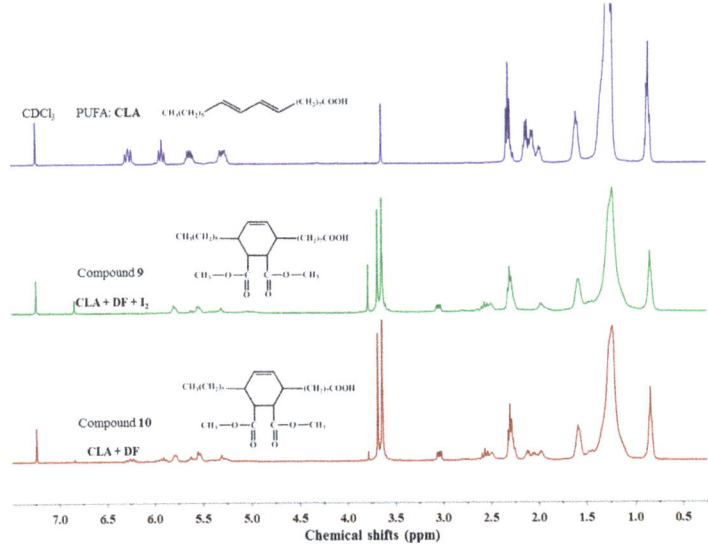

Figure 3. ^{1}H-NMR spectra of conjugated linoleic acid (CLA) and the reaction compounds with dienophiles in chloroform-d under different catalyst conditions: compound **9**, reaction products of CLA and DF with iodine catalysis; compound **10**, reaction products of CLA and DF without iodine catalysis.

Figure 4 shows the ^{1}H-NMR spectra of conjugated linoleic acid ethyl ester (CLAEE) with conjugated double bonds and the reaction compounds with dienophiles under different catalyst conditions. Compound **11** was the reaction product of CLAEE and FA with iodine catalysis; compound **12** was the reaction product of CLAEE and FA without iodine catalysis; compound **13** was the reaction product of CLAEE and DF with iodine catalysis; compound **14** was the reaction product of CLAEE and DF without iodine catalysis.

Both compound **11** and compound **12** were obtained as 22-carbon tricarboxylic acid monoethyl ester (C22TAMEE). Its molecular formula $C_{24}H_{40}O_6$ was established from ESI-MS: compound **11**, m/z 423.27524 [M − H]$^-$; compound **12**, m/z 423.27551 [M − H]$^-$. The proton signals of compound **11** at δ H 5.74 (s) and compound **12** at δ H 5.76 (m) revealed the presence of a double bond of cyclohexene; those of compound **11** at δ H 2.87 (d) and δ H 2.19–1.94 (m) and compound **12** at δ H 2.87 (d) and δ H 2.04–1.89 (m) revealed the presence of a methine of cyclohexene. The chemical shifts of compound **11** at 129.92, 44.14, and 39.53 ppm were correspondingly attributed to the carbons in the cyclohexene (Figure S27, Supplementary Materials). The chemical shifts of compound **12** at 129.85 and 44.15 ppm were correspondingly attributed to the carbons in the cyclohexene (Figure S28, Supplementary Materials). These results implied that a Diels–Alder reaction occurred between CLAEE and FA with and without iodine catalysis. Dienes containing conjugated double bonds such as CLAEE can react with dienes under heating conditions.

Both compound **13** and compound **14** were obtained as 22-carbon tricarboxylic acid dimethyl monoethyl ester (C22TADMMEE). Its molecular formula $C_{26}H_{44}O_6$ was established from ESI-MS: compound **13**, m/z 453.31989 [M + H]$^+$, 475.30170 [M + Na]$^+$, 491.27563 [M + K]$^+$; compound **14**, m/z 453.32010 [M + H]$^+$, 475.30191 [M + Na]$^+$, 491.27591 [M + K]$^+$. The proton signals of compound **13** at δ H 5.84–5.72 (m) and compound **14** at δ H 5.90 (m) revealed the presence of a double bond of cyclohexene. Those of compound **13** at δ H 3.03 (m) and δ H 2.64–2.39 (m) and compound **14** at δ H 3.04 (m) and δ H 2.64–2.39 (m) revealed the presence of a methine of cyclohexene, while the proton signals of compound **13** and compound **14** at δ H 3.72–3.61 (m) revealed the presence of a methoxy of dimethyl ester attached to cyclohexene. The chemical shifts of compound **13** at 129.10, 46.91, and 39.67 ppm were correspondingly attributed to the carbons in the cyclohexene (Figure S29, Supplementary Materials). The chemical shifts of compound **14** at 130.05, 46.91, and 39.67 ppm were correspondingly attributed to the carbons in the cyclohexene (Figure S30, Supplementary Materials). These results implied that a Diels–Alder reaction occurred between CLAEE and DF with and without iodine catalysis. Dienes containing conjugated double bonds such as CLAEE can react with dienes under heating conditions.

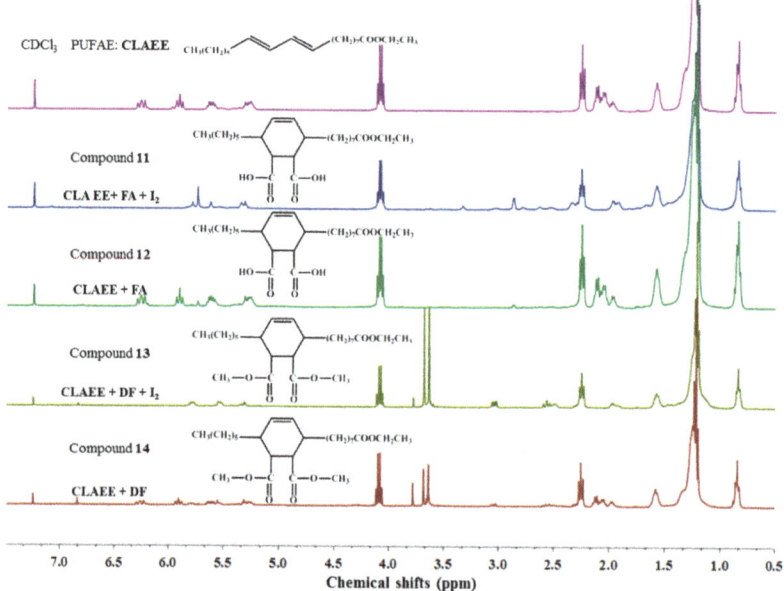

Figure 4. ^{1}H-NMR spectrum of conjugated linoleic acid ethyl ester (CLAEE) and the reaction compounds with dienophiles in chloroform-d under different catalyst conditions: compound **11**, reaction products of CLAEE and FA with iodine catalysis; compound **12**, reaction products of CLAEE and FA without iodine catalysis; compound **13**, reaction products of CLAEE and DF with iodine catalysis; compound **14**, reaction products of CLAEE and DF without iodine catalysis.

The results of ^{1}H-NMR, ^{13}C-NMR, and ESI-MS supported the successful synthesis of C22TA and C22TAE. All C22TAs and their esters, C22TADME, C22TAMME, C22TATME, C22TAMEE, and C22TADMMEE, were liquid at room temperature, which would be desirable for many applications.

2.2. The Role of Iodine in Diels–Alder Reaction

The reaction yields of C22TA and C22TAE prepared via Diels–Alder reaction under different experimental conditions, calculated according to the ^{1}H-NMR spectra, are shown in Table 1.

Under heating conditions, a Diels–Alder reaction did not take place between dienes and dienophiles in the absence of iodine catalyst, while Diels–Alder reactions occurred with iodine catalysis. These results indicated that the isomerization catalyst was necessary to transform the dienes of nonconjugated double bonds into a suitable configuration for the reaction [19]. The iodine catalyst was important for conjugated double bond formation. There are many methods of making conjugated double bonds; however, iodine was selected in this study, because iodine can minimize side reactions and unnecessary double-bond migrations [20].

In general, the Diels–Alder reactions occur with dienophiles containing conjugated double bonds [7,20–22]. Therefore, the participation of LA and its esters in the Diels–Alder reaction was due to the conversion of the nonconjugated double bonds into conjugated double bonds. The Diels–Alder reaction could occur on the conjugated double bonds at C9 and C11 for 9t,11t CLA and its ester or at C10 and C12 for 10t,12t CLA and its ester [23,24]. Herein, the reactions of PUFAs and dienophiles without catalysts were used as the reference reaction for the discussion of subsequent mechanisms with iodine catalysts and PUFA. No attempt was made to separate the isomers in this study.

Diels–Alder reactions can occur between LA and its ester with dienophiles under the catalysis of iodine [11,25]. The possible mechanism of iodine converting nonconjugated double bonds into conjugated double bonds of LA and its ester, along with the traditional alkali isomerization mechanism, is shown in Scheme 1. It was proposed that iodine turned nonconjugated double bonds into conjugated double bonds via radical processes. The mechanism of iodine isomerization of LA was possibly the same as that of photoisomerization in the presence of iodine [23]. However, the process was initiated via heat source. As shown in Scheme 1, under heating, iodine formed two I• radicals, which in turn reacted with the unsaturated molecule. When I• abstracted an allylic hydrogen, positional isomerization occurred, and the conjugated product was formed [23,26,27]. In the traditional alkali isomerization process, the main bases used include inorganic bases NaOH and KOH, and organic bases sodium ethoxide, sodium methoxide, tetraethyl ammonium hydroxide, etc. [28–30]. The essence of alkali isomerization involves the process of a carbon anion, and its reaction mechanism features the base grabbing the protons of the allylic hydrogen of LA and its ester to generate a carbon anion, followed by the carbon anion rearranging to form a stable conjugated structure, thereby achieving a conjugated product [31,32].

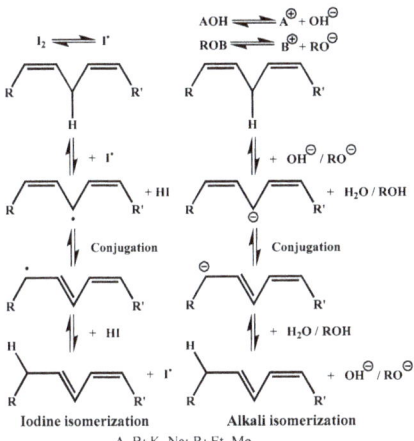

Scheme 1. Mechanism of iodine and alkali isomerization of LA.

As can be seen from the results of entries 1–8 in Table 1, it was difficult for nonconjugated dienes and dienophiles to react without a catalyst. In addition, with iodine catalysis, the substituent on dienes and dienophiles had an effect on the efficiency of the Diels–Alder reaction. The yields of C22TA (C22TAE) in the reaction of LA with FA and DA were 71.9% and 81.3%, respectively. The yield of the latter was 13.1% higher than that of the former. The yields of C22TA (C22TAE) from ML in the reaction with FA and DF were 65.4% and 76.3%, respectively. The yield of the latter was 16.7% higher than that of the former. The yield of C22TA (C22TAE) prepared via the reaction of dienes with FA was lower than that with DF. The same trend was observed for entries 11–14 in Table 1. The yield of C22TA (C22TAE) prepared via reaction with FA was lower than that with DF with and without catalyst. With catalyst, the yields of CLAEE reacting with FA and DF were 74.4% and 83.3% respectively, while the yields were 71.4% and 77.5%, respectively, without the catalyst.

This was mainly because DF, the dienophile with an electron-withdrawing group, was more favorable for the Diels–Alder reaction than FA [33]. According to frontier molecular orbital theory

(FMO), the Diels–Alder reaction is controlled by the suprafacial in-phase interaction of the highest occupied molecular orbital (HOMO) of one component and the lowest unoccupied molecular orbital (LUMO) of the other [34,35]. The reactivity of a Diels–Alder reaction depends upon the HOMO–LUMO energy gap of the components: a lower energy difference results in a lower transition state energy of the reaction. Electron-withdrawing substituents lower the energies of both HOMO and LUMO, while electron-donating groups increase their energies [36]. In general, a normal Diels–Alder reaction is accelerated by electron-donating substituents in the diene and by electron-withdrawing substituents in the dienophile [37].

As can be seen from the results of entries 9–14 in Table 1, Diels–Alder reactions between dienes and dienophiles containing conjugated double bonds could occur without a catalyst. However, the iodine catalyst had a certain positive effect on the efficiency of the Diels–Alder reaction. The yields of C22TAE were 82.0%, 79.4%, 74.4%, 71.4%, 83.3%, and 77.5%, respectively, under the conditions of CLA and DF, CLAEE and FA, and CLAEE and DF with and without catalyst. The results showed that the yields of C22TAE with iodine catalyst were higher than those without catalyst.

This was mainly because iodine is a mild Lewis acid. Lewis acids can greatly accelerate cycloaddition [38]. The catalytic effect is explained by FMO theory, considering that the coordination of the carbonyl oxygen by a Lewis acid increases the electron-withdrawing effect of the carbonyl group on the carbon–carbon double bond and lowers the LUMO dienophile energy [39]. It reduces the energy difference between the HOMO of dienes and the LUMO of dienophiles, thus facilitating the Diels–Alder reaction [37].

3. Materials and Methods

3.1. Materials

Linoleic acid (LA, 60%, Acros Organics) was purchased from Beijing Innochem Co., Ltd., Beijing, China. Fumaric acid (FA, 99%) was purchased from Shanghai Shanpu Chemical Co., Ltd., Shanghai, China. Dimethyl fumarate (DF, 97%, Accela) was purchased from Shanghai Shaoyuan Co., Ltd., Shanghai, China. Methyl alcohol (99.5%), sulfuric acid (98%), potassium hydroxide (95%), and iodine (99.5%) were purchased from Tianjin Chemical Reagent Co., Ltd., Tianjin, China. Conjugated linoleic acid (CLA, 85%) and conjugated linoleic acid ethyl ester (CLAEE, 70%) were obtained from Xinjiang Technical Institute of Physics and Chemistry (Xinjiang, China.). All chemicals were used without further purification unless notified. Deionized water was used throughout the experiments.

3.2. Synthesis of Methyl Linoleate (ML)

A typical procedure for the synthesis of ML was as follows: firstly, 100 g of LA, 70 g of methyl alcohol, and 1.4 g of sulfuric acid as catalyst were mixed in a 250-mL three-neck round-bottom flask and refluxed at 80 °C for 6 h. The flask was equipped with a mechanical stirrer, a thermometer, and a condenser. The reaction products were separated with by-product glycerol using a separating funnel for standing stratification. The supernatant was crude ML. Then, the crude ML was washed with deionized water (50 °C) to a neutral pH. The crude product was transferred into a 250-mL three-neck round-bottom flask for vacuum distillation at 60 °C to remove the residual water and methyl alcohol. Finally, the product ML was obtained.

3.3. Synthesis of 22-Carbon Tricarboxylic Acid (C22TA) and 22-Carbon Tricarboxylic Ester (C22TAE)

C22TA and C22TAE were synthesized using the method of Vijayalakshmi et al. [8]. Firstly, 1 mol of PUFA or PUFAE and 1.1 mol of dienophiles were mixed in a 250-mL three-neck round-bottom flask. To this one-step process, catalytic amounts of iodine were 0 wt.% or 0.3 wt.% based on the amounts of PUFA or PUFAE. The flask was equipped with a mechanical stirrer, a thermometer, and a condenser. The reaction mixture was heated up within 30 min to 200 °C and maintained at 2 h. The unreacted dienophile was removed by vacuum distillation in steam (<15 MPa, 200 °C). PUFAs used in this study were LA and CLA. PUFAEs used in this study were ML and CLAEE. Dienophiles used in this study were FA and DF. The typical synthesis routes of C22TA and C22TAE are shown in Scheme 2.

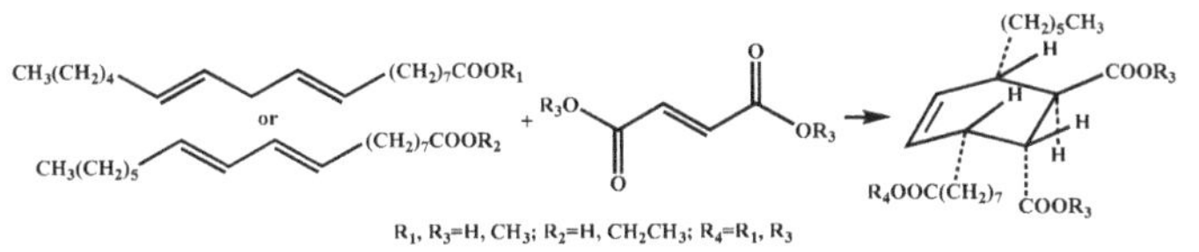

Scheme 2. The typical synthesis routes of 22-carbon tricarboxylic acid (C22TA) and its ester (C22TAE).

3.4. Characterization

NMR spectra were recorded on an NMR spectrometer (Varian, Palo Alto, CA, USA) at 400 (^{1}H) or 100 (^{13}C) MHz in deuterated chloroform (CDCl$_3$) or deuterated dimethyl sulfoxide (DMSO-d$_6$). Chemical shifts relative to that of chloroform (7.24) or DMSO (2.47) were reported. ESI-MS spectra were determined on a Waters UPLC-Quattro Premier XE UPLC–MS/MS (Ultimate 3000/Q-Exactive, Thermo Fisher Scientific., Waltham, MA, USA). The ESI flow rate of the electrospray ion source was 1–2000 µL·min^{-1}.

4. Conclusions

^{1}H-NMR, ^{13}C-NMR, and ESI-MS spectrometry were used to investigate the efficiency of Diels–Alder reactions under different reaction conditions. PUFAs (PUFAEs) such as LA, ML, CLA, and CLAEE were used as dienes, fumaric acid and dimethyl fumarate were used as dienophiles, and iodine was used as a catalyst to investigate the yield of C22TA (C22TAE) with or without catalyst. Results showed that, for dienes without conjugated double bonds to participate in a Diels–Alder reaction, iodine was necessary as the catalyst. Under heating conditions, without iodine catalyst, the nonconjugated double bonds of dienes cannot be converted effectively to conjugated double bonds. For dienes with conjugated double bonds, Diels–Alder reactions can occur with dienes under heating conditions without catalyst. However, the yield of C22TAE with iodine catalyst was higher than that without iodine catalyst. The yield of C22TA (C22TAE) prepared via the reaction of diene with DF was higher than that with FA. The mechanism of the Diels–Alder reaction for the synthesis of C22TA (C22TAE) is proposed below.

(1) The catalyst iodine plays two roles in the synthesis of C22TA (C22TAE). On the one hand, it transforms the nonconjugated double bonds of diene into conjugated double bonds via radical processes. Then, the conjugated diene participates in a Diels–Alder reaction to form C22TA (C22TAE). On the other hand, iodine can indeed improve the yield of C22TA (C22TAE), which is beneficial to the reaction. Compared with the Diels–Alder reaction system without the catalyst, iodine as a Lewis acid increases the electron-withdrawing effect of the carbonyl group on the carbon–carbon double bond and lowers the LUMO dienophile energy. It reduces the energy difference between the HOMO of diene and the LUMO of dienophile, thus facilitating the Diels–Alder reaction.

(2) The type of diene determines whether an iodine catalyst needs to be added for the synthesis of C22TA (C22TAE). For dienes without conjugated double bonds, iodine must be added as a catalyst to prepare C22TA (C22TAE). However, for dienes with conjugated double bonds, C22TA (C22TAE) can be obtained without iodine, but the product yield can be improved by adding iodine to the reaction system. However, one should keep in mind that, with the addition of iodine, the prepared product usually has a deep brown color and an extra decolorization procedure is usually needed.

(3) The type of substituent in the dienophile directly affects the yield of C22TA (C22TAE). When the dienophile has electron-withdrawing substituents, the efficiency of the Diels–Alder reaction is higher, and the yield of C22TA (C22TAE) is higher. The electron-withdrawing substituents in the dienophile reduce the HOMO–LUMO energy and accelerate the normal Diels–Alder reactions. C22TA (C22TAE) synthesized in this study has excellent biodegradability, superior properties, and great

market potential. The spectroscopic data and proposed mechanism of the Diels–Alder reaction could be useful for the design and preparation of C22TA (C22TAE) and cyclohexene derivatives.

Supplementary Materials: The following are available online at http://www.mdpi.com/2073-4344/9/12/972/s1: Figure S1: ESI-MS spectrum of compound 1 C22TA; Figure S2: ESI-MS spectrum of compound 3 C22TADME; Figure S3: ESI-MS spectrum of compound 5 C22TAMME; Figure S4: ESI-MS spectrum of compound 7 C22TATME; Figure S5: ESI-MS spectrum of compound 9 C22TADME; Figure S6: ESI-MS spectrum of compound 10 C22TADME; Figure S7: ESI-MS spectrum of compound 11 C22TAMEE; Figure S8: ESI-MS spectrum of compound 12 C22TAMEE; Figure S9: ESI-MS spectrum of compound 13 C22TADMMEE; Figure S10: ESI-MS spectrum of compound 14 C22TADMMEE; Figure S11: ^{1}H-NMR spectrum of compound 1 C22TA; Figure S12: ^{1}H-NMR spectrum of compound 3 C22TADME; Figure S13: ^{1}H-NMR spectrum of compound 5 C22TAMME; Figure S14: ^{1}H-NMR spectrum of compound 7 C22TATME; Figure S15: ^{1}H-NMR spectrum of compound 9 C22TADME; Figure S16: ^{1}H-NMR spectrum of compound 10 C22TADME; Figure S17: ^{1}H-NMR spectrum of compound 11 C22TAMEE; Figure S18: ^{1}H-NMR spectrum of compound 12 C22TAMEE; Figure S19: ^{1}H-NMR spectrum of compound 13 C22TADMMEE; Figure S20: ^{1}H-NMR spectrum of compound 14 C22TADMMEE; Figure S21: ^{13}C-NMR spectrum of compound 1 C22TA; Figure S22: ^{13}C-NMR spectrum of compound 3 C22TADME; Figure S23: ^{13}C-NMR spectrum of compound 5 C22TAMME; Figure S24: ^{13}C-NMR spectrum of compound 7 C22TATME; Figure S25: ^{13}C-NMR spectrum of compound 9 C22TADME; Figure S26: ^{13}C-NMR spectrum of compound 10 C22TADME; Figure S27: ^{13}C-NMR spectrum of compound 11 C22TAMEE; Figure S28: ^{13}C-NMR spectrum of compound 12 C22TAMEE; Figure S29: ^{13}C-NMR spectrum of compound 13 C22TADMMEE; Figure S30: ^{13}C-NMR spectrum of compound 14 C22TADMMEE.

Author Contributions: Conceptualization, Y.Z., X.Z., and Y.L.; methodology, Y.L. and L.W.; software, Y.L. and L.Z.; validation, L.W. and L.Z.; formal analysis, Y.L.; investigation, Y.L. and L.W.; resources, Y.Z., X.Z., and L.Z.; data curation, Y.L. and L.W.; writing—original draft preparation, Y.L.; writing—review and editing, Y.Z.; visualization, L.W. and L.Z.; supervision, Y.Z.; project administration, Y.Z.; funding acquisition, Y.Z. All authors contributed substantially to the work reported.

Funding: This work was financially supported by the National Natural Science Foundation of China (21464015, 21472235), the Xinjiang Tianshan Talents Program (2018xgytsyc 2-3), the One Thousand Talents Program (2019 to Zan X.), and the UESTC Talent Startup Funds (A1098 5310 2360 1208).

Conflicts of Interest: The authors declare no conflict of interest.

Table of Contents: The Role of Iodine Catalyst in the Synthesis of 22-Carbon Tricarboxylic Acid and Its Ester: A Case Study.

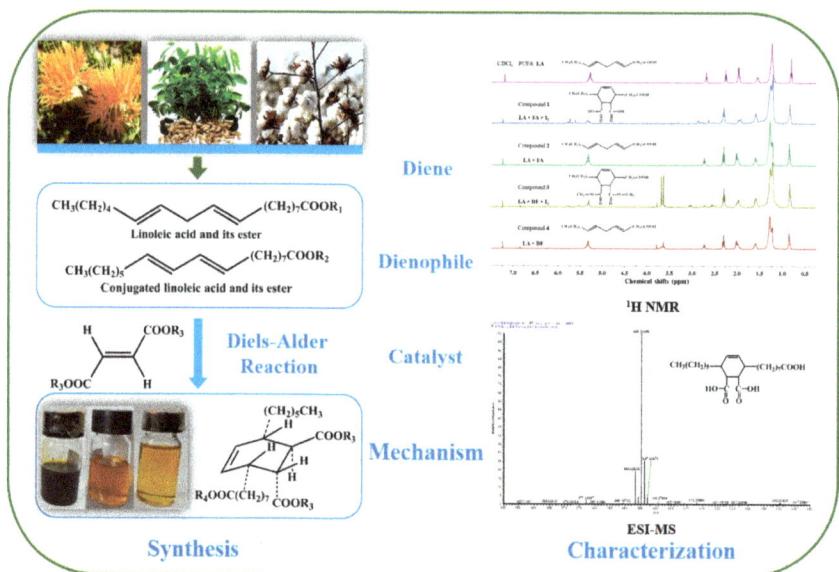

References

1. Molino, A.; Larocca, V.; Chianese, S.; Musmarra, D. Biofuels Production by Biomass Gasification: A Review. *Energies* **2018**, *11*, 811. [CrossRef]
2. Adhvaryu, A.; Biresaw, G.; Sharma, B.K.; Erhan, S.Z. Friction Behavior of Some Seed Oils: Biobased Lubricant Applications. *Ind. Eng. Chem. Res.* **2006**, *45*, 3735–3740. [CrossRef]
3. Salimon, J.; Abdullah, B.M.; Yusop, R.M.; Salih, N. Synthesis, reactivity and application studies for different biolubricants. *Chem. Cent. J.* **2014**, *8*, 16. [CrossRef] [PubMed]
4. Visentini, F.F.; Perez, A.A.; Santiago, L.G. Self-assembled nanoparticles from heat treated ovalbumin as nanocarriers for polyunsaturated fatty acids. *Food Hydrocoll.* **2019**, *93*, 242–252. [CrossRef]
5. Patterson, E.; Wall, R.; Fitzgerald, G.F.; Ross, R.P.; Stanton, C. Health Implications of High Dietary Omega-6 Polyunsaturated Fatty Acids. *J. Nutr. Metab.* **2012**, *2012*, 539426. [CrossRef]
6. Fox, N.J.; Stachowiak, G.W. Vegetable oil-based lubricants—A review of oxidation. *Tribol. Int.* **2007**, *40*, 1035–1046. [CrossRef]
7. Huang, K.; Zhang, P.; Zhang, J.; Li, S.; Li, M.; Xia, J.; Zhou, Y. Preparation of biobased epoxies using tung oil fatty acid-derived C21 diacid and C22 triacid and study of epoxy properties. *Green Chem.* **2013**, *15*, 2466–2475. [CrossRef]
8. Vijayalakshmi, P.; Subbarao, R.; Lakshminarayana, G. Preparation and surface-active properties of the sodium soaps, mono- and diethanolamides and diol and triol sulfates of cycloaliphatic C21 Di- and C22 Tricarboxylic acids. *J. Am. Oil Chem. Soc.* **1991**, *68*, 133–137. [CrossRef]
9. Fischer, E.R.; Boyd, P.G. New Water Soluble C22 Tricarboxylic Acid Amine Salts and Their Use as a Water Soluble Corrosion Inhibitors. E.P. Patent No. 0,913,384, 6 May 1999.
10. Watanabe, S.; Fujita, T.; Fukuda, S.; Hirano, K.; Sakamoto, M. Characteristic properties as cutting fluid additives of the products from the reaction of unsaturated fatty acids with maleic anhydride. *Mater. Chem. Phys.* **1986**, *15*, 89–96. [CrossRef]
11. Powers, J.R.; Pleasant, M.; Miller, E.C.; Charleston, S.C. C_{22}-Cycloaliphatic Tricarboxylic Fatty Acid Soaps. U.S. Patent No. 4,081,462, 28 March 1978.
12. Chintareddy, V.R.; Oshel, R.E.; Doll, K.M.; Yu, Z.; Wu, W.; Zhang, G.; Verkade, J.G. Investigation of Conjugated Soybean Oil as Drying Oils and CLA Sources. *J. Am. Oil Chem. Soc.* **2012**, *89*, 1749–1762. [CrossRef]
13. Jain, V.P.; Proctor, A.; Lall, R. Pilot-Scale Production of Conjugated Linoleic Acid-Rich Soy Oil by Photoirradiation. *J. Food Sci.* **2008**, *73*, 183–192. [CrossRef] [PubMed]
14. Kadamne, J.V.; Castrodale, C.L.; Proctor, A. Measurement of Conjugated Linoleic Acid (CLA) in CLA-Rich Potato Chips by ATR-FTIR Spectroscopy. *J. Agric. Food Chem.* **2011**, *59*, 2190–2196. [CrossRef] [PubMed]
15. Gangidi, R.R.; Proctor, A. Photochemical production of conjugated linoleic acid from soybean oil. *Lipids* **2004**, *39*, 577–582. [CrossRef] [PubMed]
16. Jain, V.P.; Proctor, A. Photocatalytic Production and Processing of Conjugated Linoleic Acid-Rich Soy Oil. *J. Agric. Food Chem.* **2006**, *54*, 5590–5596. [CrossRef] [PubMed]
17. Tokle, T.; Jain, V.P.; Proctor, A. Effect of Minor Oil Constituents on Soy Oil Conjugated Linoleic Acid Production. *J. Agric. Food Chem.* **2009**, *57*, 8989–8997. [CrossRef]
18. Balakrishna, R.S.; Murthy, B.G.K.; Aggarwal, J.S. Diels-Alder adducts from safflower oil fatty acids. *J. Am. Oil Chem. Soc.* **1971**, *48*, 689–692. [CrossRef]
19. Miller, W.R.; Bell, E.W.; Cowan, J.C.; Teeter, H.M. Reactions of dienophiles with vegetable oils. I. Reactions of maleic esters with sulfur dioxide catalyst. *J. Am. Oil Chem. Soc.* **1959**, *36*, 394–397. [CrossRef]
20. Danzig, M.J.; O'Donnell, J.L.; Bell, E.W.; Cowan, J.C.; Teeter, H.M. Reactions of conjugated fatty acids. V. Preparation and properties of diels-alder adducts and their esters from trans,trans-conjugated fatty acids derived from soybean oil. *J. Am. Oil Chem. Soc.* **1957**, *34*, 136–138. [CrossRef]
21. Bickford, W.G.; DuPré, E.F.; Mack, C.H.; O'Connor, R.T. The infrared spectra and the structural relationships between alpha- and beta-eleostearic acids and their maleic anhydride adducts. *J. Am. Oil Chem. Soc.* **1953**, *30*, 376–381. [CrossRef]
22. Teeter, H.M.; O'Donnell, J.L.; Schneider, W.J.; Gast, L.E.; Danzig, M.J. Reactions of Conjugated Fatty Acids. IV. Diels-Alder Adducts of 9,11-Octadecadienoic. *Acid. J. Org. Chem.* **1957**, *22*, 512–514. [CrossRef]
23. Philippaerts, A.; Goossens, S.; Jacobs, P.A.; Sels, B.F. Catalytic production of conjugated fatty acids and oils. *ChemSusChem* **2011**, *4*, 684–702. [CrossRef] [PubMed]

24. Cihelkova, K.; Schieber, A.; Lopes-Lutz, D.; Hradkova, I.; Kyselka, J.; Filip, V. Quantitative and qualitative analysis of high molecular compounds in vegetable oils formed under high temperatures in the absence of oxygen. *Eur. Food Res. Technol.* **2013**, *237*, 71–81. [CrossRef]
25. Ward, B.E.; Charleston, S.C. Selective Reaction of Fatty Acids and Their Separation. U.S. Patent No. 3,753,968, 21 August 1973.
26. Benson, S.W.; Bose, A.N. The Iodine-catalyzed, Positional Isomerization of Olefins. A New Tool for the Precise Measurement of Thermodynamic Data. *J. Am. Chem. Soc.* **1963**, *85*, 1385–1387. [CrossRef]
27. Egger, K.W.; Benson, S.W. Iodine and Nitric Oxide Catalyzed Isomerization of Olefins. V. Kinetics of the Geometrical Isomerization of 1,3-Pentadiene, a Check on the Rate of Rotation about Single Bonds, and the Allylic Resonance Energy. *J. Am. Chem. Soc.* **1965**, *87*, 3314–3319. [CrossRef]
28. Afarin, M.; Alemzadeh, I.; Yazdi, Z.K. Conjugated Linoleic Acid Production and Optimization Via Catalytic Reaction Method Using Safflower Oil. *Int. J. Eng.* **2018**, *31*, 1166–1171. [CrossRef]
29. Silva-Ramirez, A.S.; Rocha-Uribe, A.; Gonzalez-Chavez, M.M.; Gonzalez, C. Synthesis of conjugated linoleic acid by microwave-assisted alkali isomerization using propylene glycol as solvent. *Eur. J. Lipid Sci. Technol.* **2017**, *119*, 9. [CrossRef]
30. Wang, Q.Y.; Li, X.; Du, K.F.; Yao, S.; Song, H. Conjugated Linoleic Acid Production by Alkali Isomerization of Linoleic Acid from *Idesia polycarpa* Maxim. var. vestita Diels Oil. *Asian J. Chem.* **2013**, *25*, 3744–3748. [CrossRef]
31. Falkenburg, L.B.; DeJong, W.M.; Handke, D.P.; Radlove, S.B. Isomerization of drying and semi-drying oils: The use of anthraquinone as a conjugation catalyst. *J. Am. Oil Chem. Soc.* **1948**, *25*, 237–243. [CrossRef]
32. Chen, C.A.; Lu, W.; Sih, C.J. Synthesis of 9Z,11E-octadecadienoic and 10E,12Z-octadecadienoic acids, the major components of conjugated linoleic acid. *Lipids* **1999**, *34*, 879–884. [CrossRef]
33. Guerrero-Corella, A.; Asenjo-Pascual, J.; Pawar, T.J.; Díaz-Tendero, S.; Martín-Sómer, A.; Gómez, C.V.; Belmonte-Vázquez, J.L.; Ramírez-Ornelas, D.E.; Peña-Cabrera, E.; Fraile, A.; et al. BODIPY as electron withdrawing group for the activation of double bonds in asymmetric cycloaddition reactions. *Chem. Sci.* **2019**, *10*, 4346–4351. [CrossRef]
34. Houk, K.N.; Sims, J.; Duke, R.E.; Strozier, R.W.; George, J.K. Frontier molecular orbitals of 1,3 dipoles and dipolarophiles. *J. Am. Chem. Soc.* **1973**, *95*, 7287–7301. [CrossRef]
35. Sauer, J.; Sustmann, R. Mechanistic Aspects of Diels-Alder Reactions: A Critical Survey. *Angew. Chem. Int. Ed.* **1980**, *19*, 779–807. [CrossRef]
36. Kim, J.H.; Lee, H. Enhancement of efficiency in luminescent polymer by incorporation of conjugated 1,3,4-oxadiazole side chains as hole-blocker/electron-transporter. *Synth. Met.* **2004**, *143*, 13–19. [CrossRef]
37. Fringuelli, F.; Taticchi, A. Diels–Alder Reaction: General Remarks. In *The Diels–Alder Reaction*; John Wiley and Sons: New York, NY, USA, 2002; pp. 1–28. [CrossRef]
38. Fringuelli, F.; Minuti, L.; Pizzo, F.; Taticchi, A. *Reactivity and Selectivity in Lewis-Acid-Catalyzed Diels-Alder Reactions of 2-Cyclohexenones*. ChemInform; Munksgaard: Copenhagen, Denmark, 1993; p. 24. [CrossRef]
39. Zeng, X.; Yu, S.; Sun, R. Effect of functionalized multiwall carbon nanotubes on the curing kinetics and reaction mechanism of bismaleimide–triazine. *J. Therm. Anal. Calorim.* **2013**, *114*, 387–395. [CrossRef]

© 2019 by the authors. Licensee MDPI, Basel, Switzerland. This article is an open access article distributed under the terms and conditions of the Creative Commons Attribution (CC BY) license (http://creativecommons.org/licenses/by/4.0/).

Article

Benchmarking Acidic and Basic Catalysis for a Robust Production of Biofuel from Waste Cooking Oil

Claudia Carlucci [1,*], Michael Andresini [1], Leonardo Degennaro [1] and Renzo Luisi [1,2]

1 Flow Chemistry and Microreactor Technology FLAME-Lab, Department of Pharmacy-Drug Sciences, University of Bari "A. Moro" Via E. Orabona 4, Bari 70125, Italy; m.andresini1@studenti.uniba.it (M.A.); leonardo.degennaro@uniba.it (L.D.); renzo.luisi@uniba.it (R.L.)
2 Institute for the Chemistry of OrganoMetallic Compounds (ICCOM), National Research Council (CNR), Via Orabona 4, Bari 70125, Italy
* Correspondence: claudia.carlucci@uniba.it; Tel.: +39-080-544-2251

Received: 29 October 2019; Accepted: 9 December 2019; Published: 10 December 2019

Abstract: The production of biodiesel at the industrial level is mainly based on the use of basic catalysts. Otherwise, also acidic catalysis allowed high conversion and yields, as this method is not affected by the percentage of free fatty acids present in the starting sample. This work has been useful in assessing the possible catalytic pathways in the production of fatty acid methyl esters (FAMEs), starting from different cooking waste oil mixtures, exploring particularly acidic catalysis. It was possible to state that the optimal experimental conditions required concentrated sulfuric acid 20% *w/w* as a catalyst, a reaction time of twelve hours, a temperature of 85 °C and a molar ratio MeOH/oil of 6:1. The role of silica in the purification method was also explored. By evaluating the parameters, type of catalyst, temperature, reaction time and MeOH/oil molar ratios, it has been possible to develop a robust method for the production of biodiesel from real waste mixtures with conversions up to 99%.

Keywords: transesterification; catalysis; acidity; waste cooking oil; biodiesel

1. Introduction

The World Energy Forum has estimated that fossil fuels will run out in less than a century if they are not replaced by an alternative energy source. Emissions from combustion of fossil fuels, containing CO_2, CO, nitrogen and sulfur oxides, organic and particulate compounds, as well as being extremely harmful, are the main cause of the greenhouse effect and acid rainfall. Therefore, alternative energy resources such as solar energy, geothermal energy and biomass have captured the attention of the scientific community, including biodiesel, which is a widely used energy source. The valorization of waste materials is a viable alternative to traditional disposal systems, including in the field of renewable energy and bio-fuels [1]. The discovery of alternative fuels that can replace conventional resources has therefore become a goal of scientific research [2]. Biodiesel is a non-oil variant, consisting of a mixture of fatty acid methyl esters (FAMEs), obtained through the transesterification reaction of triglycerides catalyzed by acids, bases, enzymes [3–5] or by using supercritical fluids [6–9], starting from vegetable oils, fats, algae and all kinds of suitable raw materials and waste products [10], including recycled frying oil. Therefore, any source of fatty acids can be used to produce the biodiesel [11,12]. The advantages related to its use as fuel are numerous, starting from the low environmental impact, as it represents a renewable and biodegradable source of energy. The current trend is not to use edible raw materials preferring waste ones, which still must comply with the quality standards as essential prerequisites for the commercial use of any fuel [13–16]. The disposal of waste vegetable oil represents an important issue, especially due to the widespread incorrect dumping in urban wastes, which influences and increases costs for treating sewage plants and polluted waters [17–20]. Biodiesel can be produced from vegetable oils through a transesterification reaction that converted triglycerides into fatty acid methyl or ethyl esters using the low molecular weight alcohols (Scheme 1). This reaction can be carried out in different conditions and with different types of homogeneous and heterogeneous catalysts [3–5].

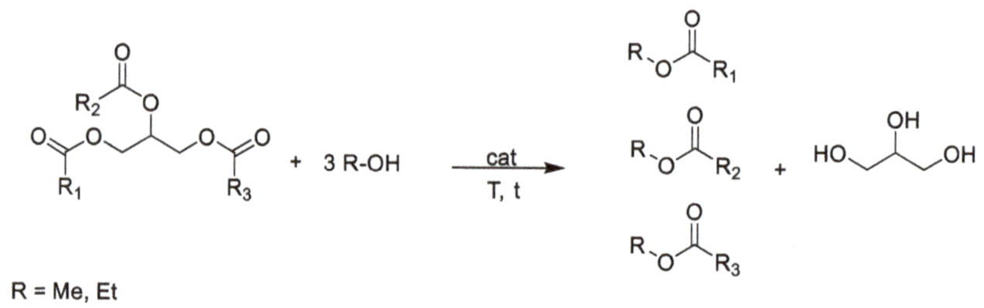

Scheme 1. Transesterification reaction.

Basic catalysis represents the most widely used industrial approach towards biodiesel production [21–23].

The advantages in using this type of catalysis concern reduced reaction times, moderate operating temperatures, the easy recovery of glycerol and the use of cost-effective catalysts. The main disadvantages relate to the saponification reaction, which entails lower yields and difficult washing of biodiesel, with excessive loss of product due to the formation of emulsions. Furthermore, an initial high content of free fatty acids (FFA) allows saponification reaction to undermine the final yield. The reaction was also conditioned by the water content of the sample, which can be reduced through an easy pretreatment such as the heating of the starting oil [24].

Otherwise, the advantage in using homogeneous acidic catalysis is related to the insensitivity to high levels of FFA, simultaneously promoting both the transesterification and the FFA esterification reactions, resulting into an improved biodiesel conversion yield [25,26].

The acidic catalysis results to be much more effective than the basic one when the vegetable oil contains more than 1% of free fatty acids. The transesterification process catalyzed by Brønsted acids, as sulfuric acid, allows very high yields in alkyl esters, however, the drawbacks are related to longer reaction times and higher temperatures [27,28].

Although the use of H_2SO_4 as a catalyst in biodiesel production was largely investigated, the transesterification of a highly pollutant starting material as waste cooking oils (WCO), catalyzed by sulfuric acid, was less explored.

Nye and coworkers reported that it is possible to produce biodiesel from exhausted frying oil using 0.1% of H_2SO_4 as a catalyst, methanol as a solvent, with a methanol/oil molar ratio of 3.6:1, at 25 °C, but the product was recovered with a yield of 72% after a reaction time of 40 h [29].

Canakci and Van Gerpen investigated the transesterification of food grade soybean oil, catalyzed by sulfuric acid, with a MeOH/oil molar ratio of 6:1, a reaction temperature of 60 °C, a catalyst amount of 3%, after a reaction time of 48 h. Nevertheless, the acid catalyst required a concentration of water lower than 0.5% to yield 88% of biodiesel [30].

Guana and Kusakabe developed the synthesis of biodiesel from waste oily sludge (WOS) with high free fatty acids (FFA) content. H_2SO_4 and $Fe_2(SO_4)_3$ were used as catalysts for the esterification of FFA, yielding biodiesel in 86% for both catalysts under suitable condition. The H_2SO_4 concentration was 3 wt% and MeOH/oil molar ratio was 10:1. The reaction rate was greatly enhanced by increasing the reaction temperature to 100 °C. However, this method required a preliminary extraction of oil from WOS containing water, using hexane [31].

In addition to homogeneous catalysis, the heterogeneous catalysis, according to the objectives of green chemistry, provides the facility of separation of the product and the catalyst, allowing the recovery and reuse of the catalyst [32]. The active catalytic material is often dispersed on the surface of a support that can be catalytically inert or can contribute to the overall catalytic activity. Typical supports include different materials such as activated carbon, black carbon and carbon nanotubes, many types of porous inorganic solids such as silica, zeolites, titanium dioxide and mesoporous materials [33].

In this work the main objective concerned the optimization of the transesterification reaction with a focus on acidic catalysis, using different waste cooking oils as starting material. In this study we offer complementary approaches towards biodiesel production from real and unknown mixtures of exhausted frying oils. The advantage of the proposed procedure consists in the robustness of the method with respect to mixtures of very different exhausted oils.

2. Results and Discussion

2.1. Basic Catalysis

We firstly explored the transesterification reaction using basic catalysis, to evaluate the composition of the fatty acids in the starting mixture. The experiments were carried out on six samples, two of which consisted exclusively of waste olive and seed oils, while four consisted of unknown waste oil mixtures. The WCO blends were heated for five hours at a temperature of 65 °C to eliminate impurities and excess of water [24].

The heated samples were then subjected to the transesterification reaction with methanol using NaOH 1% w/w as catalyst, stirring the mixtures at a temperature of 50 °C for 1 h, with a molar ratio of MeOH/oil of 6:1. All the methyl esters were obtained with very high conversion and the separation of glycerol from the methyl esters mixture easily occurred. The results are shown in Table 1.

Table 1. Optimization study of the transesterification in homogeneous basic catalysis.

Entry	Source	Crude weight [1] (g)	Conversion [2] (%)
1	WCO olive oil	68.4 [a]	94
2	WCO seed oil	65.4 [a]	>99
3	WCO mix 1	73.0 [a]	97
4	WCO mix 2	62.0 [a]	91
5	WCO mix 3	4.4 [b]	>99
6	WCO mix 4	73.0 [a]	>99

Conditions: NaOH 1% w/w, MeOH/oil ratio 6:1, 50 °C, 1h. [1] Weight of recovered biodiesel: [a] weight of the initial waste cooking oil (WCO) = 100 g; [b] weight of the initial WCO = 10 g. [2] Calculated by ^{1}H-NMR.

2.2. Acidic Catalysis

A screening was performed using concentrated H_2SO_4 (20% w/w) as a catalyst. Therefore, various reaction parameters were modified to test their influence on biodiesel yield. Table 2 showed the experiments performed using all the waste oils as the starting material.

At first the reaction was carried out on all the WCO mixtures, at a temperature of 65 °C, for 12 h, with MeOH/oil molar ratios of 6:1, reaching moderate biodiesel conversion (entries 1–6). Using WCO mix 2 as the model mixture, the temperature was raised to 85 °C and the reaction time was reduced to 6 h, while the MeOH/oil ratio was also modified. As a result, higher temperature enhanced the conversion of WCO mix 2 into biodiesel. Setting MeOH/oil ratio to 3:1 and 6:1 (entries 7–8) biodiesel was afforded with comparable conversion, which dropped when 12:1 MeOH/oil ratio was used (entry 9). By increasing the reaction time to 12 h at 85 °C, conversion has reached optimum levels up to 99% (entries 13 and 20). To investigate the scalability of the process, the reaction was scaled up using 100 g of the starting WCO mix 2 with MeOH/oil ratio of 3:1 and 6:1, and in both cases (entries 14 and 21) the conversion was found to be excellent (96% and 99% respectively). Furthermore, all the WCO mixtures were transformed into biodiesel by changing MeOH/oil ratios and temperature. The alcohol/oil molar ratio is one of the main factors that influence transesterification. An excess of the alcohol promotes the formation of the biodiesel, but the ideal alcohol/oil ratio must be established empirically, considering each individual process [30]. As a result, biodiesel was obtained with the highest conversion from all the mixtures, heating the samples at 85 °C, with a reaction time of 12 h and 6:1 MeOH/oil ratio.

Table 2. Optimization study of the transesterification in homogeneous acidic catalysis.

Entry	Source	Temperature (°C)	Time (h)	MeOH/Oil Ratio	Crude Weight [1] (g)	Conversion [2] (%)
1	WCO olive oil	65	12	6:1	8.3 [a]	66
2	WCO seed oil	65	12	6:1	6.3 [a]	63
3	WCO mix 1	65	12	6:1	7.5 [a]	45
4	WCO mix 2	65	12	6:1	6.2 [a]	48
5	WCO mix 3	65	12	6:1	7.2 [a]	34
6	WCO mix 4	65	12	6:1	7.1 [a]	40
7	WCO mix 2	85	6	3:1	7.4 [a]	75
8	WCO mix 2	85	6	6:1	4.2 [a]	72
9	WCO mix 2	85	6	12:1	9.6 [a]	42
10	WCO olive oil	85	12	3:1	5.4 [a]	79
11	WCO seed oil	85	12	3:1	6.3 [a]	74
12	WCO mix 1	85	12	3:1	9.3 [a]	68
13	WCO mix 2	85	12	3:1	7.2 [a]	98
14	WCO mix 2	85	12	3:1	72.0 [b]	96
15	WCO mix 3	85	12	3:1	7.5 [a]	69
16	WCO mix 4	85	12	3:1	7.9 [a]	85
17	WCO olive oil	85	12	6:1	6.8 [a]	92
18	WCO seed oil	85	12	6:1	8.5 [a]	85
19	WCO mix 1	85	12	6:1	8.4 [a]	91
20	WCO mix 2	85	12	6:1	6.6 [a]	99
21	WCO mix 2	85	12	6:1	88.0 [b]	99
22	WCO mix 3	85	12	6:1	9.9 [a]	86
23	WCO mix 4	85	12	6:1	8.9 [a]	91
24	WCO olive oil	85	12	12:1	8.0 [a]	61
25	WCO seed oil	85	12	12:1	7.5 [a]	63
26	WCO mix 1	85	12	12:1	8.1 [a]	35
27	WCO mix 2	85	12	12:1	7.6 [a]	70
28	WCO mix 3	85	12	12:1	7.4 [a]	37
29	WCO mix 4	85	12	12:1	7.2 [a]	60

[1] Weight of recovered biodiesel: [a] weight of the initial WCO = 10 g; [b] weight of the initial WCO = 100 g. [2] Conversion calculated by ^{1}H-NMR.

From the experimental point of view, the acid catalyzed reactions presented several problems related to the recovery of glycerol and product washing, compared to the basic catalyzed ones. The addition of concentrated H_2SO_4 to the reaction mixture caused a change of color, from yellow to black. It was hypothesized that this variation could be attributable to the formation of polymeric products [34]. In each transesterification reaction catalyzed by H_2SO_4, the formation of the dark amorphous residue was observed after some hours. This residue was insoluble in the reaction mixture and in water, whereas it was found to be soluble in acetone. Biodiesel washing represented another problem related to the use of this acid catalyst. The presence of the dark residue, in fact, makes it difficult to observe the phase separation during the work up procedure. Furthermore, an evaluation of acidic catalysis was carried out, reporting the role of silica as an element introduced directly into the reaction mixture [34]. In recent years, the development of new catalytic strategies in the production of biodiesel has witnessed an increase in the use of silica as a support for charging acid sulfonic catalysts, which have shown promising catalytic activity [35]. The use of silica was evaluated to cope with a further problem arising during the catalyzed acid transesterification reactions with H_2SO_4. These reactions were characterized by the formation of black residues, water-insoluble and hard to remove. The tests were conducted using WCO mix 2, as reference, with the addition of 50% *w/w* of silica (Table 3).

Table 3. Optimization study of the transesterification of WCO mix 2 in the presence of silica.

Entry	Catalyst	Time (h)	Crude weight [1] (g)	Conversion [2] (%)
1	SiO_2	6	-	-
2	SiO_2/Amberlyst 15 (3%)	6	-	-
3	SiO_2/HCl 4M in dioxane	6	5.3 [a]	16
4	SiO_2/H_2SO_4 (20%)	6	8.7 [a]	23
5	SiO_2/H_2SO_4 (20%)	12	8.5 [b]	92
6	SiO_2/H_2SO_4 (20%)	12	10.5 + 5.3 [b,c]	85

[1] Weight of recovered biodiesel: [a] weight of the initial WCO = 10 g; [b] weight of the initial WCO = 20 g; [c] after extraction with CH_2Cl_2. [2] Calculated by ^{1}H-NMR.

At first the reactions carried out with silica and a cation exchange resin such as Amberlyst 15 at 3% w/w (after activation with HCl 6N) were tested (entries 1 and 2), without any biodiesel conversion. Afterwards, using silica with HCl 4M in dioxane and H_2SO_4 20% w/w (entries 3 and 4) after 6 h, low conversions were observed while the best result was obtained stirring the reaction for 12 h (entry 5). However, part of the obtained biodiesel was still retained by silica, therefore supplementary extraction of the retained material on silica with CH_2Cl_2 allowed the recovery of additional 5.3 g of biodiesel (entry 6). These results were accompanied by the good ability of silica to retain the residues formed during the reaction.

2.3. Characterization

2.3.1. Acidity of WCO

The acidity of the waste oils was a fundamental parameter, as it influenced the yield of the transesterification reaction. By titration of WCO samples, it was possible to verify that free fatty acids content was lower than 3%, except of WCO mix 1 that showed an acidity of 3.16%. The analyzed samples were heated to 65 °C for 5 h and the acidity percentage was measured before and after heating. The percentage acidity values calculated for different oil mixtures are shown in Table 4.

Table 4. Percentage acidity values.

Entry	Source	Acidity % [1]	Acidity % [2]
1	WCO olive oil	1.79	1.47
2	WCO seed oil	1.73	1.61
3	WCO mix 1	3.16	2.80
4	WCO mix 2	0.68	0.84
5	WCO mix 2 [3]	0.68	0.57
6	WCO mix 3	2.05	2.22
7	WCO mix 3 [3]	2.05	2.02
8	WCO mix 4	0.34	0.31

[1] Before heating. [2] After heating. [3] Dried with anhydrous Na_2SO_4 before heating.

The acidity decreased after heating, except in two cases (entries 4 and 6) where the acidity was higher with respect to the starting sample. To justify these results, it was hypothesized that the presence of a high amount of water in the starting WCO may increase the concentration of acidic species after heating. To eliminate the excess of water, WCO mix 2 and WCO mix 3 were preliminarily dried with anhydrous Na_2SO_4. In this case, repeating the titration after the thermal treatment (entries 5 and 7), the acidity percentage decreased. Consequently, it was possible to state that water could have a significant role in lowering the transesterification reaction yield, due to the thermal promoted hydrolysis of triglycerides of WCO mixtures.

2.3.2. NMR Characterization

The starting material was analyzed by NMR spectroscopy and the analysis of ^{1}H-NMR spectra is reported in Figure 1. The signal at 5.35 ppm corresponds to the olefinic hydrogen of unsaturated fatty acids. The signal at 5.20 ppm corresponds to glycerol protons while glycerol methylene protons showed a signal at 4.19 ppm. Allylic protons showed a signal at 1.88–2.02 ppm and 2.63–2.77 ppm, signals at 2.15–2.31 and 1.47–1.62 ppm corresponded to methylene protons in α e β in the carboxy group, while the saturated groups exhibited a signal at 1.14–1.26 ppm. Signals at 0.75–0.88 and 0.95 ppm corresponded to the methyl protons of the aliphatic chains of fatty acids. A peak at a chemical shift value of 2.78 ppm, corresponding to the signal of linoleic acid and particularly abundant in seed oil, was also detected. All the mixtures were analyzed and a comparison of the composition of starting oil blends was performed, as shown in Figure 2. The transesterification reaction was verified from the analysis of the ^{1}H-NMR spectra of the corresponding biodiesel samples and the conversion was calculated according to the literature [36]. In fact, it was possible to detect the presence of a peak at a chemical shift value of 3.67 ppm (A_1), identifying the presence of the methoxy group of fatty acid esters while the signals of the methylene groups at 5.20 and 4.19 ppm (A_2) disappeared (Figure 3).

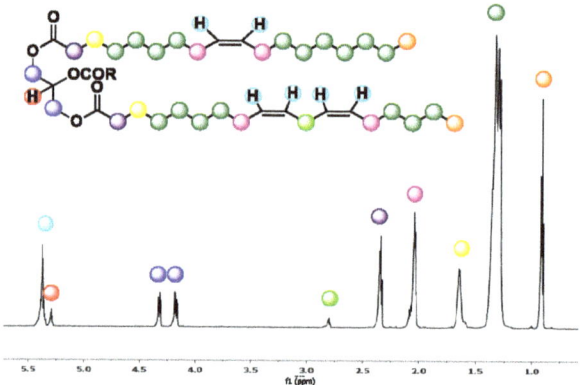

Figure 1. ^{1}H-NMR spectra of a general WCO mixture.

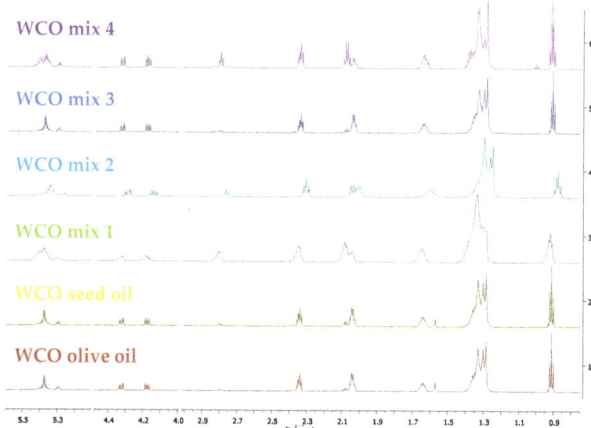

Figure 2. ^{1}H-NMR spectra of WCO mixtures.

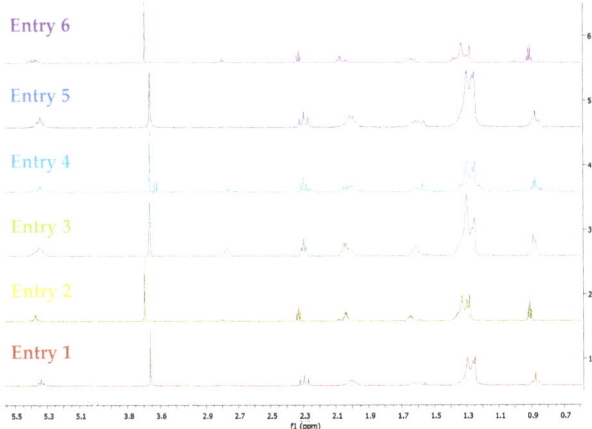

Figure 3. ^{1}H-NMR spectra of fatty acid methyl esters (FAMEs) obtained with basic catalysis (Table 1).

2.3.3. FT-IR Analysis

FT-IR spectroscopy represented a rapid and precise method for the recognition of functional groups of fatty acids deriving from WCO and biodiesel [37].

The main difference between the two FT-IR spectra was related to the transformation of the exhausted oil ester groups into the methyl esters of the produced biodiesel (Supporting Information). The exhausted vegetable oils showed an absorption that appears at 722 cm^{-1}, representative of the rocking band of the –CH$_2$ and the other at 1746 cm^{-1} typical of the stretching of the C=O of the ester. Biodiesel absorptions appeared at 1464 and 1437 cm^{-1}, representing the methyl ester groups, and at 1196 cm^{-1}, corresponding to –(C–O)– of the ester. The peak at 1163 cm^{-1} of the oil sample compared to the 1195 cm^{-1} peak of the biodiesel sample demonstrated the successful conversion.

2.3.4. Gas Chromatographic Analysis

Gas chromatography allowed the estimation of peaks obtained by the analysis of the methyl esters derived from the fatty acids after transesterification. Consequently, it was also possible to trace the composition of the fatty acids present in the triglycerides of the starting material [38].

In fact, the area underneath these peaks provided information about the related content of a specific methyl ester in the mixture (Table 5).

Table 5. Composition of FAMEs.

Entry	Source	Palmitoleate	Palmitate	Linoleate	Oleate	Stearate
1	WCO olive oil	1.441	12.284	8.701	75.255	2.319
2	WCO seed oil	1.324	12.015	13.430	70.655	2.574
3	WCO mix 1	-	6.908	53.366	35.618	4.108
4	WCO mix 2	-	7.484	37.064	52.017	3.434
5	WCO mix 3	2.400	13.046	10.163	71.785	2.515
6	WCO mix 4	-	9.027	55.759	29.864	5.350

3. Materials and Methods

3.1. Chemicals and Material

All the reactions were performed on a series of samples of waste oils for domestic use: waste olive oil, waste seed oil, mixture of different waste oils (WCO mix 1, WCO mix 2 and WCO mix 3), and mixture of different waste oils used at a constant temperature of 180 °C (WCO mix 4). All the other chemicals were commercially available and used without further purification.

3.2. General Reaction Procedures

The molecular weight of the WCO mixtures was estimated considering the composition of the lateral chains R_1, R_2 and R_3 deriving from oleic, linoleic and palmitic acid respectively, as confirmed by Gas Chromatographic analysis.

3.2.1. Homogeneous Basic Catalysis

In a 250 mL round bottom flask, 100 g of WCO (117 mmol) were refluxed at 65 °C for five hours. Of MeOH (624 mmol) 20 g was mixed with 1 g of NaOH (25 mmol), the solution was added to the oil and the reaction was refluxed at 50 °C for one hour. The organic phase was washed with NH_4Cl and distilled water, dried with Na_2SO_4, filtered and finally concentrated under vacuum.

3.2.2. Homogeneous Acidic Catalysis

In a 50 mL round bottom flask, 10 g of WCO (11.7 mmol) were mixed with 2 g of MeOH (62.4 mmol). Finally, 1.08 mL of concentrated H_2SO_4 (20 mmol) was added. The reaction was refluxed at a temperature of 85 °C for twelve hours. The organic phase was extracted with water and diethyl ether, dried with Na_2SO_4, filtered and finally concentrated under vacuum.

3.2.3. Acidic Catalysis Supported by Silica

In a 100 mL round bottom flask, 20 g of WCO mix 2 (23.3 mmol) were mixed with 4 g of MeOH (125 mmol) and 10 g of gravitational SiO_2 (166 mmol). Finally, 2.17 mL (41 mmol) of concentrated H_2SO_4 were added. The reaction was refluxed at a temperature of 85 °C for 12 h. The product was filtered under vacuum by using a Gooch filter. The organic phase was washed with CH_2Cl_2, filtered and finally concentrated under vacuum.

3.3. Acidity Percentage

The mixture of oil (10 g), ethanol 95° (100 g), diethyl ether (300 g) and phenolphthalein (2 mL) was titrated with KOH (0.1 M), until a color change from colorless to pale pink. The mL of KOH needed to reach the color change point is included in the following formula to calculate percentage acidity:

$$\%A = \frac{M \times V \times MW_{OL}}{P \times 10} \times 100 \qquad (1)$$

V = volume of KOH at the turning point;
M = KOH molarity;
MW_{OL} = molecular weight of oleic acid;
P = weight in g of oil used for titration.

3.4. Analysis Methods

The ^{1}H-NMR spectra were recorded with a VARIAN INOVA 500 MHz or Bruker 700 MHz spectrometer; all chemical shift values are reported in ppm (δ); $CDCl_3$ was used as solvent.

The conversion was calculated from the ratio between the selected areas of A_1 (methoxy group of FAME, 3.67 ppm, singlet) and A_2 (methylene group of WCO, 5.20 or 4.19 ppm, dd) according to the following equation [36]:

$$\text{Conversion } (\%) = 100 \times (2A_1/3A_2). \qquad (2)$$

FT-IR spectra were performed with a PERKIN ELMER 283 spectrophotometer. The gas chromatograms were obtained with a HP6890 plus gas chromatograph equipped with HP1 column.

4. Conclusions

Hence, it could be concluded that WCO, which is economic and easily available, is a very useful raw material for biodiesel production. The optimization of a transesterification reaction of six different waste oil mixtures was carried out. The results obtained have shown that the basic catalysis allowed us to evaluate the composition of the fatty acids in the starting mixture. The conditions of acidic catalysis have therefore been widely explored. Under optimized conditions all the mixtures show very high conversions. However, the recovery of glycerol and crude washing were difficult, due to the formation of insoluble products probably deriving from oxidation processes. To overcome this drawback, the use of silica was tested. As a result, in the presence of silica as supporting material, biodiesel was prepared with high conversion, whereas the insoluble material was retained on the silica surface. Furthermore, the role of silica facilitated the purification process because the formation of polymeric materials can contaminate the purity of the formed biodiesel.

Supplementary Materials: The following are available online at http://www.mdpi.com/2073-4344/9/12/1050/s1, Figure S1: ^{1}H-NMR spectra of FAMEs from WCO olive oil (Table 2), Figure S2: ^{1}H-NMR spectra of FAMEs from WCO seed oil (Table 2), Figure S3: ^{1}H-NMR spectra of FAMEs from WCO mix 2 (Table 2), Figure S4: ^{1}H-NMR spectra of FAMEs from WCO mix 1 (Table 2), Figure S5: ^{1}H-NMR spectra of FAMEs from WCO mix 3 (Table 2), Figure S6: ^{1}H-NMR spectra of FAMEs from WCO mix 4 (Table 2), Figure S7: ^{1}H-NMR spectra of FAMEs from WCO mix 2 (Table 3), Figure S8: ^{1}H-NMR spectrum of standard biodiesel, Figure S9: FT-IR spectrum of WCO olive oil, Figure S10: FT-IR spectrum of WCO seed oil, Figure S11: FT-IR spectrum of WCO mix 1, Figure S12: FT-IR spectrum of WCO mix 2, Figure S13: FT-IR spectrum of WCO mix 3, Figure S14: FT-IR spectrum of WCO mix 4, Figure S15: FT-IR spectrum of Entry 1 (Table 1), Figure S16: FT-IR spectrum of Entry 2 (Table 1), Figure S17: FT-IR spectrum of Entry 3 (Table 1), Figure S18: FT-IR spectrum of Entry 4 (Table 1), Figure S19: FT-IR spectrum of Entry 5 (Table 1), Figure S20: FT-IR spectrum of Entry 6 (Table 1), Figure S21: Chromatogram of Entry 1 (Table 5), Figure S22: Chromatogram of Entry 2 (Table 5), Figure S23: Chromatogram of Entry 3 (Table 5), Figure S24: Chromatogram of Entry 4 (Table 5), Figure S25: Chromatogram of Entry 5 (Table 5), Figure S26: Chromatogram of Entry 6 (Table 5), Figure S27: Chromatogram of Standard Biodiesel sample, Table S1: GC parameters.

Author Contributions: C.C. conceived, designed, performed the experiments and wrote the paper; M.A. analyzed the data and wrote the paper; R.L. and L.D. contributed reagents/materials/analysis tools.

Funding: This research received no external funding.

Acknowledgments: This work was supported by the *Intervento cofinanziato dal Fondo di Sviluppo e Coesione 2007–2013–APQ Ricerca Regione Puglia "Programma regionale a sostegno della specializzazione intelligente e della sostenibilità sociale ed ambientale-FutureInResearch"*. Saverio Cellamare and Cosimo Damiano Altomare are acknowledged for their technical and logistics support. We are also grateful to Rebecca Ostuni for the precious contribution.

Conflicts of Interest: The authors declare no conflict of interest.

References

1. Huber, G.W.; Corma, A. Synergies between Bio- and Oil Refineries for the Production of Fuels from Biomass. *Angew. Chem. Int. Ed.* **2007**, *46*, 7184–7201. [CrossRef] [PubMed]
2. Thanh, L.T.; Okitsu, K.; Van Boi, L.; Maeda, Y. Catalytic technologies for biodiesel fuel production and utilization of glycerol: A review. *Catalysts* **2012**, *2*, 191–222. [CrossRef]
3. Meher, L.C.; Vidya Sagar, D.; Naik, S.N. Technical aspects of biodiesel production by transesterification—A review. *Renew. Sustain. Energy Rev.* **2006**, *10*, 248–268. [CrossRef]
4. Marchetti, J.M.; Miguel, V.U.; Errazu, A.F. Possible methods for biodiesel production. *Renew. Sustain. Energy Rev.* **2007**, *11*, 1300–1301. [CrossRef]
5. Lam, M.K.; Lee, K.T.; Mohamed, A.R. Homogeneous, heterogeneous and enzymatic catalysis for transesterification of high free fatty acid oil (waste cooking oil) to biodiesel: A review. *Biotechnol. Adv.* **2010**, *28*, 500–518. [CrossRef] [PubMed]
6. Leone, G.P.; Balducchi, R.; Mehariya, S.; Martino, M.; Larocca, V.; Sanzo, G.D.; Iovine, A.; Casella, P.; Marino, T.; Karatza, D.; et al. Selective extraction of ω-3 fatty acids from *Nannochloropsis* sp. using supercritical CO_2 Extraction. *Molecules* **2019**, *24*, 2406. [CrossRef]
7. Machmudah, S.; Wahyu, D.; Kanda, H.; Goto, M. Supercritical fluids extraction of valuable compounds from algae: Future perspectives and challenges. *Eng. J.* **2018**, *22*, 13–30. [CrossRef]

8. Di Sanzo, G.; Mehariya, S.; Martino, M.; Larocca, V.; Casella, P.; Chianese, S.; Musmarra, D.; Balducchi, R.; Molino, A. Supercritical carbon dioxide extraction of astaxanthin, lutein, and fatty acids from haematococcus pluvialis microalgae. *Mar. Drugs* **2018**, *16*, 334. [CrossRef]
9. Catchpole, O.; Moreno, T.; Montañes, F.; Tallon, S. Perspectives on processing of high value lipids using supercritical fluids. *J. Supercrit. Fluids* **2018**, *134*, 260–268. [CrossRef]
10. Schablitzky, H.W.; Lichtscheidl, J.; Hutter, K.; Hafner, C.; Rauch, R.; Hofbauer, H. Hydroprocessing of Fischer-Tropsch biowaxes to second-generation biofuels. *Biomass Convers. Biorefinery* **2011**, *1*, 29–37. [CrossRef]
11. Canakci, M. The potential of restaurant waste lipids as biodiesel feedstocks. *Bioresour. Technol.* **2007**, *98*, 183–190. [CrossRef] [PubMed]
12. Ki-Teak, I.; Foglia, T.A. Production of alkyl ester and biodiesel from fractionated lard and restaurant grease. *J. Am. Oil Chem. Soc.* **2002**, *79*, 191–195.
13. Zhang, Y.; Dubé, M.A.; McLean, D.D.; Kates, M. Biodiesel production from waste cooking oil: Process design and technological assessment. *Bioresour. Technol.* **2003**, *89*, 1–16. [CrossRef]
14. Chhetri, A.B.; Watts, K.C.; Islam, M.R. Waste cooking oil as an alternative feedstock for biodiesel production. *Energies* **2008**, *15*, 3–18. [CrossRef]
15. Enweremadu, C.C.; Mbarawa, M.M. Technical aspects of production and analysis of biodiesel from used cooking oil—A review. *Renew. Sustain. Energy Rev.* **2009**, *13*, 2205–2224. [CrossRef]
16. Glisic, S.B.; Pajnik, J.M.; Orlovic, A.M. Process and techno-economic analysis of green diesel production from waste vegetable oil and the comparison with ester type biodiesel production. *Appl. Energy* **2016**, *170*, 176–185. [CrossRef]
17. Refaat, A.A. Different techniques for the production of biodiesel from waste vegetable oil. *Int. J. Environ. Sci. Technol.* **2010**, *7*, 183–213. [CrossRef]
18. Yaakob, Z.; Mohammada, M.; Alherbawi, M.; Alam, Z.; Sopian, K. Overview of the production of biodiesel from Waste cooking oil. *Renew. Sustain. Energy Rev.* **2013**, *18*, 184–193. [CrossRef]
19. Talebian-Kiakalaieh, A.; Amin, N.A.S.; Mazaheri, H. A review on novel processes of biodiesel production from waste cooking oil. *Appl. Energy* **2013**, *104*, 683–710. [CrossRef]
20. Glisic, S.B.; Orlović, A.M. Review of biodiesel synthesis from waste oil under elevated pressure and temperature: Phase equilibrium, reaction kinetics, process design and techno-economic study. *Renew. Sustain. Energy Rev.* **2014**, *31*, 708–725. [CrossRef]
21. Leung, D.Y.C.; Guo, Y. Transesterification of neat and used frying oil: Optimization for biodiesel production. *Fuel Process. Technol.* **2006**, *87*, 883–890. [CrossRef]
22. Demirbas, A. Biodiesel from waste cooking oil via base-catalytic and supercritical methanol transesterification. *Energy Convers. Manag.* **2009**, *50*, 923–927. [CrossRef]
23. Alcantara, R.; Amores, J.; Canoira, L.; Fidalgo, E.; Franco, M.J.; Navarro, A. Catalytic production of biodiesel from soybean oil. Used frying oil and tallow. *Biomass Bioenergy* **2000**, *18*, 515–527. [CrossRef]
24. Supple, B.; Holward-Hildige, R.; Gonzalez-Gomez, E.; Leahy, J.J. The effect of steam treating waste cooking oil on the yield of methyl ester. *J. Am. Oil Chem. Soc.* **2002**, *79*, 175–178. [CrossRef]
25. Lotero, E.; Liu, Y.; Lopez, D.E.; Suwannakarn, K.; Bruce, D.A.; Goodwin, J.G., Jr. Synthesis of biodiesel via acid catalysis. *Ind. Eng. Chem. Res.* **2005**, *44*, 5353–5363. [CrossRef]
26. Gebremariam, S.N.; Marchetti, J.M. Biodiesel production through sulfuric acid catalyzed transesterification of acidic oil: Techno economic feasibility of different process alternatives. *Energy Convers. Manag.* **2018**, *174*, 639–648. [CrossRef]
27. Schwab, A.W.; Bagby, M.O.; Freedman, B. Preparation and properties of diesel fuels from vegetable oils. *Fuel* **1987**, *66*, 1372–1378. [CrossRef]
28. Liu, K. Preparation of fatty acid methyl esters for gaschromatographic analysis of lipids in biological materials. *J. Am. Oil Chem. Soc.* **1994**, *71*, 1179–1187. [CrossRef]
29. Nye, M.J.; Williamson, T.W.; Deshpande, S.; Schrader, J.H.; Snively, W.H.; Yurkewich, T.P.; French, C.L. Conversion of used frying oil to diesel fuel by transesterification: Preliminary tests. *J. Am. Oil Chem. Soc.* **1983**, *60*, 1598–1601. [CrossRef]
30. Canakci, M.; Van Gerpen, J. Biodiesel production via acidic catalysis. *Trans. ASAE* **2009**, *42*, 1203–1210. [CrossRef]

31. Guana, G.; Kusakabe, K. Biodiesel Production from waste oily sludge by acid-catalyzed esterification. *Int. J. Biomass Renew.* **2012**, *1*, 1–5.
32. Lee, A.F.; Bennett, J.A.; Manayil, J.C.; Wilson, K. Heterogeneous catalysis for sustainable biodiesel production via esterification and transesterification. *Chem. Soc. Rev.* **2014**, *43*, 7887–7916. [CrossRef] [PubMed]
33. Carlucci, C.; Degennaro, L.; Luisi, R. Titanium dioxide as a catalyst in biodiesel production. *Catalysts* **2019**, *9*, 75. [CrossRef]
34. Isbell, T.A.; Frykman, H.B.; Abbott, T.P.; Lohr, J.E.; Drozd, J.C. Optimization of the sulfuric acid-catalyzed estolide synthesis from oleic acid. *J. Am. Oil Chem. Soc.* **1997**, *74*, 473–476. [CrossRef]
35. Yang, B.; Leclercq, L.; Clacens, J.M.; Nardello-Rataj, V. Acid/amphiphilic silica nanoparticles: New eco-friendly pickering interfacial catalysis for biodiesel production. *Green Chem.* **2017**, *19*, 4552–4562. [CrossRef]
36. Gelbard, G.; Bres, O.; Vargas, R.M.; Vielfaure, F.; Schuchardt, U.F. ^{1}H nuclear magnetic resonance determination of the yield of the transesterification of rapeseed oil with methanol. *J. Am. Oil Chem. Soc.* **1995**, *72*, 1239–1241. [CrossRef]
37. Oliveira, J.S.; Montalvão, R.; Daher, L.; Suarez, P.A.Z.; Rubim, J.C. Determination of methyl ester contents in biodiesel blends by FTIR-ATR and FTNIR spectroscopies. *Talanta* **2006**, *69*, 1278–1284. [CrossRef]
38. Issariyakul, T.; Kulkarni, M.G.; Dalai, A.K.; Bakhshi, N.N. Production of biodiesel form waste fryer grease using mixed methanol/ethanol system. *Fuel Process. Technol.* **2007**, *88*, 429–436. [CrossRef]

© 2019 by the authors. Licensee MDPI, Basel, Switzerland. This article is an open access article distributed under the terms and conditions of the Creative Commons Attribution (CC BY) license (http://creativecommons.org/licenses/by/4.0/).

Article

Contact Glow Discharge Electrolysis: Effect of Electrolyte Conductivity on Discharge Voltage

Giovanni Battista Alteri [1], Matteo Bonomo [1,2,*], Franco Decker [1] and Danilo Dini [1,*]

1. Department of Chemistry, University of Rome "La Sapienza", 00185 Piazzale Aldo Moro 5, 00178 Rome, Italy; g.alteri1@studenti.unipi.it (G.B.A.); franco.decker@uniroma1.it (F.D.)
2. Department of Chemistry, University of Turin, Via Pietro Giuria 7, 10125 Turin, Italy
* Correspondence: matteo.bonomo@unito.it (M.B.); danilo.dini@uniroma1.it (D.D.)

Received: 29 August 2020; Accepted: 21 September 2020; Published: 24 September 2020

Abstract: Contact glow discharge electrolysis (CGDE) can be exploited in environmental chemistry for the degradation of pollutants in wastewater. This study focuses on the employment of cheap materials (e.g., steel and tungsten) as electrodes for experiments of CGDE conducted in electrochemical cells with variable electrolytic composition. A clear correlation between breakdown voltage (V_B)/discharge (or midpoint) voltage (V_D) and the conductivity of the electrolyte is shown. Regardless of the chemical nature of the ionogenic species (acid, base or salt), the higher the conductivity of the solution, the lower the applied potential required for the onset of the glow discharge. Concerning practical application, these salts could be added to poorly conductive wastewaters to increase their conductivity and thus reduce the ignition potential necessary for the development of the CGDE. Such an effect could render the process of chemical waste disposal from wastewaters more economical. Moreover, it is evidenced that both V_B and V_D are practically independent on the ratio anode area to cathode area if highly conductive solutions are employed.

Keywords: contact glow discharge electrolysis; electrochemical plasma; water treatment; glycerol

1. Introduction

Plasma generated through a process of CGDE constitutes an unconventional product of electrolysis. Starting from the electrolysis in a water solution under DC conditions, it makes the development of a luminescent plasma layer at the electrode/electrolyte interface feasible [1]. Plasma can be formed either at the cathode or at the anode, depending on the operative conditions [2–4]. The glow discharge follows water electrolysis (with production of H_2 and O_2 at the cathode and the anode, respectively) and is further promoted by the formation of radical species (and their derivatives) at the interface between the plasma and the solution [5]. To date, CGDE has been proved to be a valid and cost-effective metal surface treatment technique: the method significantly and effectively increase hardness and corrosion resistance of the metal [4,6].

The CGDE phenomenon occurs for the application of a high voltage (in the order of several hundred of volts) and is characterized by a critical value of onset for the formation of the plasma [3]. Such a voltage onset mainly depends on the physical properties of the electrolyte. Once reached, a gaseous layer surrounding the electrode is formed. The gas of the layer is partly formed by water steam, due to the heating of the solution in proximity of the electrode (Joule effect) and by the products of water electrolysis [4,5]). The profile of the current intensity is characterized by the reaching of a maximum that is followed by decrease. Such a current decrease sets the end the ordinary electrolysis. The critical ignition voltage is called breakdown voltage (V_B). A further increase in the voltage leads to the stabilization of the gaseous layer around the electrode in concomitance with the continuous and slow decrease of current. The simplified mathematical model reported in ref. [1] describes the time evolution of the current in CGDE conditions through the equation:

$$I(t) = \frac{U - U_d}{R_k}\left(1 - e^{t/\tau}\right) \tag{1}$$

being R_k the total electrical resistance (mainly constituted by the ohmic drop in the interelectrode space), U_d the water decomposition potential and τ a time constant depending on the system. It is also possible to calculate the time required for the formation of the gas film around one of the two electrodes, using the following expression:

$$t_f = \frac{2\pi^2 h^2 k \rho c}{I^2} \, 2b^2(T_c - T_o) \tag{2}$$

where b is the radius of the cylindrical electrode, h the depth of the cylinder–electrode immersed in the electrolyte, I the current intensity, T_c the boiling temperature of the solvent (reached at the electrode surface) and T_o is the bulk temperature of the electrolyte having electrical conductivity k, density ρ and specific heat capacity c. This vapor film successively undergoes a transformation into a brilliant, continuous plasma which represents the peculiarity of CGDE. As previously outlined, this phenomenon occurs once the voltage is high enough to trigger the development of plasma. Plasma triggering is characterized by small and quick intermittent glow discharges that are located on the electrode surface with a spotty appearance. The electrode on which plasma is generated is indicated as active electrode [1]. The voltage required for the generation of luminescence is called midpoint voltage or discharge voltage (V_D), and the potential region between V_B and V_D is called partial CGDE. When V_D is exceeded, the current intensity starts to rise again, but with a significantly lower slope than the increase observed at voltages lower than V_B. Such a regime is called ohmic region. Complete CGDE can be formed at any valued of voltage comprised in the range delimited by the breakdown and the midpoint voltage. The occurrence of the formation of the latter event depends on the operating conditions of the cell and the polarity of the electrode. The CGDE phenomenon can occur during the electrolysis of aqueous solutions with high conductivity, non-aqueous electrolytes or melts [4,7]. In the present work, we considered prevalently aqueous electrolytes.

Recently, several studies reported on the potential application of CGDE in other sectors, such as the synthesis of tailored nanoparticles [8–10], organic compounds [11], steam generation [12], polymers [13,14] and super-adsorbent compounds [15], mineralization of water [16–32] and hydrogen production [33]. Another interesting application consists in the employment of CGDE in the synthesis of amino acids [34]. The versatility of CGDE in the fields of environmental chemistry, electrochemistry, plasma chemistry, organic and inorganic chemistry, coupled with its modest cost compared to other plasma method, fully justifies its in-depth study. Furthermore, a relatively simple experimental apparatus is required. The current–voltage characteristic curves of CGDE have a general typical shape, regardless of the electrode material or the electrolyte used [25,29,35–39].

Again, all curves have two values of fundamental practical interest: the ignition voltage, called breakdown voltage (V_B)—at which the collapse of the normal electrolysis occurs—and the midpoint voltage or discharge voltage (V_D), i.e., the voltage required for the formation of the actual luminescence, (Figure 1). Since the achievement of the V_B is always followed by a rapid decrease of the current, it is easy to identify this value as the maximum in the I–V curves of the CGDE. On the other hand, V_D is always followed by a sudden increase of the current (and it is associated with a relative minimum of current in the characteristic curve). Therefore, it is relatively handy to identify the two voltage parameters that determine the development of CGDE. In the present work, these parametric values were systematically analyzed as a function of the variables involved in each type of experiment (conductivity and concentration of the electrolyte, electrode active area, chemical nature of the electrode material, electrolyte solvent).

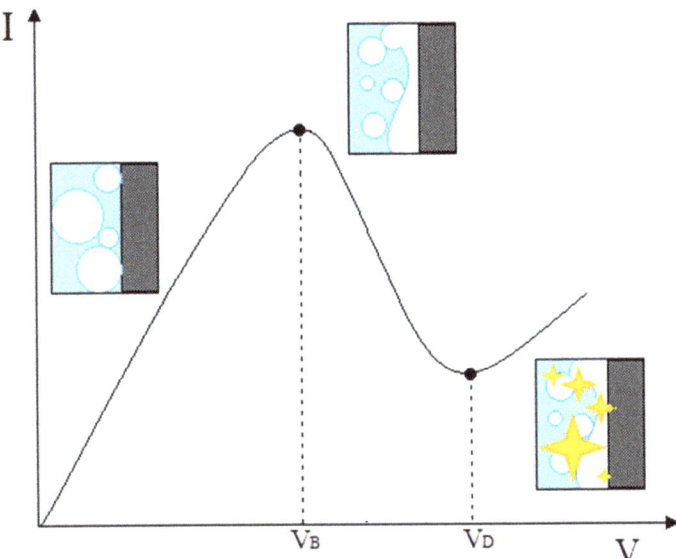

Figure 1. Individuation of V_B and V_D in a typical current–voltage characteristic curve of a contact glow discharge electrolysis (CGDE) experiment. From zero to V_B voltage, the electrolysis proceeds with the current increasing in a nonlinear fashion as expected for a faradaic regime. This initial portion of the curve is associated with the formation of gas bubbles. Once the V_B value is surpassed, the water vapor (formed because of Joule heating) plus the electrolytic gas coalesce to form a layer surrounding the active electrode. This coalescence causes a sudden drop of current intensity (I) for the increase of the electron transfer resistance at electrode–electrolyte interface. When the applied potential V_{appl} is larger than V_D this gaseous layer turns into an electrochemically generated plasma for the occurrence of a local dielectric breakdown.

Due to the complexity of CGDE phenomenon, further studies are still required to understand the mechanisms of electrolyte plasma formation. In this context, the present work focuses on the analysis of the operative conditions for the realization of CGDE. The work deals with the dependence of discharge voltage and the breakdown voltage on electrolyte conductivity and the influence that the difference of the immersed areas of anode and cathode exerts on the two characteristic voltages. As far as we are aware, there is not any study investigating both a so large range of conductivity (up to 600 mS m^{-1}) and a plethora of different electrolytic solutions, also considering a hybrid organic–aqueous environment.

2. Results and Discussion

The present study aims to define the optimal experimental conditions for the development of CGDE in both aqueous and aqueous/organic mixed solvents (namely water and glycerol mixture). Various solutes, at different concentration, were employed in order to evidence any eventual common trend of the voltage parameters with electrolyte conductivity and the area of the active electrode.

An interesting—and almost unique—feature of the CGDE is the tendency to produce molecules (originating from the plasma electrolysis), the faradaic yields of which deviate in excess from theoretical values [4,5,13,40–42]. Indeed, more than 80% of the electrolyzed molecules originate in the liquid phase at the interface close to the plasma (i.e., at the electrode surface) where $H_2O^+_{gas}$ reacts with water molecules leading to the formation of both hydroxyl and proton radicals ($H^\bullet$ e $OH^\bullet$). The latter are responsible for a higher gas evolution than expected [41]. Common reactions are briefly recalled hereafter:

$$H_2O^+_{gas} + nH_2O \rightarrow (n+1)\,H^\bullet + (n+1)\,OH^\bullet \tag{3}$$

$$OH^\bullet + OH^\bullet \to \frac{1}{2} O_2 + H_2O \tag{4}$$

$$H^\bullet + H^\bullet \to H_2 \tag{5}$$

$$OH^\bullet + OH^\bullet \to H_2O_2 \tag{6}$$

$$OH^\bullet + H_2O_2 \to HOO^\bullet + H_2O \tag{7}$$

$$HOO^\bullet + OH^\bullet \to O_2 + H_2O \tag{8}$$

The presence of a radical scavenger—such as an aliphatic alcohol—will further promote these reaction (vide infra).

2.1. CGDE in Aqueous Environment: Dependence on Solution Conductivity

In this section, the breakdown and the discharge voltage are analyzed as a function of electrolyte conductivity. Although KCl is a very well-known standard electrolyte in electrochemistry due to its large availability, low cost, negligible toxicity and high solubility in water, it presents the disadvantage of being electroactive with the anodic evolution of Cl_2 under the harsh conditions of CGDE. In fact, Cl_2 can react spontaneously with steel-based electrodes leading to the formation of iron chloride with the consequent deactivation of the electrode. For this simple reason, in the series of experiments we conducted, tungsten (W) electrodes were employed. The choice of W as electrode material was motivated by its high corrosion resistivity, the high conductivity and availability.

It has been recognized that the conductivity of the solution plays a decisive role in the determination of the value of the potential necessary for the transition from normal electrolysis to CGDE [39,43]. Lower specific conductivities of the electrolyte are associated with higher triggering voltages, while extremely high conductivities lead to a flattening of the V_D-k curve. This is observed for both V_B and V_D. Straightforwardly, we analyze V_B and V_D values obtained from different experiments using different solutions and different electrodes, as a function of the specific conductivities (mS cm^{-1}) of the fresh solutions. This approach allows to evidence any possible common trend in different starting conditions (i.e., different solutes and concentrations). In Figure 2 we reported the trend of the breakdown voltage with respect to the conductivity of the solution for cathode/anode immersed area ratio ranging from 0.05 to 0.9. In doing so we investigate the electrolytic plasma developed at the cathode surface. Under the adopted conditions, the formation of plasma was always observed at the cathode (Video S1—see Supplementary Materials).

From Figure 2 (top), one notices a rapid increase of V_B upon the decrease of k, and when electrolyte conductivity approaches the zero value there is a corresponding exponential augmentation of V_B. Highly conductive solutions, namely in the range 80–600 mS cm^{-1}, lead to a quasi-constant value of V_B with the latter showing an asymptotic behavior towards a minimum of 30 V. The latter could be set as a threshold value for the occurrence of CGDE phenomenon. This behavior is poorly dependent on the nature of the employed electrolyte, i.e., if this is acid, basic or a salt. Such a behavior could be profitably used for the optimization of the conditions under which CDGE breakdown is conducted for practical applications, e.g., wastewater purification.

The trend of V_D vs k shows a larger dispersion of the midpoint voltage with respect to the averaged profile (Figure 2, bottom). The solutions with similar conductivity display comparable V_D values. Similar to V_B (Figure 2, bottom), also V_D tends asymptotically to a minimum upon increase of the conductivity. The asymptotic value of V_D is 90 V when conductivity surpasses 80 mS cm^{-1}.

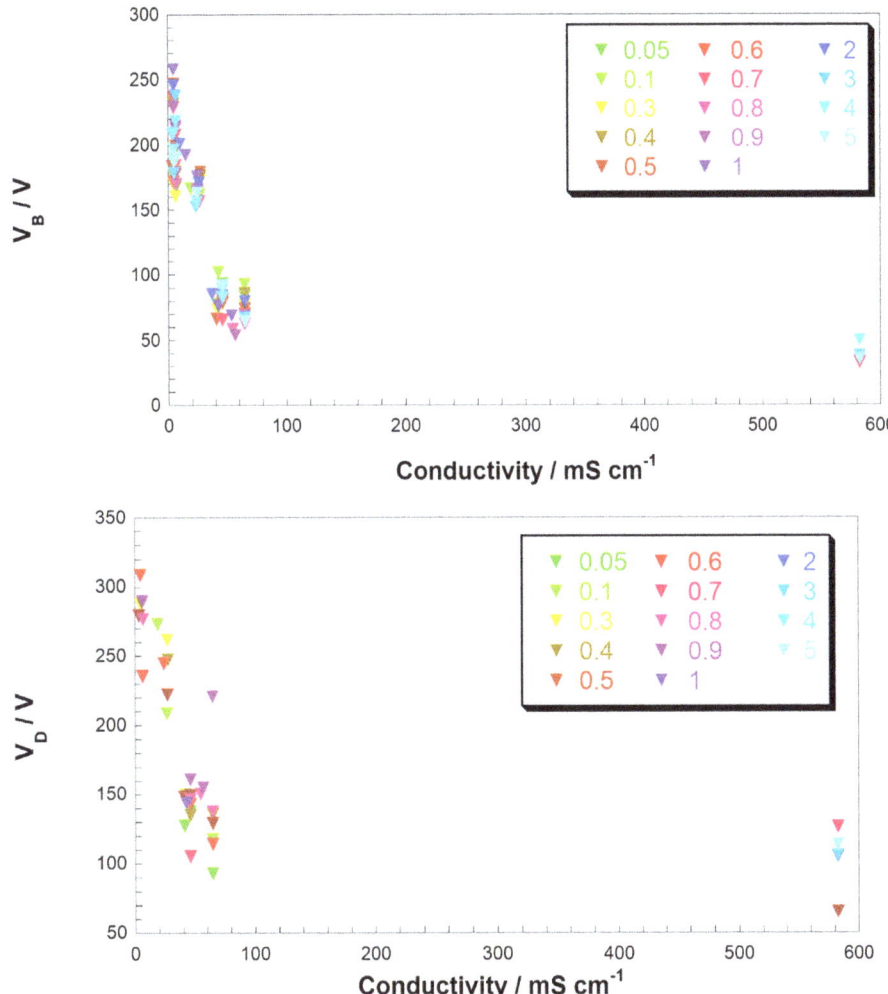

Figure 2. (**top**) Breakdown voltage V_B and (**bottom**) V_D vs. electrolyte conductivity (mS/cm). In the insets, the colored numbers represent the values of the ratio of the cathode area/anode area for each experiment.

As one can see from Figure 3 and Figure S1 (see supporting information), when the analysis is focused on the experiments of cathodic plasma formation (this is achieved when the ratio of the cathode area/anode area is lower than one, the value of V_B is dependent on the conductivity of the solution regardless the nature of the supporting electrolyte. The observed trend recalls what was previously found when only one type of electrolyte is used: the higher values of V_B are determined when k diminishes. This is a finding that can be exploited in the practical application of CGDE when the operator must choose a supporting electrolyte with low or null toxicity. For example, wastewater with a low conductivity would prevent the formation of the electrochemical plasma, but the addition of acidic, basic of salt-based supporting electrolytes will allow the application of CGDE, with possible avoidance of toxic/harmful chemicals formation. Similar trends have been recorded when anodic plasma is formed (see Figure S2—see supporting information).

The effect of the ratio of the cathode area/anode area is investigated in detail in the next section. Here, we anticipate that a clear influence of this geometric parameter on V_B and V_D values was not observed. Generally speaking, a lower/higher cathode-to-anode ratio will preferentially allow the plasma evolution at the cathode/anode being (obviously) larger the current density experimented by the electrode having the minor contact area with the electrolyte. K_2CO_3

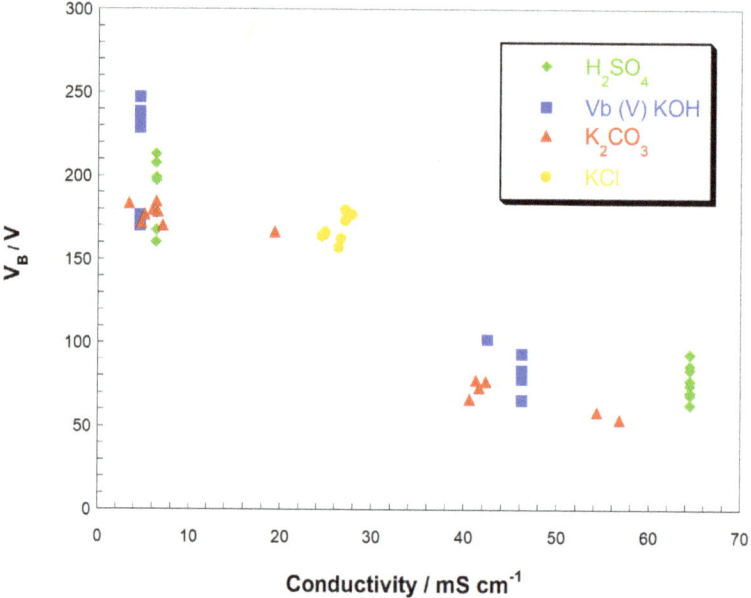

Figure 3. Breakdown voltage V_B as function of electrolyte conductivity when different supporting electrolytes are used. In this series of experiments cathodic plasma is formed. Types of electrolytes: acid (i.e., H_2SO_4, diamonds), base (KOH, squares), salt with a weak base as anion (K_2CO_3, triangles) and salt (KCl, circles).

2.2. CGDE in Aqueous Environment: Dependence on Active Electrode Area

A peculiar feature of the CGDE is its dependence on the asymmetry of the areas of the two immersed electrodes. The ratio of the cathode area/anode area determines which of the two electrodes will generate the electrolytic plasma. CGDE can take place also if the ratio between the immersed areas of the electrodes is 1:1, but this represents a special case that usually requires higher voltages with respect to a cell with asymmetrically immersed electrodes. To study this correlation, breakdown and discharge voltages are compared with respect to the electrodes immersed area ratio.

CGDE is formed preferentially at the cathode, unless the anode area is large enough [4]. In fact, when the immersed area of the anode is higher than the one of the cathode only anodic CGDE occurs (case of active anode). For comparing the two types of electrolytic plasma, we conducted also experiments in which the anode area/cathode area ratio is equal to or higher than one. In the cases of active anode (Figure 4), a quasi exponential trend is observed with a rapid increase of V_B values at low conductivities and an asymptotic decreasing behavior at higher values of conductivity. In both cases of cathodic and anodic CGDE a quasi exponential decrease of V_B with k is found. When solutions with relatively low conductivity are employed ($k <$ 10 mS cm^{-1}), the effect of the ratio of the immersed areas slightly influenced the breakdown voltage: V_B decreases if the anode experiences a larger current density (i.e., if the immersed area of the anode is smaller than the one of the cathode). On the other hand, when the solution conductivity is higher than 20 mS cm^{-1}, V_B becomes practically independent on the immersed electrode areas.

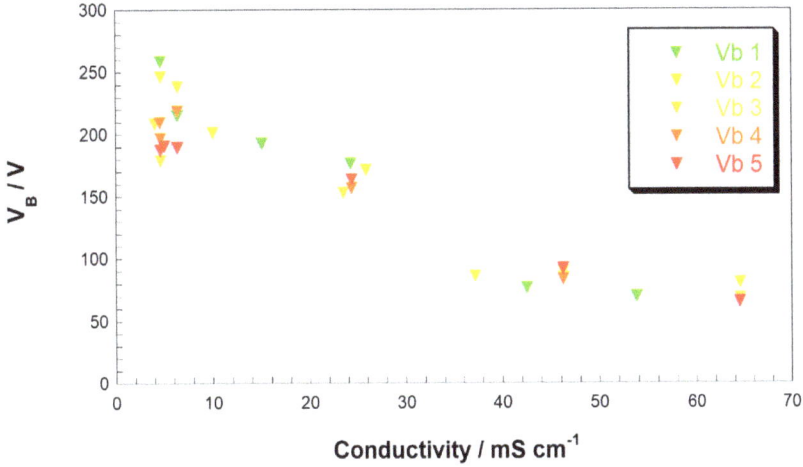

Figure 4. Breakdown voltage V_B (V) vs. conductivity (mS/cm) with anode and cathode areas-ratio that equal to or higher than 1 (case of CGDE active anode). Electrolyte is the same for all experiments.

The reported behavior can be explained considering the physical nature of the breakdown voltage in conventional electrolysis that is usually associated with the formation of a layer of vapor around the electrode [44,45]. Steam formation is caused by the local heating of the solution in proximity of the electrode experiencing the highest current density (Joule effect). The vaporization temperature of an aqueous solution poorly varies with the molar concentration of the electrolyte and, neglecting the effect of the ebullioscopic constant, it can be considered as almost constant. This can explain the similarity of V_B values in different electrolytes beyond a certain conductivity and somehow justifies the correlation between the breakdown voltage and the conductivity of the solution, which, in turn, directly affects the extent of the electrical current passing through the cell and the consequent electrode overheating. When the trend of V_D as a function of solution conductivity is analyzed (Figure 5), this results in more scattered values, as largely expected, being the voltage at which discharge occurs more influenced by the surface characteristics of the electrode/electrolyte interface.

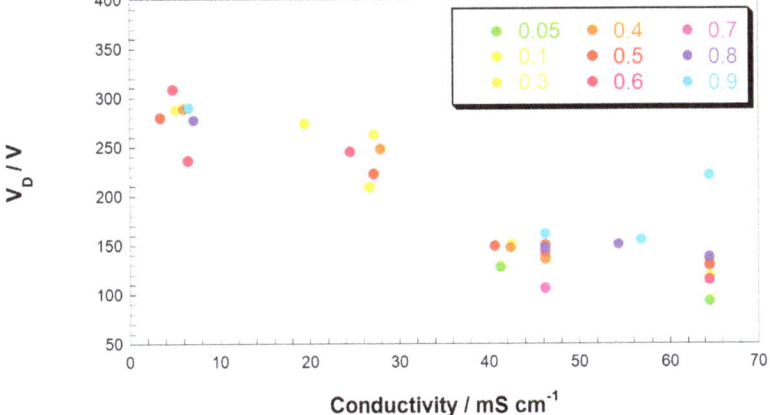

Figure 5. Discharge voltage, V_D, as a function of the conductivity of the solution. Different colors refer to different ratios of cathode-to-anode immersed area. Electrolyte was the same for all experiments.

2.3. CGDE in Aqueous Environment: Effect of the Supporting Electrolyte

In the previous sections, we show how the evolution of electrolytic plasma mainly depends on the conductivity of the electrolyte and not on its chemical composition. This implies that the choice of an electrolyte is not crucial for the occurrence of CGDE and the electrolyte can be chosen in a way that this mitigates the reactivity of a system under investigation for a CGDE treatment. The relevant example of the treatment of industrial wastewater with acidic characteristics can be tackled utilizing a basic electrolyte for avoiding the acidic corrosion of the constituents of the pipelines that carry these waste liquids. Indeed, an extremely acid or basic solution can severely damage tubes, tanks as well as the electrodes of the CGDE electrolyzer. In this context, we employed H_2SO_4, KCl, KOH and K_2CO_3 solution at different concentration to modulate both pH and conductivity. The pH values varied from 0.4 to 13 whereas k was comprised in the approximate range 5–580 mS cm^{-1}. Values of pH lower than 0.4 were avoided due to the excessive development of H_2 (an explosive gas). It must be pointed out that pH and k values refer to fresh solutions and relatively large variations of these parameters must be expected during and after plasma generation (Equations (3)–(8)).

Figure 6 resumes the trend of both V_B and V_D of H_2SO_4 (top), KOH (middle top), K_2CO_3 (middle bottom) and KCl (bottom), respectively. In general, a higher conductivity leads to a lower V_B and a to a faster kinetics of plasma production as evidenced for 1 M and 0.1 M H_2SO_4 and 0.1 M K_2CO_3 (Figure 6, left column). In the case of 1 M H_2SO_4, when the ratio of the immersed areas of cathode to anode is lower than 0.5, we were not able to record the voltage values due to the generation a huge plasma that caused severe electric interferences at the electrode surface. The experiments conducted with electrolytes having relatively low conductivity (lower than 40 mS cm^{-1}) did not produce electrolytic plasma and the determination of V_B was difficult. In these cases, characteristic curves of a conventional electrolysis are obtained, which (partially) obeys to Ohm's law, with a linear current-to-voltage trend.

As already evidenced in the former sections, the cathodic plasma is relatively easier to form with respect to anodic plasma. On the other hand, a complete anodic CGDE is formed only for 1 M H_2SO_4 solutions. For other cases, it could be observed only a faint glow that was insufficient to set a stable plasma. This means that for the occurrence of anodic CGDE a discharge voltage higher than 320 V (our detection limit) is necessary. In both diluted acid (0.01 M H_2SO_4) and basic (0.01 M KOH) solutions, a dependence of CGDE on the ratio of the cathode-to-anode-immersed areas was observed. In acidic solutions upon increase of the ratio of the cathode-to-anode-immersed areas from 0.2 to 1, a roughly linear increase of V_B is evidenced whereas a ratio larger than 1 gives a practically constant value. Both KCl and K_2CO_3-based solution showed a clear independency of the generation of the cathodic (or anodic) plasma on the ratio of the cathode-to-anode-immersed areas.

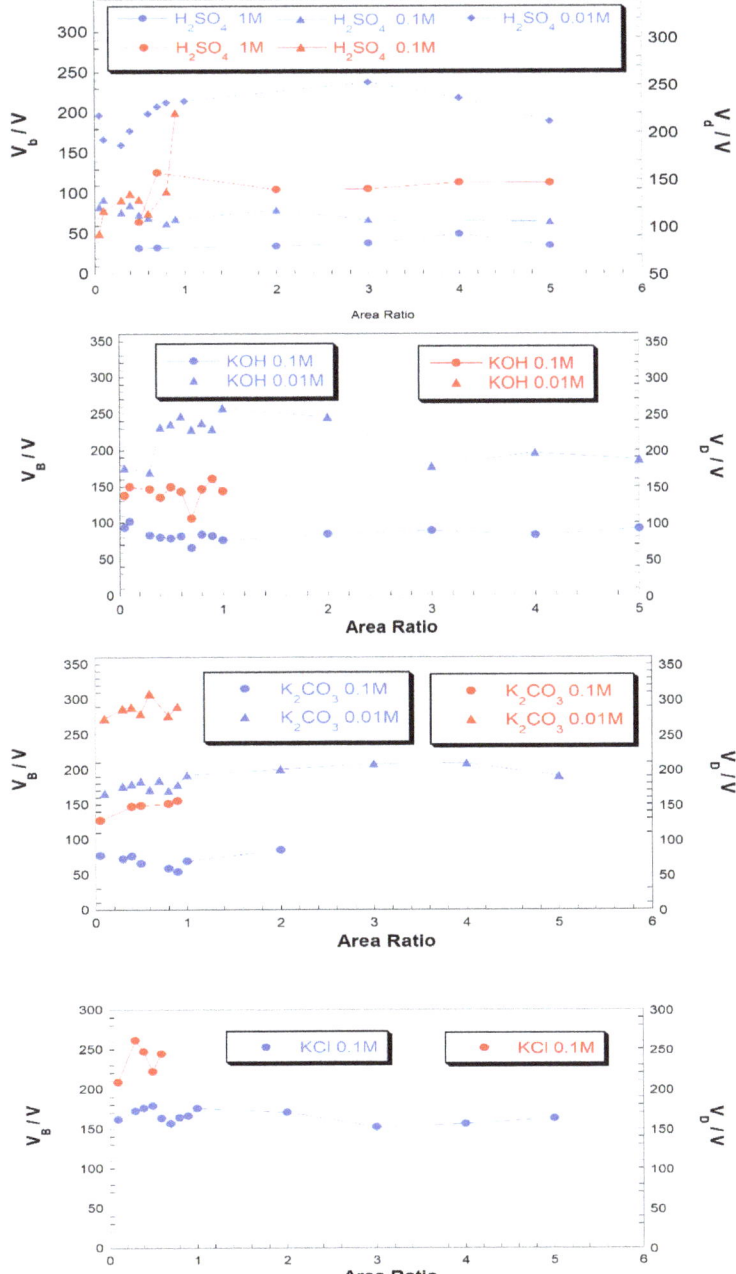

Figure 6. Trend of the characteristic CGDE parameters (VB, left y-axes in blue and VD, right y-axes in red) as a function of the cathode-to-anode ratio with different supporting electrolytes: H_2SO_4 (**top**), KOH (**middle top**), K_2CO_3 (**middle bottom**) and KCl (**bottom**). Different symbols refer to variable electrolyte concentration.

2.4. CGDE in Aqueous–Organic Environment: Water/Glycerol Mixture as Case Study

Equations (3)–(8) (vide supra) refer to a series of electrochemical and radical reactions that occur in water for generating electrolytic plasma. Similar reactions occur when an alcohol or more generally an organic hydrogen donor, is added to the aqueous media (Equations (9) and (10)) [41]:

$$RCH_2OH + H^\bullet \rightarrow RC^\bullet HOH + H_2 \tag{9}$$

$$RCH_2OH + OH^\bullet \rightarrow RC^\bullet HOH + H_2O \tag{10}$$

With respect to pure aqueous media, the addition of an organic co-solvent leads to a higher faradaic efficiency for the H_2 evolution preceding the glowing discharge. Among different (poly)alcohols, we decided to employ glycerol due to the presence of three alcoholic groups per molecule and the high miscibility in water (this avoids the formation of different phases in the same system). Glycerol/water mixture was considered a meaningful case-study since it mimics the behavior of organic-contaminated wastewater [46–48]. In order to do this, we prepared a 0.5 M glycerol solution in deionized water with 0.1 M KCl Such a solution displayed k = 23.2 mS cm^{-1} (for sake of comparison the conductivity of a 0.1 M aqueous solution of KCl is 25.6 mS cm^{-1}). The slightly lower conductivity could be ascribed to an increase of the viscosity of the solution due to the presence of glycerol. In this context, we limited our analyses on just one supporting electrolyte. Figure 7 shows that the addition of glycerol in the electrolytic solution seems does not influence the evolution of CGDE being both the V_B and V_D equal to the one experimented in pure water when the nature of the supporting electrolyte was the same and the solution had an analogous conductivity. Furthermore, the ratio of cathode to anode immersed areas is practically irrelevant in the determination of V_B. This is not longer the case when V_D is analyzed—indeed, plasma does not develop when the electrode area ratio is lower than 0.6 (i.e., for an anode with larger surface area with respect to the cathode) as verified also in case of aqueous electrolytes. This was mainly attributed to the relatively low conductivity of the solution in the adopted conditions of electrode immersion. In fact, plasma evolution at cathode-to-anode area ratios higher than 0.6 is expected also at low levels of conductivity.

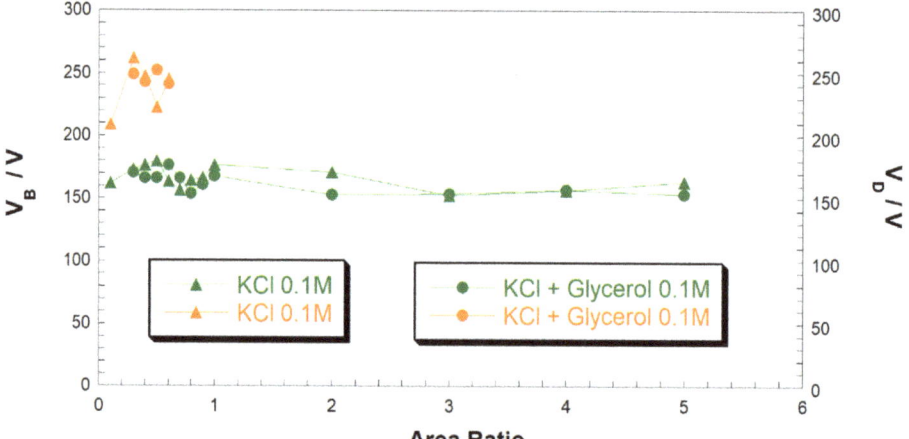

Figure 7. Breakdown voltage, V_B (in green) and discharge voltage, V_D (in orange) as a function of the ratio of the cathode-to-anode-immersed areas, with 0.1 M KCl in water (triangles) and glycerol/water mixture (circles).

3. Materials and Methods

All the reagents used were obtained from Sigma-Aldrich (Saint Louis, MO, USA) and are, respectively: glycerol (assay 86–89%), H_2SO_4 (assay 95–98%), K_2CO_3 (assay ≥ 99%), KCl (assay ≥ 99%). Every aqueous solution was prepared using distilled water. The circuital scheme depicted in Figure 8 describes the electronic circuit of the apparatus used to power the electrolysis and the subsequent CGDE. The power supplying system consisted of a current rectifier circuit connected to the laboratory electrical network, and was able to supply a voltage between 0 and 400 V, with a maximum current of 13 A. The circuit consisted of a variable autotransformer (Model Number TDGC2-3KVA, WOSN, Zhejiang, China, 50–60 Hz, 13 A, 3.51 KVA); a Graetz rectifier diode bridge (KBPC-3510, DC Components Co., Ltd., Taichung, Taiwan 400 V, 35A); six capacitors (Ducati Energia S.p.A. Bologna, Italy, 50–60 Hz, 100 µF); two HoldPeak digital multimeters; one resistance, (0.1 Ω, 100 W). The latter element was inserted in order to protect the digital multimeter which was used as an ammeter. This arrangement was motivated by the unavoidable fluctuations of current which were associated with the formation of plasma. In this way, in the frequent cased in which the plasma caused current peaks that could damage the internal resistance of the instrument, the external safety resistance acted as sacrificial element.

The electrochemical cell was made of borosilicate glass (Marbaglass S.n.c., Roma, RM, Italy), and it had the shape of a tube with a ground stopper (also in glass), equipped with a side spout having an approximate diameter of 1 cm. The latter essentially performed two functions: it allowed the gases generated by the electrolysis and the vapor produced by the overheating of the operating electrode to escape out from the reactor. This kept the pressure inside the cell practically constant and equal to the ambient value. On the other hand, it allowed filling the cell with the electrolyte with a Pasteur pipette, once closed with the upper cap (which also acted as a support for the electrode clamps). Such an experimental setup was designed to minimize the parallax error during the preparation operations of each experiment. Indeed, the cell had an indelible linear calibration mark on its external glass, the level of which coincided with the point of attachment of the terminals. Therefore, for each experiment, we could control finely the area of the immersed electrodes. The electrode support clamps were coated with a high-temperature-resistant, insulating and inert resin in order to avoid the shunting of the electrolyte and/or any parasitic contribution to the measured current. A thermometer was used to control the temperature of the solution. A high definition (HD) digital camera, was used to record each experiment and to qualitatively evaluate the behavior of the plasma (if observed) throughout the measurement time and especially during the transition range from the normal electrolysis to CGDE.

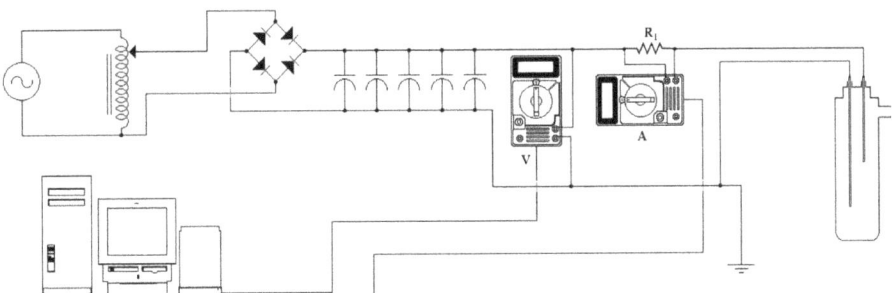

Figure 8. Diagram of the electrical circuit and the elements constituting the experimental apparatus.

For the measurements of the conductivity of the various solutions, a conductivity cell with a cell constant (K) of 10 cm^{-1} was used. This was connected to a potentiometer/galvanometer source meter Agilent E5262A (Agilent Technologies Ltd., Santa Clara, CA, USA) as a measuring instrument. The potentiometer/galvanometer, in turn, was connected to a computer on which it was installed Nova 1.9 (Metrohm Autolab B.V., Utrecht, NH) as data recording software. The conductivity of the solutions was measured both before and after each experiment in order to highlight any modification in the electronic properties.

Data on the conductivity of KOH and H_2SO_4 solutions were taken from the literature [49,50] and were not measured directly. This precaution avoided the damage of the experimental setup because of the corrosive power of these two reagents (strong base and strong acid). Each conductivity value is expressed in mS cm^{-1}, while the concentration (c) is expressed as molarity. The ratio between the electrode areas is a pure number, and it is always the relationship between the area of the cathode and the area of the anode immersed in the solution.

Since the conductivity of the solutions varied considerably with the temperature, in each measurement, the cell was equipped with a thermostat set at the plasma-triggering temperature (T_i) measured in the corresponding experiment. Temperatures higher than 80 °C were reached because the evaporation of the aqueous solution under analysis was no longer negligible: if evaporation occurs and the concentration of the electrolyte increased, the conductivity values recorded were overestimated. For this reason, when the electrolyte in the CGDE formation tests reached these values, the reference was made to this limit as the maximum temperature, and it was at this value that the conductivity of the electrolyte was measured.

The electrodes used during the experiments were of two types: rectangular plates in steel (alloy 304), 1 cm wide, 0.4 cm thick and of variable length; cylinders in pure tungsten, with a diameter of 1 mm and variable length. All the electrodes were purchased from A.G. METAL, Civita Castellana (VT) Italy.

The solutions used were: aqueous solutions of H_2SO_4 (c: 1 M, 0.1 M, 0.01 M); aqueous solutions of K_2CO_3 (c: 0.1 M, 0.01 M, 0.001 M); aqueous solutions of KOH (c: 0.1 M, 0.01 M, 0.001 M); aqueous solutions of KCl (c: 0.1 M, 0.01 M, 0.001 M); solutions 0.1, 0.001 M of KCl in mixture H_2O + $C_3H_8O_3$ (glycerol).

The reagents were chosen to study CGDE in both acidic, basic and neutral environments, with various cations and anions, in aqueous solutions and, limited to KCl—also in mixed aqueous/organic environment. Glycerol was chosen as an organic compound because of its high miscibility with water, its low cost, its ease of used and handling. Additionally, as mentioned before [5,41], the production of H_2 exploiting the non-faradaic effects of CGDE, in H_2O and alcohol solutions, was a promising energetic alternative and glycerol, having three OH alcoholic groups, was of great interest for this.

The experimental procedure executed in each test was as follows: for the preparation of the solutions, the hygroscopic salted (KCl, K_2CO_3) were first dried for one hour in an oven at 105 °C. H_2SO_4, being liquid, was added with a calibrated pipette. The electrodes were then fixed to the cell terminals: the operator checked the length of the electrodes once they were fixed. Then, the solution was injected into the cell: once the cap of the reactor was closed, the electrolytic solution was fed, by means of a Pasteur pipette, through the lateral spout. After having prepared and filled the electrolytic cell and having activated the digital multimeters, the electronic circuit was closed. The voltage was then slowly increased: initially, the voltage that passed through the cell was still practically zero—and so was the current. Thanks to the knob on the variable autotransformer, the voltage was constantly increased. Then, when possible, the plasma striking occurs as soon as V_D was reached, the development of a glow discharge took place. Once this happens, the experiment was interrupted, and the circuit was open for safety. Once the circuit was opened, the cell was opened and a small aliquot of solution was taken from the region next to the active electrode, in order to subsequently determine its conductivity. The temperature reached by the solution was also noted. It was worth mentioning that the electrode did not suffer any irreversible modification during the plasma and the replacement was due to allow a fairer comparison between the different experiments.

4. Conclusions

Throughout this article, we reported on the study of the properties and the optimal conditions necessary for the formation of the glow discharge by contact electrolysis (universally known as contact glow discharge electrolysis (CGDE)), using cheap materials and moderately high currents. These conditions were adopted to verify its applicability at both the household and the industrial level. Regardless of the chemical nature of the electrode, electrolyte or solvent, the development of the plasma glow discharge generates characteristic curves always shows the same shape, with a maximum current value associated with the breakdown voltage (V_B) that is of the electrolysis collapse, and a plasma-triggering voltage (V_D) that is associated with a sudden current resurgence. From the experiments performed, a clear dependence of the features of CGDE on the conductivity of the solution was found. Indeed, low conductivity values required very high plasma trigger voltages, even higher than 350 V. On the other hand, highly conductive solutions require extremely low potentials that could be reached with a relatively low energy costs. CGDE preferentially develops on the electrode with the highest charge density (usually coupled with the smaller surface area immersed in the electrolytic solution). When the immersed areas of cathode and the anode are the same, CDGE preferably develops at the cathode, as expected from the lower potential needed for the occurrence of the cathodic plasma. Remarkably, the dependence of V_B on the difference of the areas of the two immersed electrodes decreases upon increase of electrolyte conductivity. Finally, it was evidenced how the formation of a continuous layer of vapor around the active electrode is a necessary—but not sufficient—condition for the formation of the electrolytic plasma.

Supplementary Materials: The following are available online at http://www.mdpi.com/2073-4344/10/10/1104/s1, Figures S1–S4: V_B and V_D as a function of solution conductivity. Table S1. List of the more interesting experiments with relative values of VB, VD and Conductivity. Type of electrode, area ratio and electrolytic solution is specified as well. Video S1: Example of plasma formation.

Author Contributions: Conceptualization, G.B.A. and F.D.; methodology, G.B.A. and M.B.; software, G.B.A. and M.B.; validation, M.B., F.D. and D.D.; formal analysis, G.B.A.; investigation, G.B.A. and M.B.; resources, D.D. and F.D.; data curation, G.B.A. and M.B.; writing—original draft preparation, G.B.A. and M.B.; writing—review and editing, F.D. and D.D.; visualization, G.A. and M.B.; supervision, M.B. and D.D.; project administration, F.D. and D.D.; funding acquisition, D.D. All authors have read and agreed to the published version of the manuscript.

Funding: This research was funded by MIUR (Grant Number: PRIN-2017YH9MRK; project title: Novel Multilayered and Micro-Machined Electrode Nano-Architectures for Electrocatalytic Applications (Fuel Cells and Electrolyzers)).

Conflicts of Interest: The authors declare no conflict of interest.

References

1. Wüthrich, R.; Mandin, P. Electrochemical discharges-discovery and early applications. *Electrochim. Acta* **2009**, *54*, 4031–4035. [CrossRef]
2. Sengupta, S.K.; Singh, O.P. Contact glow discharge electrolysis: A study of its onset and location. *J. Electroanal. Chem.* **1991**, *301*, 189–197. [CrossRef]
3. Sengupta, S.K.; Srivastava, A.K.; Singh, R. Contact glow discharge electrolysis: A study on its origin in the light of the theory of hydrodynamic instabilities in local solvent vaporisation by Joule heating during electrolysis. *J. Electroanal. Chem.* **1997**, *427*, 23–27. [CrossRef]
4. Sen Gupta, S.K. Contact glow discharge electrolysis: Its origin, plasma diagnostics and non-faradaic chemical effects. *Plasma Sources Sci. Technol.* **2015**, *24*, 063001. [CrossRef]
5. Sen Gupta, S.K.; Singh, R. Cathodic contact glow discharge electrolysis: Its origin and non-faradaic chemical effects. *Plasma Sources Sci. Technol.* **2017**, *26*, 15005. [CrossRef]
6. Yerokhin, A.L.; Nie, X.; Leyland, A.; Matthews, A.; Dowey, S.J. Plasma electrolysis for surface engineering. *Surf. Coatings Technol.* **1999**, *122*, 73–93. [CrossRef]
7. Yerokhin, A.; Mukaeva, V.R.; Parfenov, E.V.; Laugel, N.; Matthews, A. Charge transfer mechanisms underlying contact glow discharge electrolysis. *Electrochim. Acta* **2019**, *312*, 441–456. [CrossRef]
8. Wüthrich, R.; Allagui, A. Building micro and nanosystems with electrochemical discharges. *Electrochim. Acta* **2010**, *55*, 8189–8196. [CrossRef]
9. Saito, G.; Nakasugi, Y.; Akiyama, T. High-speed camera observation of solution plasma during nanoparticles formation. *Appl. Phys. Lett.* **2014**, *104*, 83104.

10. Kareem, T.A.; Kaliani, A.A. Glow discharge plasma electrolysis for nanoparticles synthesis. *Ionics* **2012**, *18*, 315–327. [CrossRef]
11. Tezuka, M.; Iwasaki, M. Aromatic cyanation in contact glow discharge electrolysis of acetonitrile. *Thin Solid Films* **2002**, *407*, 169–173. [CrossRef]
12. Sharma, N.; Diaz, G.; Leal-Quirós, E. Evaluation of contact glow-discharge electrolysis as a viable method for steam generation. *Electrochim. Acta* **2013**, *108*, 330–336. [CrossRef]
13. Sengupta, S.K.; Sandhir, U.; Misra, N. A study on acrylamide polymerization by anodic contact glow-discharge electrolysis: A novel tool. *J. Polym. Sci. Part A Polym. Chem.* **2001**, *39*, 1584–1588. [CrossRef]
14. Hasanah, A.U.; Junior, A.B.; Saksono, N. Latex-starch hybrid synthesis using CGDE method with ethanol addition and air injection latex-starch hybrid synthesis using CGDE method with ethanol addition and air injection. In *AIP Conference Proceedings*; AIP Publishing LLC.: Melville, NY, USA, 2019; p. 020002.
15. Gao, J.; Wang, A.; Li, Y.; Fu, Y.; Wu, J.; Wang, Y.; Wang, Y. Synthesis and characterization of superabsorbent composite by using glow discharge electrolysis plasma. *React. Funct. Polym.* **2008**, *68*, 1377–1383. [CrossRef]
16. Liu, Y. Aqueous p-chloronitrobenzene decomposition induced by contact glow discharge electrolysis. *J. Hazard. Mater.* **2009**, *166*, 1495–1499. [CrossRef]
17. Yang, H.; Matsumoto, Y.; Tezuka, M. Exhaustive breakdown of aqueous monochlorophenols by contact glow discharge electrolysis. *J. Environ. Sci.* **2009**, *21*, S142–S145. [CrossRef]
18. Jin, X.; Bai, H.; Wang, F.; Wang, X.; Wang, X.; Ren, H. Plasma degradation of acid orange 7 with contact glow discharge electrolysis. *IEEE Trans. Plasma Sci.* **2011**, *39*, 1099–1103. [CrossRef]
19. Allagui, A.; Brazeau, N.; Alawadhi, H.; Al-momani, F.; Baranova, E.A. Cathodic contact glow discharge electrolysis for the degradation of liquid ammonia solutions. *Plasma Process. Polym.* **2015**, *12*, 25–31. [CrossRef]
20. Jin, X.; Wang, X.; Wang, Y.; Ren, H. Oxidative degradation of amoxicillin in aqueous solution with contact glow discharge electrolysis. *Ind. Eng. Chem. Res.* **2013**, *52*, 9726–9730. [CrossRef]
21. Ramjaun, S.N.; Yuan, R.; Wang, Z.; Liu, J. Degradation of reactive dyes by contact glow discharge electrolysis in the presence of Cl- ions: Kinetics and AOX formation. *Electrochim. Acta* **2011**, *58*, 364–371. [CrossRef]
22. Chen, Y.; Jin, X. Degradation of rhodamine B by contact glow discharge electrolysis with Fe3O4/BiPO4 nanocomposite as heterogeneous catalyst. *Electrochim. Acta* **2019**, *296*, 379–386. [CrossRef]
23. Zhao, C.; Yang, H.; Ju, M.; Zhao, X.; Li, L.; Wang, S.; An, B. Simultaneous degradation of aqueous trichloroacetic acid by the combined action of anodic contact glow discharge electrolysis and normal electrolytic processes at the cathode. *Plasma Chem. Plasma Process.* **2019**, *39*, 751–767. [CrossRef]
24. Jiang, B.; Hu, P.; Zheng, X.; Zheng, J.; Tan, M.; Wu, M.; Xue, Q. Rapid oxidation and immobilization of arsenic by contact glow discharge plasma in acidic solution. *Chemosphere* **2015**, *125*, 220–226. [CrossRef] [PubMed]
25. Wang, X.; Zhou, M.; Jin, X. Application of glow discharge plasma for wastewater treatment. *Electrochim. Acta* **2012**, *83*, 501–512. [CrossRef]
26. Yang, H.; Zhao, X.; Mengen, G.; Tezuka, M.; An, B.; Li, L.; Wang, S.; Ju, M. Defluorination and mineralization of difluorophenols in water by anodic contact glow discharge electrolysis. *Plasma Chem. Plasma Process.* **2016**, *36*, 993–1009. [CrossRef]
27. Gao, J.; Liu, Y.; Yang, W.; Pu, L.; Yu, J.; Lu, Q. Aqueous p-nitrotoluene oxidation induced with direct glow discharge plasma. *Cent. Eur. J. Chem.* **2005**, *3*, 377–386. [CrossRef]
28. Amano, R.; Tezuka, M. Mineralization of alkylbenzenesulfonates in water by means of contact glow discharge electrolysis. *Water Res.* **2006**, *40*, 1857–1863. [CrossRef]
29. Gai, K. Plasma-induced degradation of diphenylamine in aqueous solution. *J. Hazard. Mater.* **2007**, *146*, 249–254. [CrossRef]
30. Gao, J.; Wang, X.; Hu, Z.; Deng, H.; Hou, J.; Lu, X.; Kang, J. Plasma degradation of dyes in water with contact glow discharge electrolysis. *Water Res.* **2003**, *37*, 267–272. [CrossRef]
31. Gao, J.; Hu, Z.; Wang, X.; Hou, J.; Lu, X.; Kang, J. Oxidative degradation of acridine orange induced by plasma with contact glow discharge electrolysis. *Thin Solid Films* **2001**, *390*, 154–158. [CrossRef]
32. Saksono, N.; Seratri, R.T.; Muthia, R.; Bismo, S. Phenol degradation in wastewater with a contact glow discharge electrolysis reactor using a sodium sulfate. *Int. J. Technol.* **2015**, *6*, 1153–1163.
33. Yan, Z.C.; Li, C.; Lin, W.H. Hydrogen generation by glow discharge plasma electrolysis of methanol solutions. *Int. J. Hydrogen Energy* **2009**, *34*, 48–55. [CrossRef]

34. Munegumi, T. Chemical evolution of aminoacetonitrile to glycine under discharge onto primitive hydrosphere: Simulation experiments using glow discharge. *Asian J. Chem.* **2016**, *28*, 555–561. [CrossRef]
35. Sen Gupta, S.K. Contact glow discharge electrolysis: A novel tool for manifold applications. *Plasma Chem. Plasma Process.* **2017**, *37*, 897–945.
36. Parthasarathy, P.; Narayanan, S.K. Effect of hydrothermal carbonization reaction parameters on. *Environ. Prog. Sustain. Energy* **2014**, *33*, 676–680.
37. Allagui, A.; Elwakil, A.S. On the N-shaped conductance and hysteresis behavior of contact glow discharge electrolysis. *Electrochim. Acta* **2015**, *168*, 173–177.
38. Jin, X.; Wang, X.; Yue, J.; Cai, Y.; Zhang, H. The effect of electrolyte constituents on contact glow discharge electrolysis. *Electrochim. Acta* **2010**, *56*, 925–928.
39. Jin, X.L.; Wang, X.Y.; Zhang, H.M.; Xia, Q.; Wei, D.B.; Yue, J.J. Influence of solution conductivity on contact glow discharge electrolysis. *Plasma Chem. Plasma Process.* **2010**, *30*, 429–436. [CrossRef]
40. Sengupta, S.K.; Singh, O.P. Contact glow discharge electrolysis: A study of its chemical yields in aqueous inert-type electrolytes. *J. Electroanal. Chem.* **1994**, *369*, 113–120.
41. Gangal, U.; Srivastava, M.; Sen Gupta, S.K. Scavenging effects of aliphatic alcohols and acetone on H• radicals in anodic contact glow discharge electrolysis: Determination of the primary yield of H• radicals. *Plasma Chem. Plasma Process.* **2010**, *30*, 299–309.
42. Singh, R.; Gangal, U.; Sen Gupta, S.K. Effects of alkaline ferrocyanide on non-faradaic yields of anodic contact glow discharge electrolysis: Determination of the primary yield of OH⁻ radicals. *Plasma Chem. Plasma Process.* **2012**, *32*, 609–617. [CrossRef]
43. Saito, G.; Nakasugi, Y.; Akiyama, T. Generation of solution plasma over a large electrode surface area. *J. Appl. Phys.* **2015**, *118*, 23303. [CrossRef]
44. Allagui, A.; Wüthrich, R. Gas film formation time and gas film life time during electrochemical discharge phenomenon. *Electrochim. Acta* **2009**, *54*, 5336–5343. [CrossRef]
45. Toth, J.R.; Hawtof, R.; Matthiesen, D.; Renner, J.N.; Sankaran, R.M. On the non-faradaic hydrogen gas evolution from electrolytic reactions at the interface of a cathodic atmospheric-pressure microplasma and liquid water surface. *J. Electrochem. Soc.* **2020**, *167*, 116504. [CrossRef]
46. Santibáñez, C.; Varnero, M.T.; Bustamante, M. Residual glycerol from biodiesel Manufacturing, waste or potential source of bioenergy: A review. *Chil. J. Agric. Res.* **2011**, *71*, 469–475. [CrossRef]
47. Dashnau, J.L.; Nucci, N.V.; Sharp, K.A.; Vanderkooi, J.M. Hydrogen bonding and the cryoprotective properties of glycerol/water mixtures. *J. Phys. Chem. B* **2006**, *110*, 13670–13677. [CrossRef]
48. Mouratoglou, E.; Malliou, V.; Makris, D.P. Novel glycerol-based natural eutectic mixtures and their efficiency in the ultrasound-assisted extraction of antioxidant polyphenols from agri-food waste biomass. *Waste Biomass Valorization* **2016**, *7*, 1377–1387. [CrossRef]
49. Darling, H.E. Conductivity of sulfuric acid solutions. *J. Chem. Eng. Data* **1964**, *9*, 421–426. [CrossRef]
50. Gilliam, R.J.; Graydon, J.W.; Kirk, D.W.; Thorpe, S.J. A review of specific conductivities of potassium hydroxide solutions for various concentrations and temperatures. *Int. J. Hydrogen Energy* **2007**, *32*, 359–364. [CrossRef]

© 2020 by the authors. Licensee MDPI, Basel, Switzerland. This article is an open access article distributed under the terms and conditions of the Creative Commons Attribution (CC BY) license (http://creativecommons.org/licenses/by/4.0/).

Article

Asymmetric Cyanation of Activated Olefins with Ethyl Cyanoformate Catalyzed by Ti(IV)-Catalyst: A Theoretical Study

Zhishan Su [1], Changwei Hu [1], Nasir Shahzad [2] and Chan Kyung Kim [2,*]

[1] Key Laboratory of Green Chemistry and Technology, Ministry of Education, College of Chemistry, Sichuan University, No. 29 Wangjiang Road, Chengdu 610064, China; suzhishan@scu.edu.cn (Z.S.); changweihu@scu.edu.cn (C.H.)
[2] Department of Chemistry and Chemical Engineering, Center for Design and Applications of Molecular Catalysts, Inha University, 100 Inha-ro, Michuhol-gu, Incheon 22212, Korea; nasirchem@yahoo.com
* Correspondence: kckyung@inha.ac.kr

Received: 30 July 2020; Accepted: 15 September 2020; Published: 18 September 2020

Abstract: The reaction mechanism and origin of asymmetric induction for conjugate addition of cyanide to the C=C bond of olefin were investigated at the B3LYP-D3(BJ)/6-31+G**//B3LYP-D3(BJ)/6-31G**(SMD, toluene) theoretical level. The release of HCN from the reaction of ethyl cyanoformate (CNCOOEt) and isopropanol (HOiPr) was catalyzed by cinchona alkaloid catalyst. The cyanation reaction of olefin proceeded through a two-step mechanism, in which the C-C bond construction was followed by H-transfer to generate a cyanide adduct. For non-catalytic reaction, the activation barrier for the rate-determining C-H bond construction step was 34.2 kcal mol^{-1}, via a four-membered transition state. The self-assembly Ti(IV)-catalyst from tetraisopropyl titanate, (R)-3,3'-disubstituted biphenol, and cinchonidine accelerated the addition of cyanide to the C=C double bond by a dual activation process, in which titanium cation acted as a Lewis acid to activate the olefin and HNC was orientated by hydrogen bonding. The steric repulsion between the 9-phenanthryl at the 3,3'-position in the biphenol ligand and the Ph group in olefin raised the Pauli energy ($\Delta E^{\neq}_{Pauli}$) of reacting fragments at the *re*-face attack transition state, leading to the predominant *R*-product.

Keywords: asymmetric conjugate addition; cinchona alkaloid catalysis; cyanation reaction of olefin; self-assembly Ti(IV)-catalysis; density functional theory calculation

1. Introduction

The asymmetric catalytic cyanation of C=X bond (X = O, N or C) provides an outstanding method to obtain various optically active nitriles [1–6]. Compared to the intensively studied cyanation of aldehydes [1,2,5], ketones [1,2,5] and imine (Strecker reaction) [3], reports on the conjugate addition of cyanide to the C=C bond are limited [4,6]. Since the products from cyanide addition to C=C double bonds in the α,β-unsaturated carbonyl compounds could conveniently convert to the enantioenriched intermediates with great synthesizing value and pharmaceutical importance (e.g., γ-aminobutyric acids), developing straightforward synthetic procedures and exploring the relevant reaction mechanisms are in high demand.

Jacobsen's group reported the first catalytic asymmetric cyanation of α,β-unsaturated imides, using Al(III) complex with chiral salen ligand as a catalyst. Based on the kinetic analyses, they proposed that the reaction involved a bimetallic, dual activation process. The salen–Al(III) complex-activated cyanide was delivered to the electrophile bound as an imidate complex for highly enantioenriched cyanide adducts [7]. In a heterobimetallic system with (salen)Al and (pybox)Er complexes (pybox = 2,6-bis(2-oxazolinyl) pyridine), two catalysts operated cooperatively in the rate-determining step, promoting the conjugate addition in a highly enantioselective manner [8]. The poly(norbornene)-supported (salen)AlCl catalyst could also realize this transformation. The proximity of catalytic sites in polymeric Al-catalyst facilitated the reaction to occur via a bimetallic pathway [9]. Shibasaki et al. developed chiral gadolinium complex catalysts for the cyanation of α,β-unsaturated N-acylpyrroles [10,11] and enones [12]. Mechanistic studies suggested that the reaction is carried out through an intramolecular cyanide transfer from the gadolinium cyanide to the activated N-acylpyrrole substrate. The protic additive (e.g., HCN) efficiently facilitated both catalyst activity and enantioselectivity. Other metal complexes containing Sr(II) [13], Ru(II)/Li(I) [14,15], Mg(II) [16,17] and Li(I) [18] were also active catalysts for this kind of reaction. The bifunctional catalysis model was proposed to interpret the activation mode as well as the stereochemical outcome [15,16]. Besides, the reaction could be realized by organocatalysis [19,20] or a phase-transfer process [21,22]. The spectroscopic studies by Khan et al. verified that the N-oxide additive participated in the cyanation of nitroalkenes as a ligand and activator of trimethylsilylcyanide (TMSCN) [23]. Experiments and density functional theory (DFT) calculations by Minakata and co-workers revealed that the cyanation of the boron enolates generated from α,β-unsaturated ketones with p-toluenesulfonyl cyanide (TsCN) proceeded through a six-membered ring transition state (TS) [24]. Based on the NMR spectroscopy results, Khan et al. proposed that imidazolium cations interacted with the substrate, facilitating the attack of cyanide ions generated by the activation of acetone cyanohydrin by the acetate counter ion in 1-butyl-3-methylimidazolium (BMIM)-based ionic-liquid-catalyzed conjugate cyanation of CF_3-substituted alkylidenemalonates [25].

In 2010, Feng's group developed a modular catalyst generated in situ from cinchona alkaloid, tetraisopropyl titanate (Ti(OiPr)$_4$) and achiral 3,3'-disubstituted biphenol, achieving the efficient asymmetric cyanation of N-p-toluenesulfonyl aldimines and ketimines, as well as ketones and aldehydes [26,27]. Interestingly, this self-assembled catalyst system also exhibited excellent performance in the asymmetric cyanation of C=C bonds, using ethyl cyanoformate (CNCOOEt) as a cyanide source [28]. The enantioenriched cyanide adducts could be obtained with high yield (97%) and enantiomeric excess (ee) (up to 94%). In the reaction, the axial chirality of the biphenol ligand was induced in the formation of the complex to achieve asymmetric activation [29]. Based on previous work, they proposed that the catalytic species was (R)-biphenol, and chiral cinchonidine ligand coordinated to the Ti(IV) center simultaneously.

In this work, we employed DFT calculations to understand the mechanism for the asymmetric cyanation of activated olefin in detail (Scheme 1). The key factors controlling the enantioselectivity could be revealed to expedite the rational design of new Ti(IV)-complex catalysts.

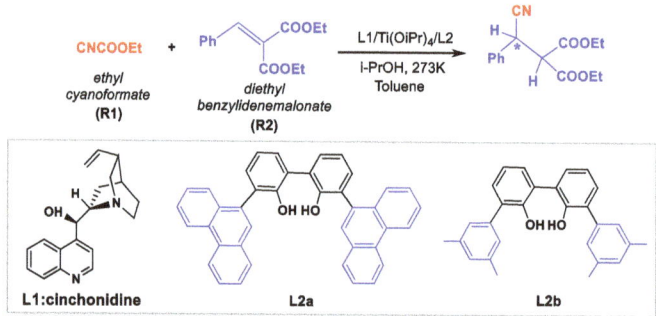

Scheme 1. Asymmetric cyanation of activated olefin catalyzed by Ti(IV)-complex.

2. Computational Details

DFT calculations were performed using the Gaussian 09 package [30] at the B3LYP-D3(BJ)/6-31G**(SMD, toluene) theoretical level. Geometries were optimized in toluene solvent and characterized by calculating the harmonic vibrational frequencies. The self-consistent reaction field (SCRF) method and SMD solvation model [31] were adopted to evaluate the effect of the solvent. The transition states were verified by the intrinsic reaction coordinate (IRC) calculation [32]. The optimized structures are summarized in the Supporting Information. Activation strain model (ASM) analysis [33–35], also known as distortion–interaction model calculation [36–39], was used to analyze the factors affecting the enantioselectivity of the catalytic reaction, in which the potential energy (ΔE) was decomposed into the distortion (ΔE_{strain}) and interaction (ΔE_{int}) energies using the Gaussian 09 program. Besides, four energy contributors (i.e., electrostatic interaction (ΔV_{elstat}), Pauli repulsion (ΔE_{Pauli}), dispersion effect (ΔE_{disp}) and orbital interaction (ΔE_{oi})) in ΔE_{int} were partitioned by energy decomposition analysis (EDA) [40] using the Amsterdam Density Functional (ADF) program [41] at the B3LYP-D3(BJ)/TZ2P level. The energy of the optimized structure was re-evaluated by single-point calculations at the B3LYP-D3(BJ)/6-31+G**(SMD, toluene) level, in which dispersion correction was also included using Grimme's D3(BJ) method [42,43]. Unless specified, the Gibbs free energies obtained at the B3LYP-D3(BJ)/6-31+G**//B3LYP-D3(BJ)/6-31G**(SMD, toluene) level at 273 K were used.

3. Results and Discussion

3.1. Release of HCN or HNC Species from CNCOOEt

The previous experimental and theoretical investigations suggested that HCN was the real cyanation reagent in the cyanation of imines with TMSCN catalyzed by Ti(IV)-complex [29]. Moreover, isopropyl alcohol (HOiPr) had a positive effect on the release of HCN from TMSCN, consequently accelerating the reaction [26,27,29,44]. Based on these results, we first studied the formation of HCN or HNC species from the reaction between HOiPr and the cyanide source, ethyl cyanoformate (CNCOOEt). Four possible pathways were considered, in which HCN was formed along paths 1 and 3, while its isomer HNC was afforded along paths 2 and 4. As shown in Scheme 2, when CNCOOEt approached HOiPr, the H atom of HOiPr transferred to the C and N atoms of CNCOOEt, yielding HCN and HNC species, respectively. The potential energy surfaces for these pathways are shown in Figure 1. The activation energies were 54.9 and 56.5 kcal mol^{-1} via 1-TS1 and 2-TS1, respectively. Besides, the reaction could also occur via other stepwise mechanisms along paths 3 and 4, in which the H atom transferred initially to the O atom of the carbonyl group in CNCOOEt having a more negative charge (−0.548 at the O atom vs. −0.501 at the CN moiety) via a four-membered ring transition state TS1. Then, this H atom transferred to the C and N atoms of the CN group through transition states 3-TS2 and 4-TS2, producing HCN (path 3) and HNC (path 4), respectively. The calculations indicated that the ΔG associated with the generation of HCN along path 3 and HNC along path 4 were 39.6 and 40.1 kcal mol^{-1}, respectively, which were lower than those along paths 1 and 2.

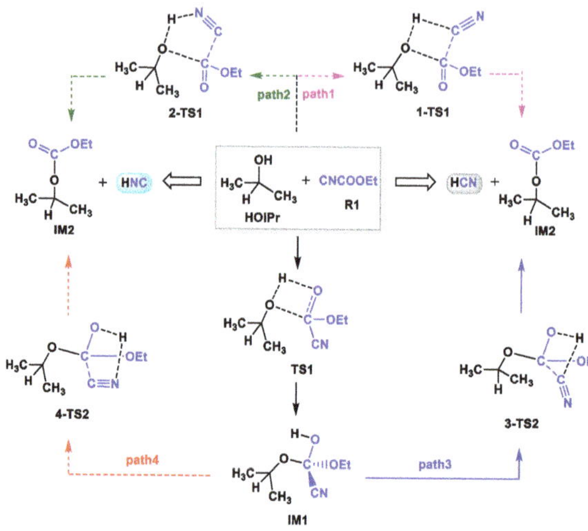

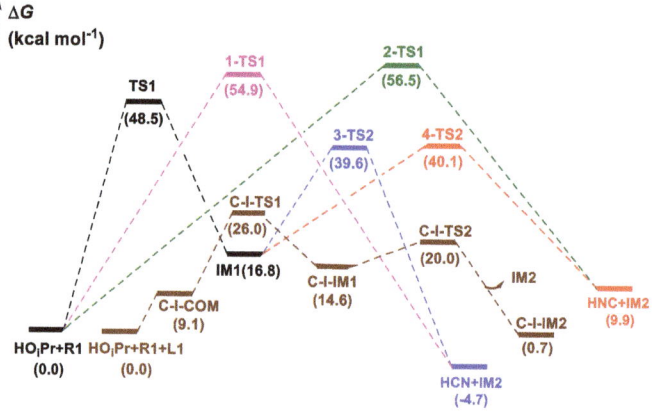

Scheme 2. Formation of HCN or HNC from the reaction between HOiPr and CNCOOEt.

Figure 1. Potential energy profiles for the formation of HCN or HNC by the reaction between HOiPr and CNCOOEt (R1) without a catalyst and in the presence of cinchona alkaloid (L1).

For comparison, the formation of HNC assisted by cinchona alkaloid (L1) was studied (Figures 1 and 2). In the initial complex (C-I-COM), the cinchona alkaloid activated HOiPr by the N atom of the tertiary amine ring and OH group simultaneously with (O)H···O and N···H distances of 1.858 and 1.834 Å, respectively. These hydrogen bonds could be verified by atoms-in-molecules (AIM) analysis, with the positive Laplacian ($\nabla^2 \rho$) on (3, −1) bond critical points (BCPs) (Figure S1). In the next step, the O atom of HOiPr approached the C atom of CNCOOEt, accompanied with H transfer from the O atom to N atom via C-I-TS1. Then, HNC was formed by breaking the C-C bond via C-I-TS2, with a ΔG of 20.0 kcal mol^{-1}. From the viewpoint of energy, cinchona alkaloid could promote the transformation of CNCOOEt to HNC by organocatalysis.

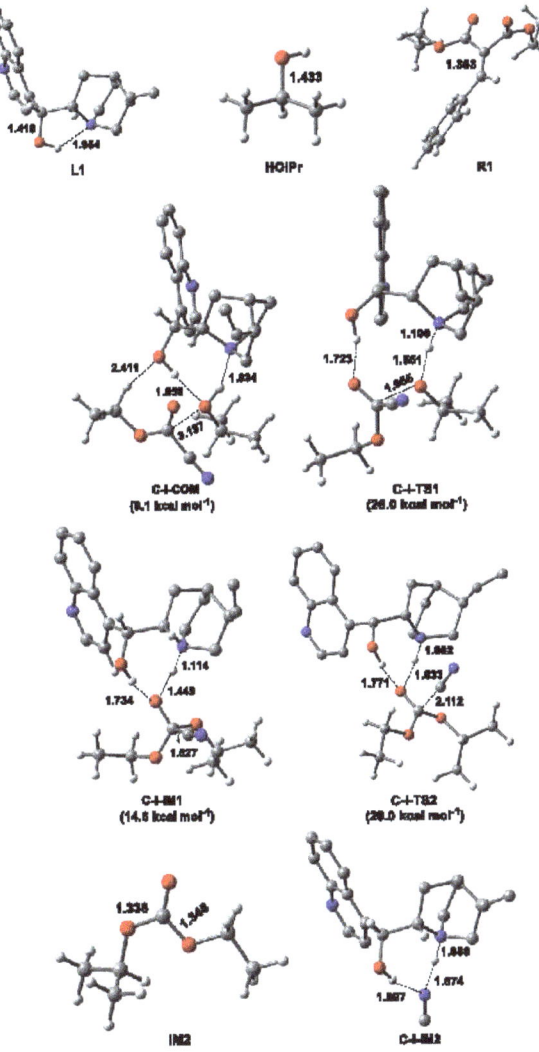

Figure 2. Optimized geometries of reactants (R1 and HOiPr), cinchona alkaloid (L1), molecular complex (C-I-COM), transition states (TSs) (C-I-TS1 and C-I-TS2) and intermediates (C-I-IM1–C-I-IM3) for the HNC formation catalyzed by cinchona alkaloid (some H atoms in cinchona alkaloid are omitted for clarity) and their relative Gibbs free energies. The intermolecular distance is in Angstroms (Å). The color definitions of atoms are red = oxygen, blue = nitrogen, gray = carbon and white = hydrogen.

3.2. Reaction Mechanism

3.2.1. Noncatalytic Reaction

The isomerization between HCN and HNC can occur quickly in the presence of cinchona alkaloid [29], establishing rapid HCN–HNC equilibrium. Then, HNC can act as an active cyanide species to construct a C-C bond by interacting with an olefin in the cyanation reaction. Based on these results, we first studied the mechanism of the noncatalytic cyanation of olefins with HNC (Figure 3). The reaction occurred through a two-step process: C-C bond formation followed by C-H bond construction. In the initial step, HNC interacted with the COOEt moiety through hydrogen bonding, forming a molecular complex b-IM1. Then, the CN group attacked the olefin to form a C-C bond via a seven-membered ring transition state b-TS1. In the final step, the product P-S was formed by shifting an H atom to a C atom (tautomerization of enol to keto), with a $\Delta G^{\neq}$ of 34.2 kcal mol^{-1}. This H-transfer step was predicted to be the rate-determining step (RDS) in the background reaction. We also located the TS involving HCN as a proton donor (b-TS1-1, Figure S2). The relative Gibbs energy of b-TS1-1 was higher than that of b-TS2 (the highest point in the energy profile) by 8.4 kcal mol^{-1}. These results indicated that the noncatalytic reaction was difficult to achieve owing to a higher barrier.

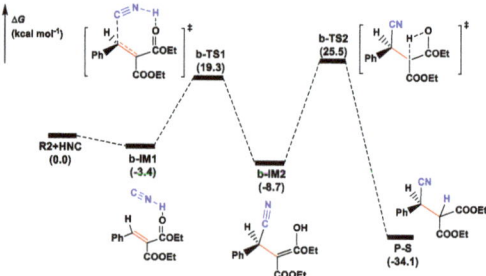

Figure 3. Potential energy profile for the noncatalytic cyanation reaction of olefin.

3.2.2. Catalytic Reaction

A previous study indicated that aldimine and HOiPr could coordinate to the Ti(IV), forming a reactive hexacoordinated Ti(IV)-complex in the Strecker reaction catalyzed by Ti(IV)-complex catalyst with cinchona alkaloid and achiral 3,3'-disubstituted 2,2'-biphenol ligands [29]. Based on experimental observation [28], a cinchonidine(L1)/Ti(IV)/(R)-biphenol (L2a) catalyst was employed as an active species in the reaction. Considering that there were two O-donors in the olefin substrate, a bidentate model was first studied in the present work, in which two carbonyl groups of olefin coordinated to the Ti(IV) center simultaneously, forming three possible hexacoordinated Ti(IV)-complexes (d-I–d-III). For comparison, the monocoordinated models (m-I–m-VI) were also investigated (Scheme 3). Nine low-energy Ti(IV)-complexes in mono- and bidentate models were located to allow the favorable *si*-face attack pathway observed in the experiment (see Table 1 and Figure S3). Table 1 shows that a bidentate Ti(IV)-complex (d-I-COM-*si*) had the lowest Gibbs free energy among nine models. In other words, this complex was the only starting species available in the reaction mixture.

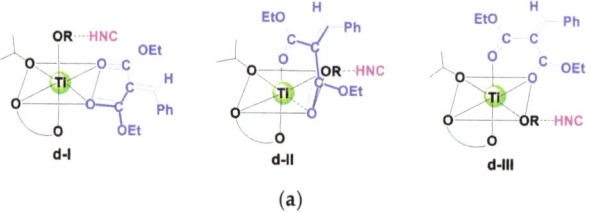

(a)

Scheme 3. *Cont.*

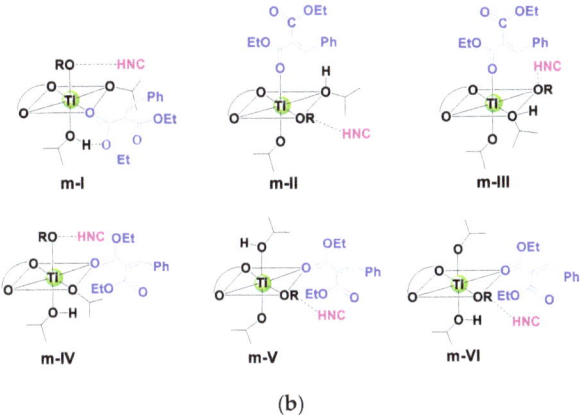

m-I   m-II   m-III

m-IV   m-V   m-VI

(b)

Scheme 3. Nine possible hexacoordinated Ti(IV)-complexes formed by coordinating olefin to Ti(IV) ion in a (**a**) bidentate and (**b**) monodentate fashion.

Table 1. Relative Gibbs free energy for the formation of hexacoordinated Ti(IV)-complexes.

	Model	Species	ΔG (kcal mol^{-1}) [1]
Monodentate	m-I	m-I-COM-si	12.2
		m-I-COM-re	7.5
	m-II	m-II-COM-si	12.6
		m-II-COM-re	6.9
	m-III	m-III-COM-si	12.6
		m-III-COM-re	16.0
	m-IV	m-IV-COM-si	14.4
		m-IV-COM-re	13.5
	m-V	m-V-COM-si	16.1
		m-V-COM-re	20.0
	m-VI	m-VI-COM-si	16.2
		m-VI-COM-re	11.0
Bidentate	d-I	d-I-COM-si	0.0
		d-I-COM-re	−2.5
	d-II	d-II-COM-si	5.5
		d-II-COM-re	−0.6
	d-III	d-III-COM-si	2.7
		d-III-COM-re	7.4

[1] The energy of d-I-COM-*si* was set to zero.

The reaction mechanisms of the cyanation of olefins in the presence of Ti(IV)-complex catalyst could be very similar for the nine coordination models, although the reaction might begin with different hexacoordinated Ti(IV)-complexes: C-C bond construction followed by H-shift. The potential energy surfaces for the *si*-face and *re*-face attack to produce *R*- and *S*-configuration enantiomers from d-I-COM-*si* and d-I-COM-*re* are shown in Figure 4.

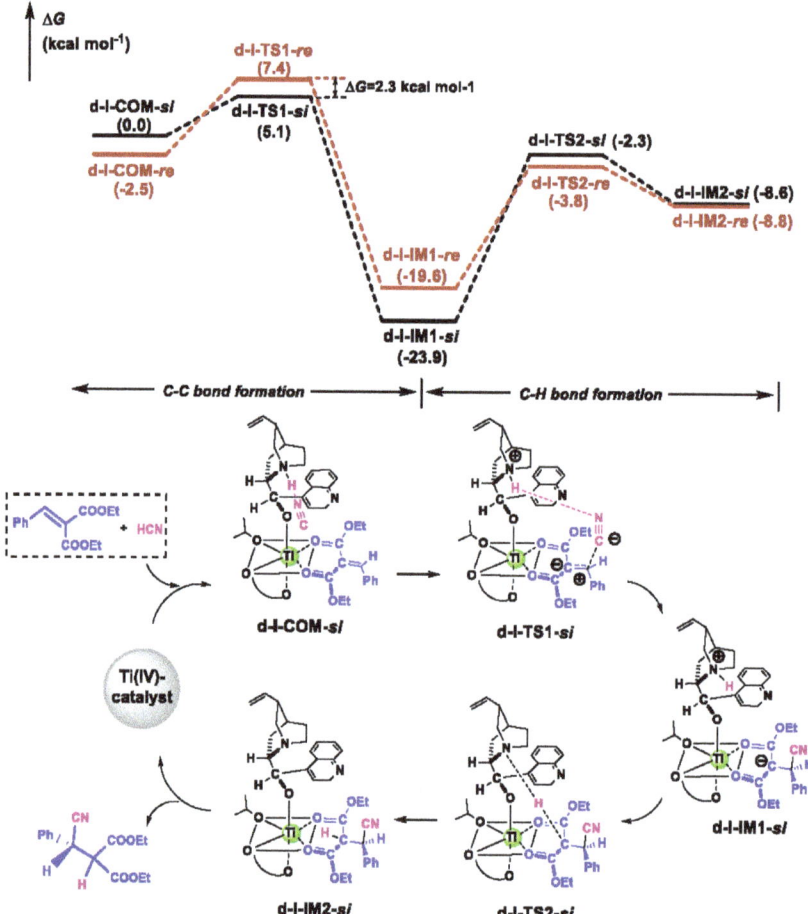

Figure 4. Gibbs free energy profiles for the catalytic cyanation of olefin mediated by Ti(IV)-complex along si- and re-face attack pathways and schematic catalytic cycle along si-face attack pathway as a representative.

Like the noncatalytic reaction, the catalytic process occurred via a stepwise mechanism. Firstly, HNC was coordinated to the tertiary amine of cinchona alkaloid through hydrogen bonding, with an H···N distance of 1.612 Å. Then, the C-C bond was constructed by the attack of the CN group to olefin, with $\Delta G^{\neq}$ of 5.1 kcal mol^{-1} via the TS d-I-TS1-si. Finally, the catalytic cycle was finished when the H atom was transferred from the N to C atom, via the TS d-I-TS2-si. A protonic reagent (e.g., HOiPr) could accelerate proton shift by hydrogen-bonding [29], and the H-shift barrier for HOiPr via the TS d-I-TS2-si-HOiPr was decreased by 3.4 kcal mol^{-1} (Figure S4). The activation barrier in the chiral-controlling step (C-C bond formation step) in the catalysis was lower than that for the noncatalytic process by 14.2 kcal mol^{-1}. We also optimized the TSs in the chiral-controlling C-C bond construction along with the si-face attack in the d-II and d-III models (Figure S5). As expected, the activation free energies via d-II-TS1-si ($\Delta G^{\neq}$ = 10.4 kcal mol^{-1}) and d-III-TS1-si ($\Delta G^{\neq}$ = 7.9 kcal mol^{-1}) were higher than that via d-I-TS1-si ($\Delta G^{\neq}$ = 5.1 kcal mol^{-1}) in the d-I model. As shown in Figure 5, the ΔG of d-I-TS1-si was lower than that of d-I-TS1-re in a chiral-controlling step by 2.3 kcal mol^{-1} at 273 K, indicating the product with R-configuration was predominant. The theoretical enantioselectivity (ee

%) was 96% using the Curtin–Hammett principle, which was close to that obtained experimentally (93% ee) [28].

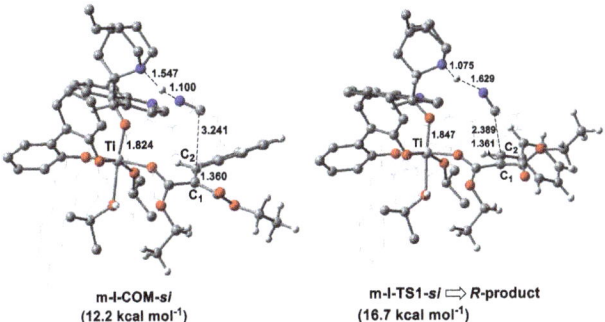

Figure 5. Optimized structures of the monodentate Ti(IV)-complex (m-I-COM-si) and TS (m-I-TS1-si) in the C-C bond construction step in the catalytic cyanation reaction of olefin along the si-face attack pathway and their relative Gibbs free energies. The intermolecular distance is in Angstroms (Å). The color definitions of atoms are red = oxygen, blue = nitrogen, gray = carbon and white = hydrogen.

We also studied the catalytic mechanism starting from the monodentate Ti(IV)-complex m-I-COM-si for comparison, and the corresponding C-C bond construction TS (m-I-TS1-si) in the chiral-controlling step was located (Figure 5). Compared with d-I-COM-si, the olefin substrate in m-I-COM-si was slightly weakened, with a large Wiberg bond index of 1.664 for the C1=C2 bond. Accordingly, m-I-TS1-si was less stable than d-I-TS1-si by 11.6 kcal mol^{-1}. Thus, olefin tended to participate in the cyanation reaction in a bidentate fashion.

3.3. Origin of Stereoselectivity

The hexacoordinated Ti(IV)-complexes d-I-COM-si and d-I-COM-re had a pocket-like chiral cavity, with the dihedral angles formed by the 9-phenanthryl groups at the 3,3'-position in L2a of 93.7° and 87.3°, respectively. The topographic steric maps [45–47] of d-I-COM-si and d-I-COM-re are shown in Scheme 4, which characterizes the surface that the ligand L2a offers to the olefin substrate. The percentage of buried volume (%V_{Bur}) quantified the first coordination sphere around the Ti center occupied by L2a ligand. For d-I-COM-re, %V_{Bur} was 39.7, which was larger than that of d-I-COM-si (35.8). Importantly, the 9-phenanthryl groups provided stronger hindrance in the north-western and south-eastern quadrants (yellow colored area). Consequently, the unfavorable steric repulsion between the Ph group of olefin and 9-phenanthryl group of ligand L2a (in the south-eastern quadrant) became more significant in d-I-COM-re as well as in the corresponding C-C bond formation TS.

Then, we further analyzed the structures of d-I-TS1-si and d-I-TS1-re (Figure 6). The Ph group of the olefin moiety in d-I-TS1-re was in proximity to the neighboring 3-substituted group in the biphenol ligand, with a Ph···Ph distance of about 2.5 Å. Accordingly, the steric repulsion raised the Pauli energy ($\Delta E^{\neq}_{Pauli}$) of the reacting fragments at d-I-TS1-re (139.0 vs. 130.4 kcal mol^{-1}) (Table 2). Consequently, the ΔG of d-I-TS1-re was higher than that of d-I-TS1-si (7.4 vs. 5.1 kcal mol^{-1}). This steric repulsion was also visualized by noncovalent interaction analysis using Multiwfn software [48], in which the larger yellow area between the Ph of olefin and phenanthryl group of L2a (Figure S6) in d-I-TS1-re was observed. In contrast, this unfavorable steric interaction was avoided efficiently in d-I-TS1-si because the Ph groups were located far away. Consequently, two reacting fragments could interact easily, with a more stabilizing $\Delta E^{\neq}_{int}$ (−5.6 kcal mol^{-1}) and low reaction barrier of 5.1 kcal mol^{-1} in the C-C bond formation step.

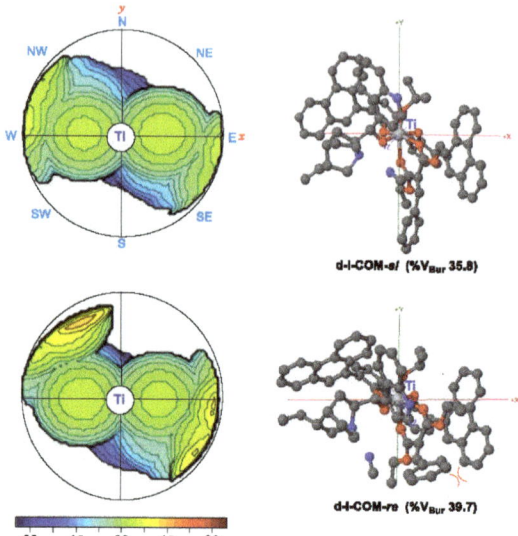

Scheme 4. Topographic steric maps for d-I-COM-*si* and d-I-COM-*re*. %V_{Bur} is the percentage of buried volume, obtained by SambVca 2.1 software [45–47].

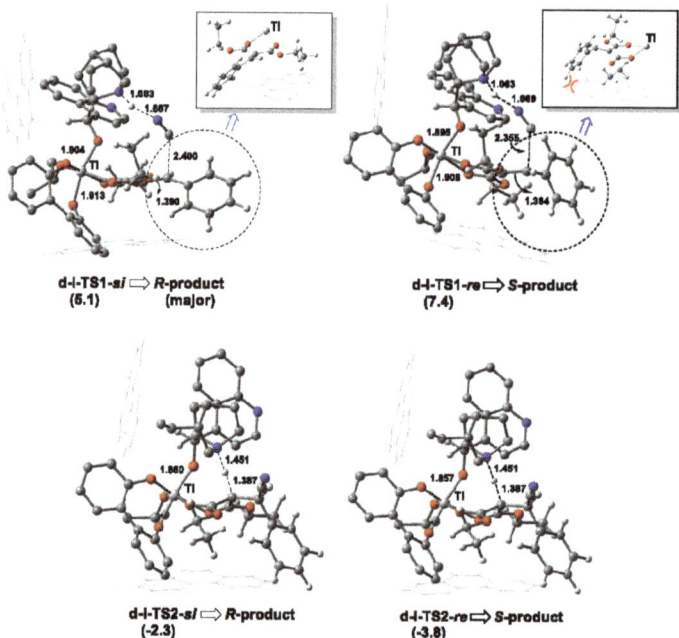

Figure 6. Optimized geometries of TSs in the C-C bond formation step (d-I-TS1-*si* and d-I-TS1-*re*) and H-transfer step (d-I-TS2-*si* and d-I-TS2-*re*) in the cyanation reaction of olefins catalyzed by Ti(IV)-complex along two competing pathways for *R*- and *S*-configuration products, respectively, associated with the relative Gibbs free energies (kcal mol^{-1}). The intermolecular distance is in Angstroms (Å). The color definitions of atoms are red = oxygen, blue = nitrogen, gray = carbon and white = hydrogen.

Table 2. Activation strain model (ASM) [1] analysis and energy decomposition analysis (EDA) [2] for the catalytic cyanation reaction of olefins via d-I-TS1-*si* and d-I-TS1-*re*. The energies are given in kcal mol^{-1}.

TS	$\Delta E^{\neq}_{strain}$	$\Delta E^{\neq}_{int}$	$\Delta E^{\neq}_{Pauli}$	$\Delta E^{\neq}_{oi}$	$\Delta V^{\neq}_{elstat}$	$\Delta E^{\neq}_{disp}$
d-I-TS1-*si*	8.8	−5.6	130.4	−77.9	−99.9	−28.7
d-I-TS1-*re*	8.6	−4.9	139.0	−82.6	−104.0	−29.4

[1] ASM calculations were done at the B3LYP-D3(BJ)/6-31+G**(SMD, toluene) level of theory; [2] EDA calculations were done at the B3LYP-D3(BJ)/TZ2P level of theory.

The influence of the 3,3′-substitute of biphenol on the stereoselectivity was further studied. When the 9-phenanthryl group in L2a was replaced by a 3,5-dimethyl phenyl group (L2b), an opening chiral pocket in d-I-COM-*si*-L2b was observed, with the dihedral angle of the two substituent groups at the 3,3′-position of 109.6°. The relative energy of the two competing TSs (d-I-TS1-*si*-L2b and d-I-TS1-*re*-L2b) in the chiral-controlling step (i.e., C-C bond construction step) was comparable (7.3 vs. 7.2 kcal mol^{-1}), affording racemic products (Figure 7). These results indicated that the bulky substituent at the 3,3′-position of the biphenol ligand was essential for asymmetric induction in the Ti(IV)-complex-catalyzed cyanation of activated olefins.

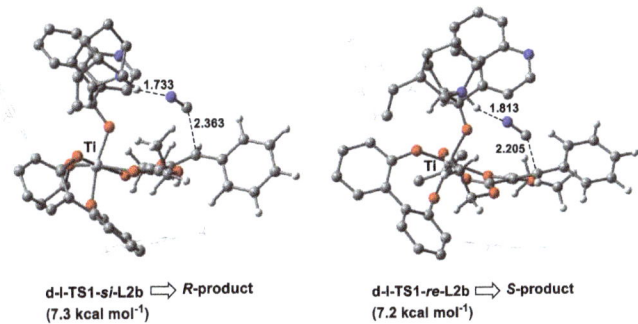

Figure 7. Optimized geometries of TSs in the C-C bond formation step (d-I-TS1-*si*-L2b and d-I-TS1-*re*-L2b) in the cyanation reaction of olefins catalyzed by Ti(IV)-complex with L2b along two competing pathways for R- and S-configuration products, respectively, associated with the relative Gibbs free energies. The intermolecular distance is in Angstroms (Å). The color definitions of atoms are red = oxygen, blue = nitrogen, gray = carbon and white = hydrogen.

4. Conclusions

DFT calculations on the reaction mechanism of asymmetric cyanation of activated olefin-catalyzed Ti(IV)-complexes revealed the following results:

(i). Cinchona alkaloid facilitated the reaction between HOiPr and ethyl cyanoformate (CNCOOEt) to release the reacting species HCN (or HNC) by organocatalysis with free energy barrier of 26.0 kcal mol^{-1}.

(ii). The cyanation reaction of olefin proceeded via a two-step mechanism, in which the C-C bond construction was followed by H-transfer to generate a cyanide adduct. For noncatalytic reaction, the $\Delta G^{\neq}$ for the rate-determining C-H bond construction step was up to 34.2 kcal mol^{-1}, through a four-membered TS. In the catalytic reaction, the olefin coordinated to the self-assembly cinchonidine/Ti(IV)/(R)-3,3′-disubstituted biphenol catalyst in the bidentate model, forming a highly reactive hexacoordinated Ti(IV)-complex. The HNC activated by the quinuclidine tertiary amine moiety of cinchonidine ligand performed a nucleophilic attack towards the activated C=C bond of olefin, generating a cyanide adduct. The catalytic reaction required about 19.9 kcal mol^{-1} lower energy barrier compared to the noncatalytic reaction.

(iii). EDA showed that the steric repulsion between the bulky group (e.g., 9-phenanthryl substituent) at the 3-position in the biphenol ligand and the phenyl group in olefin raised the Pauli energy ($\Delta E^{\neq}_{Pauli}$) of the reacting fragments at the *re*-face attack TS, leading to the predominant *R*-product through the *si*-face attack, as observed in the experiment.

Supplementary Materials: The following are available online at http://www.mdpi.com/2073-4344/10/9/1079/s1, Figure S1: Laplacian ($\nabla^2 \rho$) and electronic density (ρ, in parentheses) of selected bond critical points (BCP) for molecular complex C-I-COM were obtained by AIM analysis, using Multiwfn software. Figure S2: Optimized geometry of transition state b-TS1-1. The intermolecular distance is in Angstroms (Å). The color definitions of atoms are Red = oxygen, blue = nitrogen, gray = carbon, and white = hydrogen. Figure S3: Optimized geometries of low-energy hexacoordinated Ti(IV)-complexes formed by coordinating olefin to metal center in (a) bidentate (b) monodentate fashion along *si*-face attack pathway. The intermolecular distance is in Angstroms (Å). The color definitions of atoms are Red = oxygen, blue = nitrogen, gray = carbon, and white = hydrogen. Figure S4: Optimized geometries of hexacoordinated Ti(IV)-complexes and H-shift transition state in the presence of HOiPr along *si*-face attack pathway. The intermolecular distance is in Angstroms (Å). The color definitions of atoms are Red = oxygen, blue = nitrogen, gray = carbon, and white = hydrogen. Figure S5: Optimized geometries of two competing transition states in C-C bond formation step in d-II and d-III models as well as their relative Gibbs free energies (in kcal mol^{-1}). The intermolecular distance is in Angstroms (Å). The color definitions of atoms are Red = oxygen, blue = nitrogen, gray = carbon, and white = hydrogen. Figure S6: Visualization of the main noncovalent interaction described by contour plots of the reduced density gradient isosurfaces (density cutoff of 0.7 au) for transition states d-I-TS1-*si* and d-I-TS1-*re*. The surface color code is blue for strongly attractive, green for weakly attractive, and red for strongly repulsive interactions.

Author Contributions: Conceptualization, Z.S., C.H. and C.K.K.; methodology, Z.S., C.H. and C.K.K.; software, C.H.; validation, Z.S. and C.K.K.; formal analysis, Z.S. and N.S.; investigation, Z.S.; resources, C.H. and C.K.K.; data curation, N.S.; writing—original draft preparation, Z.S.; writing—review and editing, Z.S., C.H., N.S. and C.K.K.; visualization, Z.S.; supervision, C.H. and C.K.K.; project administration, C.K.K.; funding acquisition, Z.S. and C.K.K. All authors have read and agreed to the published version of the manuscript.

Funding: This research was funded by the Framework of International Cooperation Program managed by the National Research Foundation of Korea (2019K2A9A2A06023069, FY2019) and National Natural Science Foundation of China (21911540465).

Conflicts of Interest: The authors declare no conflict of interest.

References

1. Khan, N.H.; Kureshy, R.I.; Abdi, S.H.R.; Agrawal, S.; Jasra, R.V. Metal catalyzed asymmetric cyanation reactions. *Coord. Chem. Rev.* **2008**, *252*, 593–623. [CrossRef]
2. North, M.; Usanov, D.L.; Young, C. Lewis acid catalyzed asymmetric cyanohydrin synthesis. *Chem. Rev.* **2008**, *108*, 5146–5226. [CrossRef] [PubMed]
3. Wang, J.; Liu, X.H.; Feng, X.M. Asymmetric strecker reactions. *Chem. Rev.* **2011**, *111*, 6947–6983. [CrossRef] [PubMed]
4. Kurono, N.; Ohkuma, T. Catalytic Asymmetric Cyanation Reactions. *ACS Catal.* **2016**, *6*, 989–1023. [CrossRef]
5. Zeng, X.P.; Sun, J.C.; Liu, C.; Ji, C.B.; Peng, Y.Y. Catalytic Asymmetric Cyanation Reactions of Aldehydes and Ketones in Total Synthesis. *Adv. Synth. Catal.* **2019**, *361*, 3281–3305. [CrossRef]
6. Kouznetsov, V.V.; Galvis, C.E.P. Strecker reaction and α-amino nitriles: Recent advances in their chemistry, synthesis, and biological properties. *Tetrahedron* **2018**, *74*, 773–810. [CrossRef]
7. Mazet, C.; Jacobsen, E.N. Dinuclear {(salen)Al} complexes display expanded scope in the conjugate cyanation of α,β-unsaturated imides. *Angew. Chem. Int. Ed.* **2008**, *47*, 1762–1765. [CrossRef]
8. Sammis, G.M.; Danjo, H.; Jacobsen, E.N. Cooperative dual catalysis: Application to the highly enantioselective conjugate cyanation of unsaturated imides. *J. Am. Chem. Soc.* **2004**, *126*, 9928–9929. [CrossRef]
9. Madhavan, N.; Weck, M. Highly Active Polymer-Supported (Salen)Al Catalysts for the Enantioselective Addition of Cyanide to α,β-Unsaturated Imides. *Adv. Synth. Catal.* **2008**, *350*, 419–425. [CrossRef]
10. Mita, T.; Sasaki, K.; Kanai, M.; Shibasaki, M. Catalytic enantioselective conjugate addition of cyanide to α,β-unsaturated N-acylpyrroles. *J. Am. Chem. Soc.* **2005**, *127*, 514–515. [CrossRef]
11. Fujimori, I.; Mita, T.; Maki, K.; Shiro, M.; Sato, A.; Furusho, S.; Kanai, M.; Shibasaki, M. Toward a rational design of the assembly structure of polymetallic asymmetric catalysts: Design, synthesis, and evaluation of new chiral ligands for catalytic asymmetric cyanation reactions. *Tetrahedron* **2007**, *63*, 5820–5831. [CrossRef]

12. Tanaka, Y.; Kanai, M.; Shibasaki, M. A catalytic enantioselective conjugate addition of cyanide to enones. *J. Am. Chem. Soc.* **2008**, *130*, 6072–6073. [CrossRef] [PubMed]
13. Tanaka, Y.; Kanai, M.; Shibasaki, M. Catalytic Enantioselective Construction of β-Quaternary Carbons via a Conjugate Addition of Cyanide to β,β-Disubstituted α,β-Unsaturated Carbonyl Compounds. *J. Am. Chem. Soc.* **2010**, *132*, 8862–8863. [CrossRef] [PubMed]
14. Kurono, N.; Nii, N.; Sakaguchi, Y.; Uemura, M.; Ohkuma, T. Asymmetric hydrocyanation of alpha, beta-unsaturated ketones into beta-cyano ketones with the [Ru(phgly)$_2$(binap)]/C$_6$H$_5$OLi catalyst system. *Angew. Chem. Int. Ed.* **2011**, *50*, 5541–5544. [CrossRef]
15. Sakaguchi, Y.; Kurono, N.; Yamauchi, K.; Ohkuma, T. Asymmetric conjugate hydrocyanation of alpha, beta-unsaturated N-acylpyrroles with the Ru(phgly)$_2$(binap)-CH$_3$OLi catalyst system. *Org. Lett.* **2014**, *16*, 808–811. [CrossRef]
16. Zhang, J.L.; Liu, X.H.; Wang, R. Magnesium complexes as highly effective catalysts for conjugate cyanation of α,β-unsaturated amides and ketones. *Chem. Eur. J.* **2014**, *20*, 4911–4915. [CrossRef]
17. Dong, C.; Song, T.; Bai, X.F.; Cui, Y.M.; Xu, Z.; Xu, L.W. Enantioselective conjugate addition of cyanide to chalcones catalyzed by a magnesium-Py-BINMOL complex. *Catal. Sci. Technol.* **2015**, *5*, 4755–4759. [CrossRef]
18. Hatano, M.; Yamakawa, K.; Ishihara, K. Enantioselective Conjugate Hydrocyanation of α,β-Unsaturated N-Acylpyrroles Catalyzed by Chiral Lithium(I) Phosphoryl Phenoxide. *ACS Catal.* **2017**, *7*, 6686–6690. [CrossRef]
19. Kawai, H.; Okusu, S.; Tokunaga, E.; Sato, H.; Shiro, M.; Shibata, N. Organocatalytic asymmetric synthesis of trifluoromethyl-substituted diarylpyrrolines: Enantioselective conjugate cyanation of beta-aryl-beta-trifluoromethyl-disubstituted enones. *Angew. Chem. Int. Ed.* **2012**, *51*, 4959–4962. [CrossRef]
20. Wang, Y.F.; Zeng, W.; Sohail, M.; Guo, J.; Wu, S.; Chen, F.X. Highly Efficient Asymmetric Conjugate Hydrocyanation of Aromatic Enones by an Anionic Chiral Phosphate Catalyst. *Eur. J. Org. Chem.* **2013**, *2013*, 4624–4633. [CrossRef]
21. Liu, Y.; Shirakawa, S.; Maruoka, K. Phase-Transfer-Catalyzed Asymmetric Conjugate Cyanation of Alkylidenemalonates with KCN in the Presence of a Brønsted Acid Additive. *Org. Lett.* **2013**, *15*, 1230–1233. [CrossRef] [PubMed]
22. Provencher, B.A.; Bartelson, K.J.; Liu, Y.; Foxman, B.M.; Deng, L. Structural study-guided development of versatile phase-transfer catalysts for asymmetric conjugate additions of cyanide. *Angew. Chem. Int. Ed.* **2011**, *50*, 10565–10569. [CrossRef] [PubMed]
23. Jakhar, A.; Sadhukhan, A.; Khan, N.H.; Saravanan, S.; Kureshy, R.I.; Abdi, S.H.R.; Bajaj, H.C. Asymmetric Hydrocyanation of Nitroolefins Catalyzed by an Aluminum(III) Salen Complex. *ChemCatChem* **2014**, *6*, 2656–2661. [CrossRef]
24. Nagata, T.; Tamaki, A.; Kiyokawa, K.; Tsutsumi, R.; Yamanaka, M.; Minakata, S. Enantioselective Electrophilic Cyanation of Boron Enolates: Scope and Mechanistic Studies. *Chem. Eur. J.* **2018**, *24*, 17027–17032. [CrossRef]
25. Jakhar, A.; Ansari, A.; Nandi, S.; Gupta, N.; Khan, N.H.; Kureshy, R.I. Bronsted Basic Ionic Liquid as Catalytic and Reusable Media for Conjugate Cyanation of CF3-Substituted Alkylidenemalonates Using Acetone Cyanohydrin. *ChemistrySelect* **2017**, *2*, 11346–11351. [CrossRef]
26. Wang, J.; Hu, X.L.; Jiang, J.; Gou, S.H.; Huang, X.; Liu, X.H.; Feng, X.M. Asymmetric Activation of Tropos 2,2′-Biphenol with Cinchonine Generates an Effective Catalyst for the Asymmetric Strecker Reaction of N-Tosyl-Protected Aldimines and Ketoimines. *Angew. Chem. Int. Ed.* **2007**, *46*, 8468–8470. [CrossRef]
27. Wang, J.; Wang, W.T.; Li, W.; Hu, X.L.; Shen, K.; Tan, C.; Liu, X.H.; Feng, X.M. Asymmetric cyanation of aldehydes, ketones, aldimines, and ketimines catalyzed by a versatile catalyst generated from cinchona alkaloid, achiral substituted 2,2′-biphenol and tetraisopropyl titanate. *Chem. Eur. J.* **2009**, *15*, 11642–11659. [CrossRef]
28. Wang, J.; Li, W.; Liu, Y.L.; Chu, Y.Y.; Lin, L.L.; Liu, X.H.; Feng, X.M. Asymmetric Cyanation of Activated Olefiins with Ethyl Cyanoformate Catalyzed by a Modular Titanium Catalyst. *Org. Lett.* **2010**, *12*, 1280–1283. [CrossRef]
29. Su, Z.S.; Li, W.Y.; Wang, J.; Hu, C.W.; Feng, X.M. A theoretical investigation on the Strecker reaction catalyzed by a Ti(IV)-complex catalyst generated from a cinchona alkaloid, achiral substituted 2,2′-biphenol, and tetraisopropyl titanate. *Chem. Eur. J.* **2013**, *19*, 1637–1646. [CrossRef]

30. Frisch, J.; Schlegel, H.B.; Scuseria, G.E.; Robb, M.A.; Cheeseman, J.R.; Montgomery, J.A., Jr.; Vreven, T.; Kudin, K.N.; Burant, J.C.; Millam, J.M.; et al. *Pople, Gaussian 09 (Revision D.01)*; Gaussian, Inc.: Wallingford, CT, USA, 2013.
31. Marenich, A.V.; Cramer, C.J.; Truhlar, D.G. Universal solvation model based on solute electron density and on a continuum model of the solvent defined by the bulk dielectric constant and atomic surface tensions. *J. Phys. Chem. B* **2009**, *113*, 6378–6396. [CrossRef]
32. Gonzalez, C.; Schlegel, H.B. An improved algorithm for reaction path following. *J. Chem. Phys.* **1989**, *90*, 2154. [CrossRef]
33. Fernandez, I. Combined activation strain model and energy decomposition analysis methods: A new way to understand pericyclic reactions. *Phys. Chem. Chem. Phys.* **2014**, *16*, 7662–7671. [CrossRef]
34. Zeist, W.J.; Bickelhaupt, F.M. The activation strain model of chemical reactivity. *Org. Biomol. Chem.* **2010**, *8*, 3118–3127. [CrossRef] [PubMed]
35. Fernandez, I.; Bickelhaupt, F.M. The activation strain model and molecular orbital theory: Understanding and designing chemical reactions. *Chem. Soc. Rev.* **2014**, *43*, 4953–4967. [CrossRef] [PubMed]
36. Ess, D.H.; Houk, K.N. Distortion/Interaction Energy Control of 1,3-Dipolar Cycloaddition Reactivity. *J. Am. Chem. Soc.* **2007**, *129*, 10646–10647. [CrossRef]
37. Thomas, B.E.; Loncharich, R.J.; Houk, K.N. Force Field Modeling of Transition Structures of Intramolecular Ene Reactions and ab Initio Transition Structures for an Activated Enophile. *J. Org. Chem.* **1992**, *57*, 1354–1362. [CrossRef]
38. Hong, X.; Liang, Y.; Griffith, A.K.; Lambert, T.H.; Houk, K.N. Distortion-accelerated cycloadditions and strain-release-promoted cycloreversions in the organocatalytic carbonyl-olefin metathesis. *Chem. Sci.* **2014**, *5*, 471–475. [CrossRef]
39. Houk, K.N.; Beno, B.R.; Nendel, M.; Black, K.; Yoo, H.Y.; Wilsey, S.; Lee, J.K. Exploration of pericyclic reaction transition structures by quantum mechanical methods: Competing concerted and stepwise mechanisms. *J. Mol. Struct. Theochem.* **1997**, *398*, 169–179. [CrossRef]
40. Hopffgarten, M.; Frenking, G. Energy decomposition analysis. *WIREs Comput. Mol. Sci.* **2012**, *2*, 43–62. [CrossRef]
41. Baerends, J.; Autschbach, J.; Bashford, D.; Bérces, A.; Bickelhaupt, F.M.; Bo, C.; Boerrigter, P.M.; Cavallo, L.; Chong, D.P.; Deng, L.; et al. *ADF2016, Theoretical Chemistry*; Vrije Universiteit: Amsterdam, The Netherlands, 2016. Available online: http://www.scm.com (accessed on 10 July 2020).
42. Grimme, S.; Antony, J.; Ehrlich, S.; Krieg, H. A consistent and accurate ab initio parametrization of density functional dispersion correction (DFT-D) for the 94 elements H-Pu. *J. Chem. Phys.* **2010**, *132*, 154104. [CrossRef]
43. Grimme, S.; Ehrlich, S.; Goerigk, L. Effect of the damping function in dispersion corrected density functional theory. *J. Comput. Chem.* **2011**, *32*, 1456–1465. [CrossRef] [PubMed]
44. Sammis, G.M.; Jacobsen, E.N. Highly Enantioselective, Catalytic Conjugate Addition of Cyanide to α,β-Unsaturated Imides. *J. Am. Chem. Soc.* **2003**, *125*, 4442–4443. [CrossRef] [PubMed]
45. Falivene, L.; Cao, Z.; Petta, A.; Serra, L.; Poater, A.; Oliva, R.; Scarano, V.; Cavallo, L. Towards the online computer-aided design of catalytic pockets. *Nat. Chem.* **2019**, *11*, 872–879. [CrossRef] [PubMed]
46. Poater, A.; Ragone, F.; Giudice, S.; Costabile, C.; Dorta, R.; Nolan, S.P.; Cavallo, L. Thermodynamics of N-Heterocyclic Carbene Dimerization: The Balance of Sterics and Electronics. *J. Am. Chem. Soc.* **2008**, *27*, 2679–2681. [CrossRef]
47. Poater, A.; Ragone, F.; Mariz, R.; Dorta, R.; Cavallo, L. Comparing the enantioselective power of steric and electrostatic effects in transition-metal-catalyzed asymmetric synthesis. *Chem. Eur. J.* **2010**, *16*, 14348–14353. [CrossRef]
48. Lu, T.; Chen, F.W. Multiwfn: A multifunctional wavefunction analyzer. *J. Comput. Chem.* **2012**, *33*, 580–592. [CrossRef]

© 2020 by the authors. Licensee MDPI, Basel, Switzerland. This article is an open access article distributed under the terms and conditions of the Creative Commons Attribution (CC BY) license (http://creativecommons.org/licenses/by/4.0/).

Article

In Silico Acetylene [2+2+2] Cycloadditions Catalyzed by Rh/Cr Indenyl Fragments

Shah Masood Ahmad, Marco Dalla Tiezza and Laura Orian *

Dipartimento di Scienze Chimiche Universita' degli Studi di Padova, Via Marzolo 1, 35131 Padova, Italy
* Correspondence: laura.orian@unipd.it; Tel.: +39-0-498-275-140

Received: 20 July 2019; Accepted: 8 August 2019; Published: 9 August 2019

Abstract: Metal-catalyzed alkyne [2+2+2] cycloadditions provide a variety of substantial aromatic compounds of interest in the chemical and pharmaceutical industries. Herein, the mechanistic aspects of the acetylene [2+2+2] cycloaddition mediated by bimetallic half-sandwich catalysts [Cr(CO)$_3$IndRh] (Ind = (C$_9$H$_7$)$^-$, indenyl anion) are investigated. A detailed exploration of the potential energy surfaces (PESs) was carried out to identify the intermediates and transition states, using a relativistic density functional theory (DFT) approach. For comparison, monometallic parent systems, i.e., CpRh (Cp = (C$_5$H$_5$)$^-$, cyclopentadienyl anion) and IndRh, were included in the analysis. The active center is the rhodium nucleus, where the [2+2+2] cycloaddition occurs. The coordination of the Cr(CO)$_3$ group, which may be in syn or anti conformation, affects the energetics of the catalytic cycle as well as the mechanism. The reaction and activation energies and the turnover frequency (TOF) of the catalytic cycles are rationalized, and, in agreement with the experimental findings, our computational analysis reveals that the presence of the second metal favors the catalysis.

Keywords: acetylene [2+2+2] cycloadditions; DFT calculations; rhodium; chromium; half-sandwich catalysts; turnover frequency (TOF); activation strain analysis; indenyl effect; metal slippage; slippage span model

---

1. Introduction

The [2+2+2] cycloadditions of small unsaturated molecules, such as alkynes and nitriles, afford a variety of aromatic, heterocyclic and polycyclic compounds of paramount importance in the chemical and pharmaceutical industries [1]. In 1867, the synthesis of benzene by thermal cyclotrimerization of acetylene was reported for the first time [2]. Despite the reaction being highly exothermic, it is strongly disfavored by entropic factors, and this limits its synthetic utility. Reppe and Schweckendieck demonstrated that low valent metal nuclei catalyze these [2+2+2] cycloadditions [1]. Further studies assessed that a wide variety of metals such as Ti, Zr, Ru, Co, Rh, Ir, Ni, and Pd may play an important role as catalysts in the synthesis of benzene and its derivatives and, in the past decades, experimental results and theoretical insight have been reported to establish the correct mechanism and tune the efficiency and regioselectivity [3–7].

One important class of catalysts for alkyne [2+2+2] cycloadditions are the half-sandwich complexes, i.e., metal-cyclopentadienyl (CpM) or metal-ligand complexes in which the metal is coordinated to the Cp moiety of larger polycyclic ligands. These compounds possess peculiar structural features and reactivity properties; for this reason, and for their synthetic versatility, they are largely used [8–13]. They are denoted here with the general formula Cp'ML$_n$, where Cp' = Cp, Ind, and L$_n$ are the ancillary ligands coordinated to the metal center (M), so that the 18-electrons rule is satisfied. Group 9 metals, i.e., cobalt, rhodium, and iridium, have been largely employed and, particularly, cobalt and rhodium have revealed significant catalytic efficiency [14–17]. The bonding mode of the metal to the Cp moiety (hapticity) is not perfectly symmetric (η^5), but typically exhibits a distortion toward allylic (η^3)

coordination [18,19], and eventually to an extreme structure in which a σ metal-carbon bond forms (η^1). This phenomenon is called metal slippage, and the different bonding modes are shown in Scheme 1a.

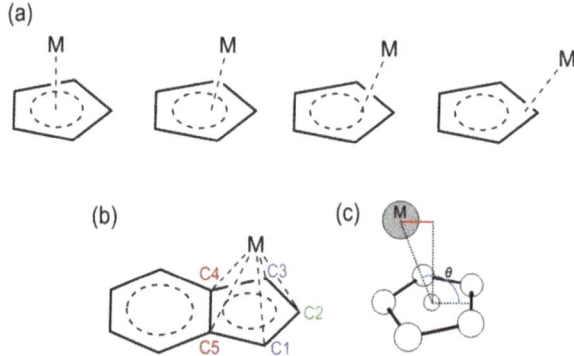

Scheme 1. (a) Different coordination modes (hapticities) of a metal (M) to a Cp ring. (b) Labelling scheme used in the definition of Δ and LISP (Equations (1) and (2)). (c). Definition of θ angle for LISP calculation.

To quantify the slippage, Basolo and coworkers [20,21] introduced the geometrical parameter Δ (Equation (1)).

$$\Delta = \frac{(\text{M-C4} + \text{M-C5}) - (\text{M-C1} + \text{M-C3})}{2} \quad (1)$$

As shown in Scheme 1b, M-C4 and M-C5 are the longest distances between M and two adjacent C atoms of the Cp ring, and M-C3 and M-C1 are the distances between M and the C atoms adjacent to C4 and C5, respectively.

The slippage variations occurring during the catalytic cycle can be quantified referring to the value of Δ, which changes from 0 Å (η^5) to nearly 0.3 Å (η^3) till 0.6 Å or even higher values (η^1). Some of us have recently pointed out that this definition suitably applies to symmetric systems [22], and thus have introduced another descriptor, i.e., the label independent slippage parameter (LISP) (Equation (2)).

$$\text{LISP}(\text{Å}) = \frac{d}{N} \sum_{i=1}^{N} \left| \sin\left(\theta_i - \frac{\pi}{2}\right) \right| \quad (2)$$

LISP is actually the sum of the five average minimum distances from a normal vector passing through the centroid and the metal; d is the distance between the metal and the centroid of the ring of N atoms (Scheme 1c). Importantly, LISP is also suitable for describing non-symmetric displacements. In the same paper [22], a relationship between the catalytic activity of several half-sandwich group 9 metal complexes for alkyne [2+2+2] cycloadditions and the slippage span expressed in terms of LISP (computed as the difference between the maximum and the minimum value of LISP along the catalytic cycle) was established. The slippage span ΔLISP was then related to the turn-over frequency (TOF) values, calculated with the energy span model [23], and it emerged that the lower the ΔLISP is, the higher the TOF is. Finally, in order to improve the sensitivity of ΔLISP, ΔLISP* was introduced:

$$\Delta \text{LISP}^* (\text{Å}) = \sum_{i=1}^{N-1} |\text{LISP}_1 - \text{LISP}_{i+1}| + \sum_{i=1}^{N-1} |\text{LISP}_i - \text{LISP}_{i+1}| + |\text{LISP}_N - \text{LISP}_1| \quad (3)$$

In this descriptor (Equation (3)), the first term indicates how far/close each intermediate/transition state of the catalytic cycle is from/to the starting point. The second term accounts for the slippage difference between two consecutive states along the whole catalytic cycle. The last term includes the slippage variation between the last intermediate and the initial state. The availability of a flexible parameter to quantify the metal slippage, which intuitively influences the catalytic activity, and of a relationship between the metal slippage and the turn-over frequency is a valuable tool. So far, the slippage span model has been applied to monometallic Co and Rh half-sandwich catalysts for alkyne [2+2+2] cycloadditions [22]; in this work, we apply it to bimetallic Rh/Cr indenyl catalysts to assess its general validity.

Different strategies have been developed to tune the regioselectivity and the efficiency of the half-sandwich catalysts. Ingrosso et al. [24,25] experimentally studied the influence of organic moieties, i.e., cyclopentadienyl (Cp), indenyl (Ind) and fluorenyl (FN), in Rh(I) half-sandwich catalysts on alkyne [2+2+2] cycloadditions. Booth et al. [26] reported that IndRh initiates the reaction ten times faster as compared to CpRh, and this was related to the so-called *indenyl effect* (a phenomenon firstly reported by Adam J. Hart-Davis and Roger J. Mawby in 1969 [27], it was thoroughly explored and named by Fred Basolo [28]; it consists in an enhancement of the rate of the substitution reactions at the metal when indenyl is used instead of cyclopentadienyl aromatic ligand). It was also observed that the presence of an electron withdrawing group in the Cp ligand reduces the catalytic activity at low temperature [29,30]. Based on these studies, it emerged that structural and electronic modifications to the aromatic moiety of the half-sandwich catalyst influence its efficiency.

Changing the metal also plays a role. For example, considering group 9 metals, Co is highly preferred when compared to Rh and Ir [31,32].

Finally, another modification to a half-sandwich catalyst is the coordination of a second metal to form a bimetallic complex [1] when the aromatic ligand is polycyclic; the second metal can be in syn or anti position. A very nice example, reported by Ceccon et al. [33–35], is [Cr(CO)$_3$IndRh]L$_2$ (Scheme 2). The idea behind the design of these compounds was that the presence of two metal centers within the same molecule may profoundly affect both the physical properties and the reactivity of the catalyst. In fact, it was found that the presence of a second metal in the anti position strongly enhances the reactivity as compared to monometallic complexes (*extra-indenyl effect*) [36].

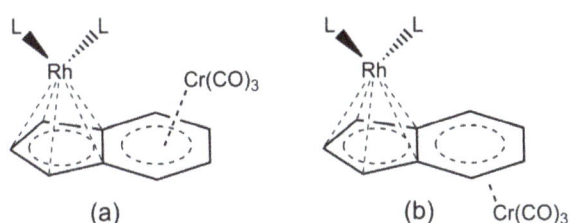

Scheme 2. Anti-(a) and syn-(b) [Cr(CO)$_3$IndRhL$_2$].

Nowadays, DFT computational methodologies make it possible to investigate the mechanistic details of catalytic reactions, defining with accuracy their thermodynamic (reaction energies) as well as their kinetic (activation energies) features. After the pioneering work by Albright and co-workers on CpCo catalyzed acetylene [2+2+2] cycloaddition to benzene [37], several important computational studies were carried out by Calhorda and Kirchner on CpRuCl [38–40], Orian and Bickelhaupt on CpRh and IndRh [41–43] and analogous heteroaromatic Rh(I) catalysts [44], Koga and co-workers on CpCo [31,45,46], and Hapke et al. on CpIr [45]. To the best of our knowledge, no theoretical mechanistic investigation on the use of bimetallic half-sandwich catalysts has been reported so far. The use of bimetallic complexes as catalysts in organic synthesis is interesting, because the reaction rate and selectivity can be tuned via possible inter-metal cooperative effects. Ceccon et al. have provided rather complete information on anti-and syn-[Cr(CO)$_3$IndRhL$_2$]. Particularly, they studied the cyclotrimerization of methylpropriolate (MP) and dimethyl-acetylenecarboxylate (DMAD) with mono and bimetallic catalysts, i.e., IndRh(COD), p-NO$_2$-IndRh(COD) and bimetallic anti-[Cr(CO)$_3$IndRh(COD)]. They found that the Rh/Cr catalyst leads to a greatly enhanced catalytic efficiency compared to the monometallic one. This increase of catalytic activity was ascribed to a synergic or "cooperative" interaction between the two metals in activating the substrate of interest [33–35].

The main goal of this study is the detailed investigation of the mechanism of acetylene [2+2+2] cycloaddition catalyzed by Rh/Cr indenyl fragments, particularly focusing on (i) the presence of the second metal, i.e., Cr, on the mechanism and energetics; (ii) the relationship between rhodium slippage and catalytic activity, and (iii) the outline of general guidelines for the design of Rh(I) half-sandwich catalysts based on the slippage span model. Points (i) and (ii) will be presented in the Results section, while point (iii) will be treated in the Discussion.

2. Results

The computational mechanistic investigation of acetylene [2+2+2] cycloaddition to benzene catalyzed by the monometallic catalysts CpRh and IndRh was first reported in 2006 [41]. In Scheme 3a (Path I), the well-known and widely accepted mechanism, proposed by Albright for CpCo catalysis [47,48] is shown. The catalytic cycle begins with the replacement of the ancillary ligands L of the catalyst precursor, i.e., Cp-or Ind-RhL$_2$ (L = CO, PPh$_3$ or COD (1,5-cyclooctadiene)) by two molecules of acetylene, leading to the bis-acetylene complex **Z-1**. The coordinated acetylene molecules undergo oxidative coupling and the unsaturated 16-electrons rhodacycle **Z-2** forms. This elementary step typically has the highest activation energy and was recently discussed in detail for group 9 metal-Cp fragments [49]. The subsequent coordination of a third acetylene molecule occurs without an appreciable activation energy and leading to **Z-3**, and, after its addition to the π-electron system of the rhodacycle, the intermediate **Z-4** is obtained, which is characterized by an unsaturated bent six-membered ring. By further stepwise addition of two acetylene molecules, the intermediate **Z-5** first forms and then the initial catalyst is regenerated with the cleavage of benzene.

An alternative mechanism was postulated by Booth and co-workers [26] on the basis of their experimental findings, which implies that a ligand of the catalyst precursor remains bonded to the metal center throughout the whole catalytic cycle; this is shown in Scheme 3b (Path II). This mechanism was rationalized by Orian et al. [43] in a recent systematic study on the *indenyl effect* and its connection to metal slippage. The presence of an ancillary ligand imposes strong hapticity variations in both CpRh and IndRh catalysis. In addition, the bicyclic **CO-Z-b**, and the heptacyclic **CO-Z-h** were located along this catalytic path, which resembles the one described for CpRuCl catalysis [38].

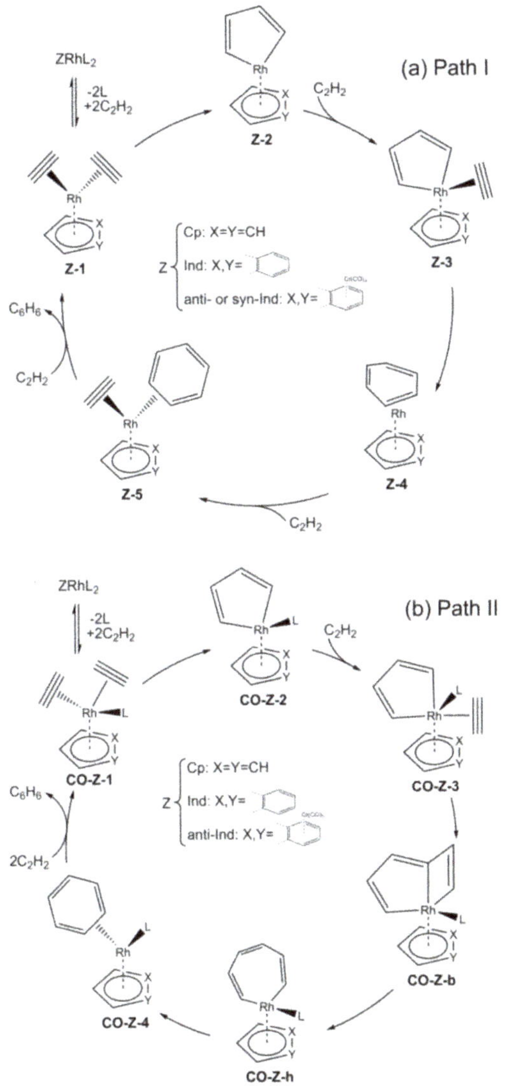

Scheme 3. Mechanism of acetylene [2+2+2] cycloaddition to benzene catalyzed by ZRh (Z = Cp, Ind and anti or syn-Ind) (Path I) (**a**) and by CO-ZRh fragments (Z = Cp, Ind and anti-Ind; L = CO) (Path II) (**b**).

Inspired by the experimental work by Ceccon et al. [33–35], we chose the bimetallic catalyst [Cr(CO)$_3$ IndRhL$_2$], where the second metal group Cr(CO)$_3$ can be coordinated both in anti and syn conformations and examined both paths I and II of Scheme 3. The optimized molecular structures (ZORA-BLYP/TZ2P) of anti- and syn-[Cr(CO)$_3$IndRh(CO)$_2$] (Figure S1) are in good agreement with the X-ray crystallographic structures (labelled as HEXPOP [50] and HAPPOD [34] in the Cambridge database (CSD) [51]). Significant geometry parameters are compared in Table S1.

For simplicity, in the ongoing discussion, anti-and syn-[Cr(CO)$_3$IndRh] fragments are abbreviated as anti-or syn-IndRh (Scheme 3).

2.1. Acetylene [2+2+2] Cycloaddition Catalyzed by Anti-[Cr(CO)₃IndRh] Fragment: Reaction Mechanism and PES (Path I)

The intermediates and transition states found for acetylene [2+2+2] cycloaddition catalyzed by the bimetallic anti-IndRh along Path I (Scheme 3a), are shown in Figure 1. Those found on the PESs of the parent monometallic catalysts, i.e., CpRh and IndRh, are in Figures S2 and S3, respectively. The computed energy profile for anti-RhInd is shown in Figure 2a.

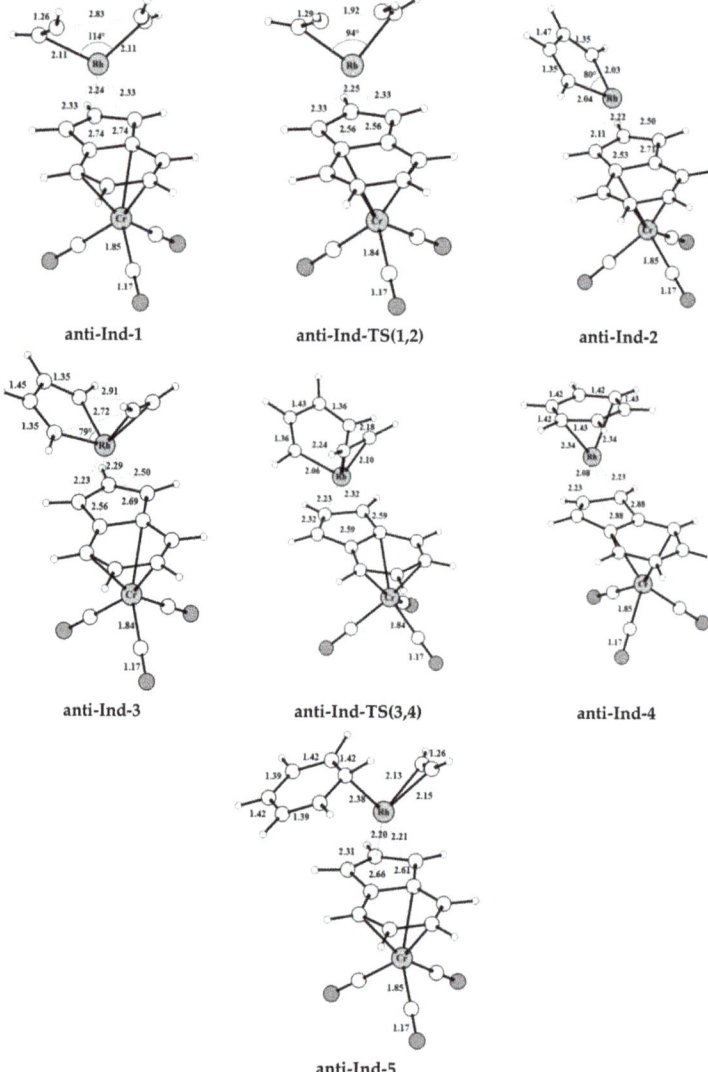

Figure 1. Optimized structures with selected interatomic distances (Å) and angles (deg) of the intermediates and transition states located on the PES of the anti-IndRh catalyzed acetylene [2+2+2] cycloaddition to benzene (Path I, Scheme 2a). Level of theory: ZORA-BLYP/TZ2P.

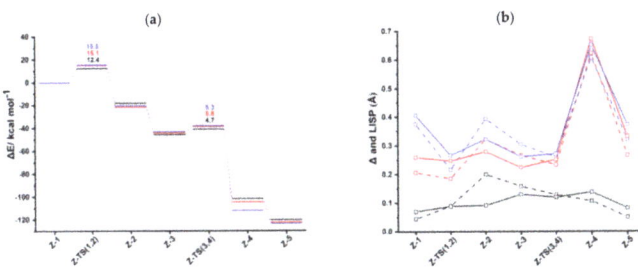

Figure 2. (a) Energy profiles of acetylene [2+2+2] cycloaddition to benzene catalyzed by CpRh (black), IndRh (red), and bimetallic anti-IndRh (blue) (Scheme 3a, Path I). (b) Profiles of the slippage parameters Δ (dashed line) and LISP (solid line) for the acetylene [2+2+2] cycloaddition cycles catalyzed by CpRh (black), IndRh (red), and bimetallic anti-IndRh (blue) along Path I (Scheme 3a). Level of theory: ZORA-BLYP/TZ2P.

The catalytic cycle mediated by anti-IndRh is very similar to the cycle described for the monometallic parent catalysts CpRh and IndRh [41] since the Cr(CO)$_3$ group is coordinated in anti, and thus there is no steric effect. It begins with replacement of the ancillary ligands L by two acetylene molecules to form a bis-acetylene complex labelled **anti-Ind-1** (Scheme 3a). This process usually occurs experimentally by thermal or photochemical activation and might be dissociative or associative depending on the nature of the metal, on the electrophilicity of the ligands and on the substituents on the Cp ring [21,28,52]. In **anti-Ind-1**, the acetylene molecules are slanted with respect to the plane of the indenyl ring and the C-C bond length is 1.26 Å. The Rh-C$_\alpha$ and Rh-C$_\beta$ bond lengths are 2.11 Å and 2.13 Å, respectively; they are shorter as compared to those of **Ind-1** (2.13 Å and 2.16 Å). This suggests that acetylene is more tightly bonded, likely due to the electron withdrawing effect of the second metal group Cr(CO)$_3$. Additionally, the Rh-Cp coordination is more distorted in **anti-Ind-1** than in the parent **Cp-1** and **Ind-1**, as also quantified by the metal slippage parameters Δ and LISP, which were calculated for the intermediates and the transition states along the whole catalytic cycle (Figure 2b).

By inspecting the frontier molecular orbitals of **Cp-1**, **Ind-1** and **anti-Ind-1** shown in Figure 3, the π-antibonding character between Cp'-π system and valence d orbitals of Rh is found to increase in the order **Cp-1** < **Ind-1** < **anti-Ind-1**, leading to a corresponding increase of metal slippage. Calhorda et al. have reported the same observation in their pioneering work on the nature of indenyl effect, which was related to the nodal characteristics of Cp'-π orbitals of CpRh and IndRh [19,44]. Herein, it is found that the metal-π anti-bonding nature with Ind-π system is further enhanced in the presence of Cr(CO)$_3$ in **anti-Ind-1** compared to the parent **Ind-1**, leading to extra slippage of rhodium in the former.

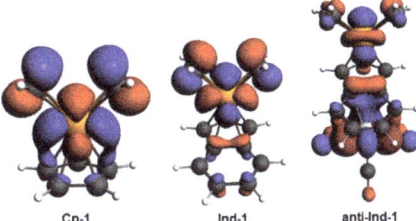

Figure 3. Kohn-Sham HOMOs of **Cp-1**, **Ind-1**, and **anti-Ind-1**; level of theory: ZORA-BLYP/TZ2P. The isodensity value is 0.03.

The oxidative coupling of the two coordinated acetylene molecules in **anti-Ind-1** leads to the 16-electrons unsaturated rhodacyclic intermediate **anti-Ind-2** (Figure 2), crossing an activation barrier of 15.5 kcal mol^{-1}, which is higher than those computed for the formation of **Ind-2**

(15.1 kcal mol^{-1}) and **Cp-2** (12.4 kcal mol^{-1}). This reaction step is exothermic by 20.6 kcal mol^{-1}. These metallacycles are generally described by two resonating structures, i.e., a metallacyclopentadiene, as found for CpCo-(C$_4$H$_4$) [37,49] and CpRh-(C$_4$H$_4$) [41,49], and a metallacyclopentatriene, as for CpRuCl-(C$_4$H$_4$) [38]. In **anti-Ind-2**, the C$_\alpha$-C$_\beta$ and C$_\beta$-C$_\beta'$ distances are 1.35 Å and 1.47 Å, respectively, which are rather well matched to the length of the ethylene double bond and to the length of the σ-bond between two carbon atoms, respectively. This suggests the character of rhodacyclopentadiene of **anti-Ind-2**.

A third acetylene molecule easily coordinates to **anti-Ind-2**, which is converted into **anti-Ind-3**. This step is barrierless and exothermic by 22.8 kcal mol^{-1}, about 5 kcal mol^{-1} less exothermic than the formation of **CpRh-3**; this value is almost identical to the reaction energy computed for the formation of **IndRh-3**. Subsequently, by Diels-Alder-like [4+2] addition of the coordinated acetylene to the rhodacycle, **anti-Ind-4** forms, with an activation energy of 5.3 kcal mol^{-1} (Figure 2a). This step is strongly exothermic by 68.6 kcal mol^{-1}, 8.4 kcal mol^{-1} more exothermic than the same step in the IndRh catalytic cycle; conversely, the energy barriers are very similar. Structurally, **anti-Ind-4** (Figure 1) is characterized by the presence of a six-carbon arene ring coordinated to rhodium in η^6 fashion, while η^3 coordination is found in the Cp-Rh moiety. A similar bonding mode of rhodium is found in **Ind-4**; conversely, in **Cp-4**, η^4 hapticity is observed with the arene ring and $\eta^3 + \eta^2$ coordination to Cp moiety. This pronounced slippage in **Ind-4** and **anti-Ind-4** explains the spikes in the profiles of Δ and LISP (Figure 2b). The coordination of another acetylene leads to the formation of **anti-Ind-5** (Figure 2a), accompanied by a variation of hapticity from η^3 to $\eta^3 + \eta^2$ in the anti-IndRh fragment and by the release of 11.5 kcal mol^{-1}. The cleavage of benzene by the incoming second acetylene completes the cycle with the regeneration of the catalyst. The released energy is 16.4 kcal mol^{-1}. Summarizing, the $\eta^3 + \eta^2$ coordination is found along the whole cycle catalyzed by **anti-IndRh** except in intermediate **anti-Ind-4**, where η^3 coordination is predicted.

Consistently with CpRh and IndRh catalysis, the first step **Z-1 → Z-2**, that is the oxidative coupling of the acetylene molecules, has the highest energy barrier along the cycle (Figure 2a). To gain insight on the origin of this barrier, an activation strain analysis (ASA) has been carried out and compared to those already reported for CpRh and IndRh [49]. For this purpose, the complexes were divided into two fragments, i.e., Cp'Rh (Cp' = Cp, Ind, and anti-[Cr(CO)$_3$Ind]) and the C$_4$H$_4$ moiety. Being an intramolecular reaction, the activation energy $\Delta E^\ddagger$ is conveneintly given as the change, upon going from the reactant to the TS, in strain within the two fragments plus the change, upon going from the reactant to the TS, in the interaction between these two fragments [53].

$$\Delta E^\ddagger = \Delta\Delta E_{strain} + \Delta\Delta E_{int} \qquad (4)$$

The results are shown in Table 1. The $\Delta\Delta E_{strain}$ contributions increase from CpRh to IndRh and to anti-IndRh and are very similar for the Cp'Rh fragments (ranging from 2.4 to 2.7 kcal mol^{-1}), but increase significantly for the C$_4$H$_4$ fragment going from 33.2 to 37.0 and 37.3 kcal mol^{-1}, respectively. Since $\Delta\Delta E_{int}$ are very similar, varying from -23.2 kcal mol^{-1} (CpRh) to −24.6 kcal mol^{-1} (IndRh) and −24.5 kcal mol^{-1} (anti-IndRh), the increase of the barrier in the indenyl catalysts is mainly due to the strain effects localized on the bis-acetylene moiety. Based on the identical $\Delta\Delta E_{int}$ for IndRh and anti-IndRh, no influence of the second metal of the latter is found in the barrier of this oxidative coupling.

Table 1. Activation strain analysis for the oxidative coupling **Z-1 → Z-2** (Path I); all values are in kcal mol^{-1}. The fragments are Cp'Rh (Cp' = Cp, Ind, and anti-[Cr(CO)$_3$Ind]) and the C$_4$H$_4$ moiety (the reference are two acetylene molecules).

	$\Delta\Delta E_{strain}$			$\Delta\Delta E_{int}$	$\Delta E^{\neq}$
	2(C$_2$H$_2$)	Cp'Rh	Total		
Cp-1/Cp-TS (1,2)	33.21	2.36	35.57	−23.16	12.41
Ind-1/Ind-TS (1,2)	37.02	2.66	39.68	−24.59	15.09
anti-Ind-1/anti-Ind-TS (1,2)	37.31	2.72	40.03	−24.52	15.51
syn-Ind-1/syn-Ind-TS (1,2)	39.67	46.98	86.65	−71.47	15.18

We also examined **Z-4**, which was found to be more stable in the case of anti-IndRh than in the cases of IndRh and CpRh. ASA was carried out on **Z-4**, by considering these two fragments: Cp'Rh, where (Cp' = Cp, Ind, and anti-[Cr(CO)$_3$Ind]) and the C$_6$H$_6$ moiety; the results are listed in Table 2. A very high total ΔE_{strain} is found in **Cp-4** compared to **Ind-4** and **anti-Ind-4**, which comes out from the benzene fragment and reflects the structural differences. In fact, benzene is bent in **Cp-4** with η^4 coordination (Figure S2), while in **Ind-4** and **anti-Ind-4**, the ring is almost planar and coordinated to rhodium in η^6 fashion and η^3 coordination is found for the Rh coordination to the Cp ring. On the other hand, the large ΔE_{strain} of **Cp-4** is well balanced by a large negative ΔE_{int}, which leads to larger ΔE by about 12.2 kcal/mol than **Ind-4** and **anti-Ind-4**.

Table 2. Activation strain analysis for **Cp-4**, **Ind-4**, and **anti-Ind-4**; all values are in kcal mol^{-1}. The fragments are Cp'Rh (Cp' = Cp, Ind, and anti-[Cr(CO)$_3$Ind]) and C$_6$H$_6$ moiety (the reference is benzene).

	ΔE_{strain}			ΔE_{int}	ΔE
	C$_6$H$_6$	Cp'Rh	Total		
Cp-4	39.14	4.95	44.09	−83.44	−39.35
Ind-4	1.86	1.45	3.31	−30.44	−27.13
anti-Ind-4	1.64	1.26	2.90	−31.60	−28.70

Although the catalytic center is rhodium, we also investigated the coordination of Cr(CO)$_3$ in all the intermediates and transition states of the anti-IndRh catalyzed cycle. ASA was carried out using as fragments Cr(CO)$_3$ and IndRhX$_i$ (Table 3). Significantly larger ΔE_{strain} values are found for **anti-Ind-TS (1,2)** and **anti-Ind-TS (3,4)**; the biggest contribution comes from the IndRhL$_n$ fragment, reflecting the changes occurring at the Rh center. The ΔE_{strain} of Cr(CO)$_3$ fragment remains almost equal along the path, reflecting the fact that no important structural changes occur within the Cr(CO)$_3$ moiety. ΔE_{int} values fluctuate in the range~47–54 kcal mol^{-1}. In **anti-Ind-4**, a sudden increase of ΔE_{int} is computed. In fact, on the opposite side of the indenyl ligand, Rh-Cp coordination is highly slipped (Figure 2). Therefore, in order to compensate for the weakening of Rh-Cp coordination, Cr(CO)$_3$ binds more tightly the benzene moiety of the aromatic ligand. In **anti-Ind-4**, the increased ΔE_{int} is due to a larger electrostatic contribution which is not sufficiently counterbalanced by an increase of ΔE_{Pauli}.

Table 3. Activation strain analysis (ASA) of all the intermediates and transition states of the anti-IndRh catalyzed cycle (Path I); all the values are in kcal mol^{-1}. The fragments are Cr(CO)$_3$ and IndRhX$_i$.

	ΔE_{strain}			ΔV_{elstat}	ΔE_{Pauli}	ΔE_{oi}	ΔE_{int}	ΔE
	Cr(CO)$_3$	IndRhX$_i$	Total					
anti-Ind-1	1.23	2.29	3.52	−76.47	129.17	−101.20	−48.50	−44.98
anti-Ind-TS(1,2)	1.26	17.17	18.43	−73.37	124.90	−99.39	−47.86	−29.43
anti-Ind-2	1.67	1.78	3.45	−73.03	130.93	−105.13	−47.24	−43.79
anti-Ind-3	1.36	1.85	3.21	−73.99	129.02	−102.09	−47.06	−43.85
anti-Ind-TS(3,4)	1.28	7.77	9.05	−73.38	125.83	−100.03	−47.58	−38.53
anti-Ind-4	1.23	1.17	2.40	−80.45	128.03	−101.74	−54.16	−51.76
anti-Ind-5	1.25	1.65	2.90	−73.39	123.39	−97.75	−47.75	−44.85

2.2. Acetylene [2+2+2] Cycloaddition Catalyzed by Syn-[Cr(CO)$_3$IndRh] Fragment: Reaction Mechanism and PES (Path I)

As stated above, the coordination of the Cr(CO)$_3$ group to the benzene moiety of IndRh catalyst may occur in anti or syn conformation [54]. Ceccon and co-workers have illuminated the stereochemistry of the syn catalyst, i.e., syn-[Cr(CO)$_3$IndRh], and found that it is kinetically more stable because of a Rh-Cr interaction.

First, we investigated the acetylene [2+2+2] cycloaddition to benzene catalyzed by syn-IndRh catalyst along Path I (Scheme 3a). The structures of the intermediates and transition states found on the PES with their relevant parameters are shown in Figure S4, while the computed energy profile is in Figure 4a. Like for anti-IndRh catalysis, at the beginning of the cycle, the ancillary ligands L are replaced by two molecules of acetylene, leading to the formation of the bis-acetylene complex **syn-Ind-1** (Figure S4). In **syn-Ind-1**, the Rh-C$_\alpha$ bond length is 2.15 Å, the distance C$_\beta$-C$_\beta$' of two acetylenes coordinated to rhodium is 2.91 Å, and the angle C$_\alpha$-Rh-C$_\alpha$' is 91°. To quantify the Rh-Cp slippage along the catalytic cycle, Δ and LISP parameters were computed for all the intermediates and transition states and are shown in Figure 4b. In **syn-Ind-1**, their values are 0.40 Å and 0.37 Å, respectively, with no appreciable difference with respect to **anti-Ind-1**; this might be ascribed to Rh-Cr interaction, despite high steric effects are present.

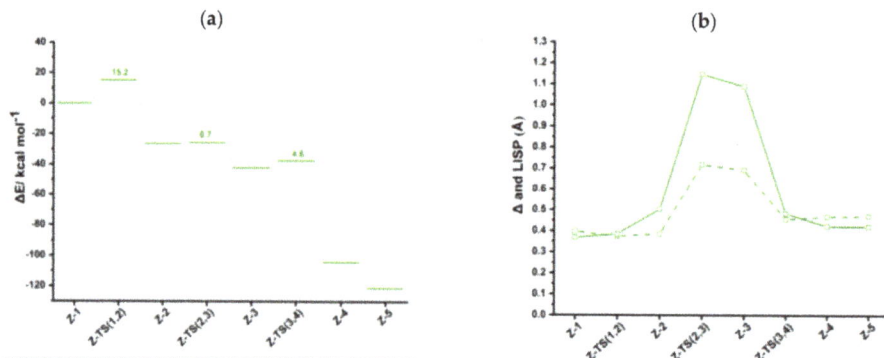

Figure 4. (a) Energy profile of acetylene [2+2+2] cycloaddition to benzene catalyzed by syn-IndRh (Scheme 3a, Path I). (b) Profiles of the slippage parameters (dashed line) and LISP (solid line) along the acetylene [2+2+2] cycloaddition cycle catalyzed by syn-IndRh (Scheme 3a, Path I). Level of theory: ZORA-BLYP/TZ2P.

The oxidative coupling in **syn-Ind-1** leads to the formation of the five-membered ring rhodacycle **syn-Ind-2** with an activation energy of 15.2 kcal mol^{-1}, nearly the same as found for anti-IndRh. Notably, in **syn-Ind-2** complex, one CO of Cr(CO)$_3$ interacts with Rh. In fact, the distance between the CO ligand and rhodium is 2.30 Å, and the distance between Cr and Rh center is smaller by 0.34 Å than in **syn-Ind-1**, implying stabilization of **syn-Ind-2**. In this situation, the rhodacycle **syn-Ind-2** is a nearly 18-electron saturated complex, in contrast to **anti-Ind-2** and the parent **Ind-2** and **Cp-2**. The conversion of **syn-Ind-1** into **syn-Ind-2** is exothermic by 26.4 kcal mol^{-1}, about 6 kcal mol^{-1} more than in the anti-catalyst and in the parent IndRh (Figure 3a). Finally, in this step, an hapticity shift from $\eta^3+\eta^2$ to η^3 occurs (Figure 4b).

The coordination of the third acetylene to **syn-Ind-2** occurs from the upper side with a low barrier of 0.7 kcal mol^{-1}, via **syn-Ind-TS (2,3)** (Figure S4) and a large slippage variation (LISP changes by approximately 0.6 Å) leading to almost η^1 coordination. The conversion **syn-Ind-2** → **syn-Ind-3** occurs only with a slight modification in the carbon-carbon bonds of the rhodacycle. The reaction is exothermic by 15.6 kcal mol^{-1}. In the next step, **syn-Ind-4** forms, crossing a barrier of 4.6 kcal mol^{-1}. The LISP value drops from 1.09 Å to 0.42 Å (Figure 4b). **syn-Ind-4** has a six-carbon arene ring coordinated to rhodium in η^4 fashion, while in Rh-Cp the coordination is $\eta^3 + \eta^2$. Thus, the bonding mode of rhodium is different from **anti-Ind-4** (Figure 2), where η^6 coordination and η^3 coordination are found between the six-carbon arene ring and rhodium and the Cp ring and rhodium, respectively. However, **syn-Ind-4** resembles the case of the parent **Cp-4** [41]. This step is exothermic by 62.3 kcal mol^{-1}.

Finally, the coordination of another acetylene leads to the formation of **syn-Ind-5** with no appreciable activation energy and with the release of 16.8 kcal mol^{-1} (Figure 4a). It is less exothermic than the formation of **Ind-5** but more exothermic by 5 kcal/mol than the formation of **anti-Ind-5**. The cleavage of benzene promoted by another acetylene completes the cycle and leads to the regeneration of the catalyst. The energy released in this last step is 18.8 kcal mol^{-1}. During the catalysis, the Rh-Cr distance varies in the range 3.05–3.51 Å, but remains close to the crystallographic value of 3.1 Å [34].

The hapticity variations along the catalytic cycle are definitively more pronounced in syn-IndRh catalysis than in the anti-IndRh one.

By inspecting the energy profile (Figure 4a), in the syn-IndRh catalyzed process the oxidative coupling **Z-1** → **Z-2** has the highest energy barrier. From ASA (Table 1), a very high $\Delta\Delta E_{strain}$ contribution to the barrier $\Delta E^{\neq}$ has been found compared to anti-IndRh and also to the parent catalysts IndRh and CpRh. However, $\Delta\Delta E_{int}$ is a pretty strongly stabilizing term and compensates for the strain, resulting in a lowering of $\Delta E^{\neq}$. In contrast to anti-IndRh, the strain arising from the Cp'Rh fragment is very high (Table 1) because of the steric effects and because Cr(CO)$_3$ undergoes deformation to interact with the Rh center.

To further assess the interaction and the role of Cr(CO)$_3$ in the syn-IndRh catalyzed process, ASA was carried out for the intermediates and transition states of the whole catalytic cycle, considering Cr(CO)$_3$ and IndRhX$_i$ fragments; the results are reported in Table 4. ΔE_{int} are in the range ~49–58 kcal mol^{-1}, larger than those computed for the anti-IndRh molecular species (Table 3), suggesting Rh-Cr stabilizing interaction in the former case. On the other hand, ΔE_{strain} is also larger. This is mainly ascribed to steric factors. In particular, the contribution of Cr(CO)$_3$ to ΔE_{strain} in **syn-Ind-2, syn-Ind-TS (2,3)** and **syn-Ind-3** is higher, suggesting some structural differences from the anti-analogous species. In fact, in these structures one carbonyl ligand is coordinated to Rh (Figure S4). This feature also leads to larger ΔE_{int} in syn-Ind-2.

Table 4. Activation strain analysis (ASA) of all the intermediates and transition states along the syn-IndRh catalyzed cycle (Path I); all the values are in kcal mol^{-1}. The fragments are Cr(CO)$_3$ and IndRhX$_i$.

	ΔE_{strain}						ΔE_{int}	ΔE
	Cr(CO)$_3$	IndRhX$_i$	Total	ΔV_{elstat}	ΔE_{Pauli}	ΔE_{oi}		
syn-Ind-1	1.65	5.02	6.67	−80.49	132.33	−102.88	−51.04	−44.37
syn-Ind-TS (1,2)	1.62	20.29	21.91	−79.52	130.84	−102.42	−51.10	−29.19
syn-Ind-2	2.25	4.31	6.56	−117.69	195.36	−133.21	−55.54	−48.98
syn-Ind-TZ (2,3)	2.57	7.56	10.13	−117.3	188.69	−129.83	−58.45	−48.32
syn-Ind-3	2.52	11.78	14.30	−87.65	141.15	−109.59	−56.09	−41.79
syn-Ind-TS (3,4)	1.82	12.54	14.36	−81.06	135.13	−105.64	−51.57	−37.21
syn-Ind-4	1.15	4.50	5.65	−79.61	129.57	−99.01	−49.04	−43.39
syn-Ind-5	1.56	5.28	6.84	−78.6	128.78	−99.76	−49.58	−42.74

As an example of an existing favorable inter-metal interaction, in Figure 5, we show HOMO-3 of **syn-Ind-1**, which is formed by the contributions of the HOMOs of Cr(CO)$_3$ and IndRh(C$_2$H$_2$)$_2$ fragments, whose metal d-based molecular orbitals (MOs) indicate stabilizing d-d bonding between Rh-Cr.

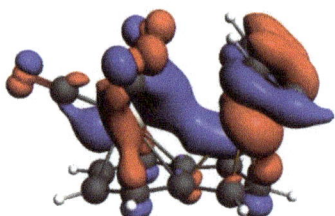

Figure 5. HOMO-3 of **syn-Ind-1**; the isodensity value is 0.03.

2.3. Acetylene [2+2+2] Cycloaddition Catalyzed by Anti-[Cr(CO)$_3$IndRh] Fragment: Reaction Mechanism and PES (Path II)

According to the alternative mechanistic path by Booth et al. [26], an ancillary ligand of the catalyst precursor remains bonded to rhodium during the catalytic cycle; this mechanism is denoted Path II (Scheme 3b). For CpRh(CO) and IndRh(CO), this path has been thoroughly explored, making it possible to give an interpretation of the higher efficiency of the indenyl catalyst observed in the experiment [35]. Thus, here we considered the same hypothesis for anti-IndRh(CO) catalyzed acetylene [2+2+2] cycloaddition. The structures of the intermediates and transition states found on the PES with their relevant parameters are in Figures S5–S7. The energy profile of the process is shown in Figure 6a.

CO-anti-Ind-1 is characterized by η^1 coordination which is due to the presence of two acetylenes and the CO ligand, as observed for CpRh(CO) and IndRh(CO) catalysis [43]. Δ and LISP values are 0.92 Å and 1.60 Å, respectively (Figure 6b). The initial oxidative coupling leads to the formation of **CO-anti-Ind-2**, in which rhodium is coordinatively saturated due to the presence of CO. The energy barrier required to cross **CO-anti-Ind-TS (1,2)** is 13.9 kcal mol^{-1} slightly higher if compared to those computed for CpRh(CO) and IndRh(CO) catalysis (by about 1–2 kcal mol^{-1}). This is the same energy trend observed along Path I. The conversion of **CO-anti-Ind-1** into **CO-anti-Ind-2** is accompanied by the hapticity change from η^1 to distorted η^5 and is exothermic by 43.1 kcal mol^{-1} (Figure 6a), 4.5 kcal mol^{-1} and 23.0 kcal mol^{-1} less than the analogous step in IndRh(CO) and CpRh(CO) catalysis, respectively. The addition of the third acetylene leads to the formation of the η^1 **CO-anti-Ind-3** with an activation energy of 9.8 kcal mol^{-1}. The barriers for this step are much higher in the cases of IndRh(CO) and CpRh(CO) catalysis, i.e., 28.5 and 43.6 kcal mol^{-1}, respectively. Thus, this step is kinetically favored with the bimetallic anti-IndRh(CO). In addition, the formation of **CO-anti-Ind-3** is endothermic by 4.4 kcal mol^{-1}, that is overall less endothermic if compared to the same step in IndRh(CO) and

CpRh(CO) catalysis, for which 9.3 kcal mol^{-1} and 22.5 kcal mol^{-1} are computed, respectively. Thus, the bimetallic catalyst has also a thermodynamic advantage. In the next step, the activation energy of 2.4 kcal mol^{-1} is necessary to cross **CO-anti-Ind-TS (3,b)** and generate the bicyclic intermediate **CO-anti-Ind-b** (Figure S7), with negligible difference from the IndRh(CO) catalyzed step and lower by 1.2 kcal mol^{-1} than in CpRh(CO) catalyzed step. The formation of **CO-anti-Ind-b** is accompanied by the release of 19.1 kcal mol^{-1}, being 3.8 kcal mol^{-1} more exothermic than the formation of the parent **CO-Ind-b** but 1.8 kcal mol^{-1} less exothermic than **CO-Cp-b**. One can notice that the bicyclic intermediate in CpRh(CO) catalysis has an higher energy while in bimetallic anti-IndRh(CO) and IndRh(CO) catalysis it lies at a lower energy (Figure 6a), suggesting a better catalytic efficiency in these latter cases. By crossing a modest barrier of 1.2 kcal mol^{-1}, the bicyclic **CO-anti-Ind-b** readily transforms into the heptacyclic intermediate **CO-Ind-h**; this step is accompanied by the haptotropic shift $\eta^1 \rightarrow$ distorted η^5 and by the release of 28.0 kcal mol^{-1}, a value lower than those computed for the parent catalysts, i.e., −37.4 kcal mol^{-1} in the case of IndRh(CO) and −37.8 kcal mol^{-1} in the case of CpRh(CO), respectively. **CO-anti-Ind-h** undergoes reductive elimination with an activation energy of 1.7 kcal mol^{-1}. **CO-anti-Ind-4** forms and 46.4 kcal mol^{-1} are released. Finally, benzene is cleaved from **CO-anti-Ind-4** by stepwise addition of two acetylene molecules and the catalyst is regenerated.

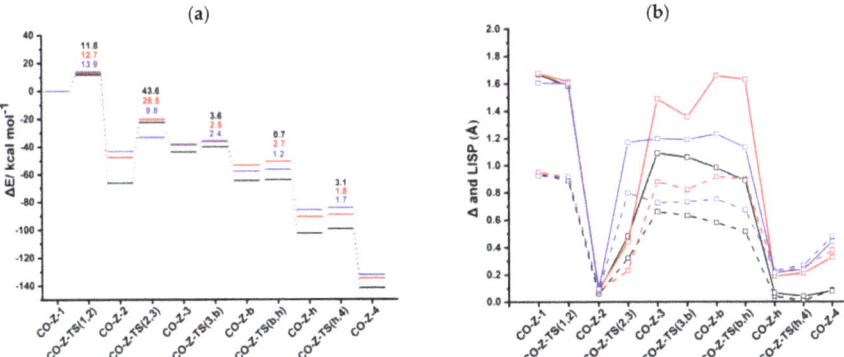

Figure 6. (**a**) Energy profiles of acetylene [2+2+2] cycloaddition to benzene catalyzed by CpRh(CO) (black), IndRh(CO) (red), and anti-IndRh(CO) (blue) (Scheme 3b, Path II). (**b**) Profiles of the slippage parameters Δ (dashed line) and LISP (solid line) along the acetylene [2+2+2] cycloaddition cycles catalyzed by CpRh(CO) (black), IndRh(CO) (red), and anti-IndRh(CO) (blue) (Scheme 3b, Path II). Level of theory: ZORA-BLYP/TZ2P.

With the formation of the 18-electron intermediate **CO-Z-2**, a rather flat portion of the PES begins (Figure 6a). ASA was carried out on **CO-Z-2** considering CO-Cp'Rh (Cp' = Cp, Ind and anti-[Cr(CO)$_3$Ind]) and the C$_4$H$_4$ moiety as fragments; the results are listed in Table 5. The similar strain in fragment C$_4$H$_4$ computed for **CO-Cp-2**, **CO-Ind-2**, and **CO-anti-Ind-2** (Table 5) revealed that the structure of this moiety is very similar in the three catalysts (see also Figures S5–S7). The strain contribution of the fragment CO-Cp'Rh is much higher in **CO-anti-Ind-2** and **CO-Ind-2** than in **CO-Cp-2**. This is related to the least slippage predicted for this last species. The stronger interaction ΔE_{int} in **CO-Ind-2** and **CO-anti-Ind-2** does not counterbalance their strain term and thus **CO-Cp-2** results the most stabilized among the three.

Table 5. Activation strain analysis (ASA) of **CO-Cp-2**, **CO-Ind-2**, and **CO-anti-Ind-2**; all values are in kcal mol^{-1}. The fragments are: CO-Cp'Rh (Cp' = Cp, Ind, and anti-[Cr(CO)$_3$Ind]) and the C$_4$H$_4$ moiety (the reference are two acetylene molecules).

	ΔE_{strain}			ΔE_{int}	ΔE
	C$_4$H$_4$	CO-Cp'RhL$_n$	Total		
CO-Cp-2	50.63	6.08	56.71	−138.15	−81.44
CO-Ind-2	50.96	34.04	85.00	−153.65	−68.65
CO-anti-Ind-2	50.89	36.06	86.95	−149.23	−62.28

In the subsequent step, i.e., the addition of third acetylene to the 18-electrons rhodacycle **CO-Z-2**, an energy barrier is found along the catalytic Path II (Figure 6a). **CO-Z-TS (2,3)** was divided into two fragments, i.e., CO-Cp'Rh(C$_4$H$_4$) (Cp' = Cp, Ind and anti-[Cr(CO)$_3$Ind]) and acetylene and ASA was performed. The results are shown in Table 6. The acetylene in all cases is only slightly deformed, whereas CO-Cp'Rh(C$_4$H$_4$) is highly strained.

Table 6. Activation strain analysis (ASA) of the transition states **CO-Cp-TS (2,3)**, **CO-Ind-TS (2,3)** and **CO-anti-Ind-TS (2,3)**; all values are in kcal mol^{-1}. The fragments are CO-Cp'RhL$_n$ (Cp' = Cp, Ind, and anti-[Cr(CO)$_3$Ind]) and acetylene.

	ΔE_{strain}			ΔE_{int}	ΔE
	C$_2$H$_2$	CO-Cp'RhL$_n$	Total		
CO-Cp-TS (2,3)	0.18	43.96	44.14	−0.50	43.64
CO-Ind-TS (2,3)	0.06	28.83	28.89	−0.34	28.55
CO-anti-Ind-TS (2,3)	0.12	14.27	14.39	−4.58	9.81

To summarize, we found that most of the steps are kinetically as well as thermodynamically more favored in the bimetallic anti-IndRh(CO) catalysis. At a glance, this can be seen also from the energy profile (Figure 6a), which is flatter in the case of anti-IndRh(CO) than in the parent IndRh(CO) and CpRh(CO) catalyzed processes. Alternatively, the slippage variations quantified with Δ and LISP (Figure 6b) are less pronounced in anti-IndRh(CO) than those computed for the parent monometallic catalysts. Thus, the smaller slippage variations and the flatter potential energy profile of the bimetallic anti-IndRh(CO) suggest an higher catalytic efficiency than the monometallic IndRh(CO) and CpRh(CO), which is consistent with the experimental findings [26]. The phenomena of *indenyl effect* for IndRh and *extra-indenyl effect* for the bimetallic anti-IndRh can be fully explained when considering the mechanism of path II.

Notably, path II has been excluded for syn-IndRh(CO) because in this particular case the rhodium center would be too much crowded.

3. Discussion

To quantify the catalytic efficiency, we calculated the turnover frequencies (TOFs) for the studied catalytic cycles by using the energy span model [23,55]. Calculations were run at standard room temperature (298.15 K), as well as at the reflux temperature of toluene, i.e., 383.65 K, which is occasionally used in alkyne [2+2+2] cycloadditions as solvent [25,26,29]; the values are listed in Table 7.

Following the catalytic Path I (Scheme 3a), the TOFs are in the order CpRh > IndRh > anti-IndRh, suggesting that the bimetallic catalyst anti-IndRh is worse than the monometallic parent catalysts IndRh and CpRh. This is in contrast with the experimental results [36]. Also, no appreciable difference in terms of TOF is found between IndRh and anti-IndRh, suggesting that the influence of the second metal group, i.e., Cr(CO)$_3$, on the efficiency of the catalyst is negligible. In all these cases, the TOF determining intermediate (TDI) is the bis-acetylene intermediate **Z-1** and the TOF determining transition state (TDTS) is the subsequent transition state **Z-TS (1,2)** (Figure 2a).

Conversely, along the catalytic Path II (Scheme 3b), in terms of TOF, a significant enhancement of the catalytic activity of the bimetallic anti-IndRh(CO) is found compared to those of the parent catalysts IndRh(CO) and CpRh(CO). On the basis of the TOF values reported in Table 7, the following trend can be established anti-IndRh(CO) > IndRh(CO) > CpRh(CO), which is consistent with the experimental observation [26]. In this case, the intermediate **CO-Z-2** is the TDI and the subsequent transition state **CO-Z-TS (2,3)** is the TDTS (Figure 6a). We can thus confirm that the coordination of $Cr(CO)_3$ favors the catalytic activity.

As recently reported [22], the slippage variations in intermediates and transition states along the catalytic cycle may be related to the chemical activity of the half-metallocene catalysts in this class of reactions. For this purpose, the slippage span $\Delta LISP^*$ was calculated. Along Path I (Scheme 3a), $\Delta LISP^*$ values follow the trend anti-IndRh > IndRh > CpRh. Since TOFs' trend is reverse, i.e., anti-IndRh < IndRh < CpRh, this indicates that the higher the slippage span along the catalytic cycle is, the lower the performance of the catalyst is [22]. When following Path II, $\Delta LISP^*$ values follow the trend anti-IndRh(CO) < IndRh(CO) < CpRh(CO), while TOFs follow the reverse trend anti-IndRh(CO) > IndRh(CO) > CpRh(CO), in agreement with the presence of extra indenyl and indenyl effect in the bimetallic and monometallic Ind catalysts, respectively.

The mechanism for syn-IndRh catalyzed acetylene [2+2+2] cycloaddition is different, and so we cannot directly compare the TOF values. However, the higher energy barriers and higher slippage variations certainly do not favor the syn catalyst.

Table 7. Calculated TOF (s^{-1}) and slippage span $\Delta LISP^*$ (Å) for the catalytic Path I and II of the [2+2+2] cycloaddition of acetylene to benzene.

	$TOF_{298.15\,K}$ (s^{-1})	$Ratio_{298.15\,K}$	$TOF_{383.65\,K}$ (s^{-1})	$Ratio_{383.65\,K}$	$\Delta LISP^*$ (Å)
Path I					
CpRh	4.83×10^3	1.82×10^2	6.66×10^5	5.79×10^1	0.85
IndRh	5.23×10	1.97	2.00×10^4	1.74	1.75
anti-IndRh	2.66×10	1	1.15×10^4	1	1.81
syn-IndRh	4.57×10	–	1.78×10^4	–	3.42
Path II					
CpRh(CO)	6.10×10^{-20}	1	1.07×10^{-12}	1	15.59
IndRh(CO)	6.41×10^{-8}	1.05×10^{12}	2.35×10^{-3}	2.20×10^9	14.11
anti-IndRh(CO)	3.97×10^2	6.51×10^{21}	9.55×10^4	8.93×10^{16}	10.54

4. Materials and Methods

All the equilibrium and transition state geometries were fully optimized (i.e., without any constraint) in gas-phase using density functional theory (DFT) approach as implemented in the Amsterdam Density Functional (ADF 2016, SCM, Vrije Universiteit: Amsterdam, The Netherlands, 2016) program [56,57]. The BLYP [58–61] function in combination with the TZ2P basis set was applied for all elements. The scalar relativistic effects were accounted for within the zeroth-order regular approximation (ZORA), which is an excellent approximation of the relativistic Dirac equation [62–64]. The TZ2P basis set [65] is a large uncontracted set of Slater-type orbitals (STOs) of triple-ζ quality and has been augmented with two sets of polarization functions on each atom: 2p and 3d for H, 3d, and 4f for C and O, 4p and 4f for Cr, and 5p and 4f for Rh. The frozen-core approximation was adopted for core electrons: up to 1s for C and O, up to 2p for Cr and up to 3d for Rh. This level of theory has been applied with success in previous studies [22,41–44,49].

Frequency calculations were computed to confirm that all the intermediates have positive frequencies, whereas the transition states have one imaginary frequency. The character of the normal mode associated with this imaginary frequency was carefully examined to verify that the correct transition state was found.

Activation strain analyses (ASA) were performed on selected geometries [66,67]. ASA is an approach based on molecule fragmentation, useful to understand the properties of the chemical bonding. For this purpose, the total energy of a complex can be decomposed as the sum of a strain contribution (ΔE_{strain}) and an interaction contribution (ΔE_{int}) (Equation (5)):

$$\Delta E = \Delta E_{strain} + \Delta E_{int} \quad (5)$$

ΔE_{strain} is the energy required for the geometrical deformation of the fragments when they are brought from infinite distance to the geometry they acquire in the complex, while ΔE_{int} is the actual energy change when the deformed fragments are combined to form the overall complex. ΔE_{int} can be further decomposed into electrostatic interaction (ΔV_{elstat}), Pauli repulsion (ΔE_{Pauli}) and orbital interactions (ΔE_{oi}) in the framework of Kohn-Sham molecular orbital (MO) theory (Equation (6)):

$$\Delta E_{int} = \Delta V_{elstat} + \Delta E_{Pauli} + \Delta E_{oi} \quad (6)$$

According to Fernandez et al. [53], who proposed an extension to the activation strain model for unimolecular reaction steps, the activation barrier can be given as the change, upon going from the reactant to the TS, in strain within the two defined fragments plus the change, upon going from the reactant to the TS, in the interaction between the same two fragments:

$$\Delta E^{\ddagger} = \Delta\Delta E_{strain} + \Delta\Delta E_{int} \quad (7)$$

This situation is encountered in the present study, when the five-membered rhodacycle is formed from the bis-acetylene precursor.

The turnover frequency (TOF) was calculated by using the energy span model proposed by Kozuch and Shaik [23,55]. The expression is:

$$TOF = \frac{k_B T}{h} \frac{e^{\frac{-\Delta G_r}{RT}} - 1}{\sum_{i,j=1}^{N} e^{(TS_i - I_j - \delta G_{i,j})/RT}} \quad (8)$$

where ΔG_r is reaction Gibbs free energy, T_i and I_j are Gibbs free energies of the ith transition state and intermediate, respectively. $\delta G_{i,j}$, called the energy span, is equal to ΔG_r if $i > j$ or to 0 if $i \leq j$. Instead of Gibbs free energies, electronic energies were used, as it was demonstrated that there are no significant differences in the energy profiles and corresponding TOF ratios for analogous catalytic cycles [44].

5. Conclusions

We performed a theoretical investigation of the bimetallic system [Cr(CO)$_3$IndRh]-mediated [2+2+2] cycloaddition of acetylene to benzene. Through a detailed exploration of the potential energy surfaces (PESs), the intermediates and transition states were located using density functional theory (DFT) methods following two mechanistic paths, i.e., Path I and II (Scheme 3a,b). The bimetallic catalysts anti- and syn-[Cr(CO)$_3$IndRh] were tested in silico and compared to the monometallic parent catalysts CpRh and IndRh. The anti or syn coordination of Cr(CO)$_3$ affects the energetics of the cycle and also to the mechanism. The reaction energies and barriers, the turn over frequency (TOF) and the change of the slippage parameters along the catalytic cycles are discussed.

Considering Path I, the established trend for the slippage span ΔLISP* is anti-IndRh > IndRh > CpRh while TOF values follow the opposite order, i.e., CpRh > IndRh > anti-IndRh. This leads to the conclusion that the lower slippage span along the catalytic cycle, the higher the catalytic performance. In any case, this does not explain the highest catalytic efficiency of the bimetallic Rh/Cr compound (extra-indenyl effect).

Conversely, if we follow the catalytic cycle along Path II, a dramatic TOF enhancement of the bimetallic system anti-IndRh(CO) relative to the parent CpRh(CO) and IndRh(CO) is found. On the basis of the TOF, the following trend of catalytic efficiency can be established anti-IndRh(CO) > IndRh(CO) > CpRh(CO), in agreement with experimental findings. In this case, the slippage span is inverted, too, i.e., CpRh(CO) > IndRh(CO) > anti-IndRh(CO), which again leads us to conclude that the lower the slippage span is, the higher the catalytic efficiency is. We can thus conclude that the coordination of $Cr(CO)_3$ in the bimetallic indenyl catalyst improves the efficiency along Path II.

The hapticity variations of intermediates and transition states along the catalytic cycle are highly pronounced in the syn bimetallic conformer, implying low TOF values and leading to the conclusion that the syn-[$Cr(CO)_3$IndRh] catalyzed process is not favored.

Supplementary Materials: The following are available online at http://www.mdpi.com/2073-4344/9/8/679/s1: Figure S1: Optimized geometries of anti- and syn-[$Cr(CO)_3$IndRh$(CO)_2$]. Level of theory: ZORA-BLYP/TZ2P; Figure S2: Optimized structures with relevant geometric parameters, i.e., bond lengths (Å) and angles (deg) of the intermediates and transition states along Path (I) for acetylene [2+2+2] cycloaddition catalyzed by CpRh fragment; Figure S3: Optimized structures with relevant geometric parameters, i.e., bond lengths (Å) and angles (deg) of the intermediates and transition states along Path (I) for acetylene [2+2+2] cycloaddition catalyzed by IndRh fragment; Figure S4: Optimized structures with relevant geometric parameters, i.e., bond lengths (Å) and angles (deg) of the intermediates and transition states along Path (I) for acetylene [2+2+2] cycloaddition catalyzed by syn-IndRh fragment; Figure S5: Optimized structures with relevant geometric parameters, i.e., bond lengths (Å) and angles (deg) of the intermediates and transition states along Path (II) for acetylene [2+2+2] cycloaddition catalyzed by CpRh(CO) fragment; Figure S6: Optimized structures with relevant geometric parameters, i.e., bond lengths (Å) and angles (deg) of the intermediates and transition states along Path (II) for acetylene [2+2+2] cycloaddition catalyzed by IndRh(CO) fragment; Figure S7: Optimized structures with relevant geometric parameters, i.e., bond lengths (Å) and angles (deg) of the intermediates and transition states along Path (II) for acetylene [2+2+2] cycloaddition catalyzed by anti-IndRh(CO) fragment, Table S1: Selected geometric parameters of the anti- and syn-[$Cr(CO)_3$IndRh$(CO)_2$], computed at ZORA-BLYP/TZ2P level of theory, Table S2: Cartesian coordinates (in Å) and ADF total energies (in kcal mol^{-1}) of all the intermediates and the transition states along Path I catalyzed by anti-[$Cr(CO)_3$IndRh] fragment, computed at ZORA-BLYP/TZ2P level of theory, Table S3: Cartesian coordinates (in Å) and ADF total energies (in kcal mol^{-1}) of all the intermediates and the transition states along Path I catalyzed by syn-[$Cr(CO)_3$IndRh] fragment, computed at ZORA-BLYP/TZ2P level of theory, Table S4: Cartesian coordinates (in Å) and ADF total energies (in kcal mol^{-1}) of all the intermediates and the transition states along Path I catalyzed by anti-[$Cr(CO)_3$IndRh] fragment, computed at ZORA-BLYP/TZ2P level of theory.

Author Contributions: Conceptualization, L.O.; investigation, S.M.A. and M.D.T.; data curation, S.M.A. and M.D.T.; writing—original draft preparation, S.M.A. and L.O.; writing—review and editing, L.O.; visualization, M.D.T.; supervision, L.O.; funding acquisition, L.O.

Funding: This research was funded by CINECA (Casalecchio di Reno, Italy) ISCRA Grant STREGA (Filling the STructure-REactivity GAp: in silico multiscale approaches to rationalize the design of molecular catalysts), who provided generous allocation of computational time on Galileo; P. I.: L. O. The APC was waived by the journal.

Acknowledgments: M.D.T. is grateful to Fondazione CARIPARO for financial support (PhD grant). This work has been inspired by the thorough synthetic and reactivity studies of heterobimetallic half-sandwich compounds, carried out in the past three decades by Alberto Ceccon and his co-workers, whom are all gratefully acknowledged.

Conflicts of Interest: The authors declare no conflict of interest. The funders had no role in the design of the study; in the collection, analyses, or interpretation of data; in the writing of the manuscript, or in the decision to publish the results.

References

1. Jones, G. Pyridines and their Benzo Derivatives: (V) Synthesis. *Compr. Heterocycl. Chem.* **1984**, *2*, 395–510.
2. Berthelot, M. Ueber die Polymeren des Acetylens. *Ann. Chem. Pharm.* **1867**, *141*, 173–184.
3. Reppe, W.; Schlichting, O.; Klager, K.; Toepel, T. Cyclisierende Polymerisation von Acetylen I Über Cyclooctatetraen. *Justus Liebigs Ann. Chem.* **1948**, *560*, 1–92. [CrossRef]
4. Wakatsuki, Y.; Yamazaki, H. Novel synthesis of heterocyclic compounds from acetylenes. *J. Chem. Soc. Chem. Commun.* **1973**, 280a. [CrossRef]
5. Wakatsuki, Y.; Yamazaki, H. Improved Catalytic Activity of Cyclopentadienylcobalt in the Preparation of Pyridines from Acetylenes and Nitrites. *Bull. Chem. Soc. Jpn.* **1985**, *58*, 2715–2716. [CrossRef]
6. Suzuki, D.; Urabe, H.; Sato, F. Metalative Reppe Reaction. Organized Assembly of Acetylene Molecules on Titanium Template Leading to a New Style of Acetylene Cyclotrimerization. *J. Am. Chem. Soc.* **2001**, *123*, 7925–7926. [CrossRef]

7. Takahashi, T.; Li, Y.; Stepnicka, P.; Kitamura, M.; Liu, Y.; Nakajima, K.; Kotora, M. Coupling Reaction of Zirconacyclopentadienes with Dihalonaphthalenes and Dihalopyridines: A New Procedure for the Preparation of Substituted Anthracenes, Quinolines, and Isoquinolines. *J. Am. Chem. Soc.* **2002**, *124*, 576–582. [CrossRef]
8. Werner, H. Electron-Rich Half-Sandwich Complexes?Metal Basespar excellence. *Angew. Chem. Int. Ed. Engl.* **1983**, *22*, 927–949. [CrossRef]
9. Consiglio, G.; Morandini, F. Half-sandwich chiral ruthenium complexes. *Chem. Rev.* **1987**, *87*, 761–778. [CrossRef]
10. Bauer, E.B. Chiral-at-metal complexes and their catalytic applications in organic synthesis. *Chem. Soc. Rev.* **2012**, *41*, 3153–3167. [CrossRef]
11. Saito, S.; Yamamoto, Y. Recent Advances in the Transition-Metal-Catalyzed Regioselective Approaches to Polysubstituted Benzene Derivatives. *Chem. Rev.* **2000**, *100*, 2901–2916. [CrossRef]
12. Collman, J.P. Priciples and Applications of Organotransition Metal Chemistry. *Univ. Sci. Books* **1987**, 324.
13. Varela, J.A.; Saá, C. Construction of Pyridine Rings by Metal-Mediated [2 + 2 + 2] Cycloaddition. *Chem. Rev.* **2003**, *103*, 3787–3802. [CrossRef]
14. Bönnemann, H. Cobalt-Catalyzed Pyridine Syntheses from Alkynes and Nitriles. *Angew. Chem. Int. Ed. Engl.* **1978**, *17*, 505–515. [CrossRef]
15. Bönnemann, H. Organocobalt Compounds in the Synthesis of Pyridines-An Example of Structure-Effectivity Relationships in Homogeneous Catalysis. *Angew. Chem. Int. Ed. Engl.* **1985**, *24*, 248–262. [CrossRef]
16. Schmid, G.; Schütz, M. 1, 2-Azaborolyl complexes. Cobalt 1, 2-azaborolyl diene complexes. *Organometallics* **1992**, *11*, 1789–1792. [CrossRef]
17. Koga, N.; Morokuma, K. Ab initio molecular orbital studies of catalytic elementary reactions and catalytic cycles of transition-metal complexes. *Chem. Rev.* **1991**, *91*, 823–842. [CrossRef]
18. Calhorda, M.J.; Veiros, L.F. Ring slippage in indenyl complexes: Structure and bonding. *Coord. Chem. Rev.* **1999**, *185–186*, 37–51. [CrossRef]
19. Calhorda, M.J.; Romão, C.C.; Veiros, L.F. The nature of the indenyl effect. *Chem. Eur. J.* **2002**, *8*, 868–875. [CrossRef]
20. Basolo, F.; Pearson, R.G. *Mechanisms of Inorganic Chemistry*; John Wiley Sons Inc.: Hoboken, NJ, USA, 1967.
21. Schuster-Woldan, H.G.; Basolo, F. Kinetics and Mechanism of Substitution Reactions of π-Cyclopentadienyldicarbonylrhodium. *J. Am. Chem. Soc.* **1966**, *88*, 1657–1663. [CrossRef]
22. Dalla-Tiezza, M.; Bickelhaupt, F.M.; Orian, L. Half-Sandwich Metal-Catalyzed Alkyne [2+2+2] Cycloadditions and the Slippage Span Model. *Chem. Open* **2018**, 143–154. [CrossRef]
23. Kozuch, S.; Shaik, S. How to Conceptualize Catalytic Cycles? The Energetic Span Model. *Acc. Chem. Res.* **2011**, *44*, 101–110. [CrossRef]
24. Diversi, P.; Ingrosso, G.; Lucherini, A.; Porzio, W.; Zocchi, M. Synthesis and x-ray structures of cobalta-, rhodia-, and iridiacycloalkanes. Observation of novel structural features in the metallocyclopentane rings. *Inorg. Chem.* **1980**, *19*, 3590–3597. [CrossRef]
25. Borrini, A.; Diversi, P.; Ingrosso, G.; Lucherini, A.; Serra, G. Highly active rhodium catalysts for the [2+2+2] cycloaddition of acetylenes. *J. Mol. Catal.* **1985**, *30*, 181–195. [CrossRef]
26. Abdulla, K.; Booth, B.L.; Stacey, C. Cyclotrimerization of acetylenes catalyzed by (η^5-cyclopentadienyl)rhodium complexes. *J. Organomet. Chem.* **1985**, *293*, 103–114. [CrossRef]
27. Hart-Davis, A.J.; Mawby, R.J. Reactions of π-indenyl complexes of transition metals. Part I. Kinetics and mechanisms of reactions of tricarbonyl-π-indenylmethylmolybdenum with phosphorus(III) ligands. *J. Am. Chem. Soc.* **1969**, *0*, 2403–2407. [CrossRef]
28. Rerek, M.E.; Ji, L.-N.; Basolo, F. The indenyl ligand effect on the rate of substitution reactions of Rh(η-C$_9$H$_7$)(CO)$_2$ and Mn(η-C$_9$H$_7$)(CO)$_3$. *J. Chem. Soc. Chem. Commun.* **1983**, 1208–1209. [CrossRef]
29. Cioni, P.; Diversi, P.; Ingrosso, G.; Lucherini, A.; Ronca, P. Rhodium-catalyzed synthesis of pyridines from alkynes and nitriles. *J. Mol. Catal.* **1987**, *40*, 337–357. [CrossRef]
30. Diversi, P.; Ermini, L.; Ingrosso, G.; Lucherini, A. Electronic and steric effects in the rhodium-complex catalysed co-cyclization of alkynes and nitriles to pyridine derivatives. *J. Organomet. Chem.* **1993**, *447*, 291–298. [CrossRef]

31. Dahy, A.A.; Yamada, K.; Koga, N. Theoretical Study on the Reaction Mechanism for the Formation of 2-Methylpyridine Cobalt(I) Complex from Cobaltacyclopentadiene and Acetonitrile. *Organometallics* **2009**, *28*, 3636–3649. [CrossRef]
32. Trost, B.M.; Ryan, M.C. Indenylmetal Catalysis in Organic Synthesis. *Angew. Chem. Int. Ed.* **2017**, *56*, 2862–2879. [CrossRef]
33. Bonifaci, C.; Carta, G.; Ceccon, A.; Gambaro, A.; Santi, S.; Venzo, A. Heterobimetallic Indenyl Complexes. Kinetics and Mechanism of Substitution and Exchange Reactions of trans -[Cr(CO)$_3$-indenyl-Rh(CO)$_2$] with Olefins. *Organometallics* **1996**, *15*, 1630–1636. [CrossRef]
34. Bonifaci, C.; Ceccon, A.; Gambaro, A.; Ganis, P.; Santi, S.; Valle, G.; Venzo, A. Heterobimetallic indenyl complexes. Synthesis and structure of cis-[Cr(CO)3(indenyl)RhL2 (L2 = norbornadiene, (CO)2). *Organometallics* **1993**, *12*, 4211–4214. [CrossRef]
35. Ceccon, A.; Santi, S.; Orian, L.; Bisello, A. Electronic communication in heterobinuclear organometallic complexes through unsaturated hydrocarbon bridges. *Coord. Chem. Rev.* **2004**, *248*, 683–724. [CrossRef]
36. Ceccon, A.; Gambaro, A.; Santi, S.; Venzo, A. On different chemical and catalytic behavior of (η-indenyl)-Rh(η^4-COD) and Cr(CO)$_3$(μ-η:η-indenyl)Rh(η^4-COD) complexes. *J. Mol. Catal.* **1991**, *69*, L1–L6. [CrossRef]
37. Hardesty, J.H.; Koerner, J.B.; Albright, T.A.; Lee, G.-Y. Theoretical Study of the Acetylene Trimerization with CpCo. *J. Am. Chem. Soc.* **1999**, *121*, 6055–6067. [CrossRef]
38. Kirchner, K.; Calhorda, M.J.; Schmid, R.; Veiros, L.F. Mechanism for the Cyclotrimerization of Alkynes and Related Reactions Catalyzed by CpRuCl. *J. Am. Chem. Soc.* **2003**, *125*, 11721–11729. [CrossRef]
39. Calhorda, M.J.; Costa, P.J.; Kirchner, K.A. Benzene and heterocyclic rings formation in cycloaddition reactions catalyzed by RuCp derivatives: DFT studies. *Inorganica Chim. Acta* **2011**, *374*, 24–35. [CrossRef]
40. Dazinger, G.; Torres-Rodrigues, M.; Kirchner, K.; Calhorda, M.J.; Costa, P.J. Formation of pyridine from acetylenes and nitriles catalyzed by RuCpCl, CoCp, and RhCp derivatives—A computational mechanistic study. *J. Organomet. Chem.* **2006**, *691*, 4434–4445. [CrossRef]
41. Orian, L.; van Stralen, J.N.P.; Bickelhaupt, F.M. Cyclotrimerization Reactions Catalyzed by Rhodium(I) Half-Sandwich Complexes: A Mechanistic Density Functional Study. *Organometallics* **2007**, *26*, 3816–3830. [CrossRef]
42. Orian, L.; van Zeist, W.J.; Bickelhaupt, F.M. Linkage Isomerism of Nitriles in Rhodium Half-Sandwich Metallacycles. *Organometallics* **2008**, *27*, 4028–4030. [CrossRef]
43. Orian, L.; Swart, M.; Bickelhaupt, F.M. Indenyl Effect Due to Metal Slippage? Computational Exploration of Rhodium-Catalyzed Acetylene [2+2+2] Cyclotrimerization. *Chem. Phys. Chem.* **2014**, *15*, 219–228. [CrossRef]
44. Orian, L.; Wolters, L.P.; Bickelhaupt, F.M. In Silico Design of Heteroaromatic Half-Sandwich Rh(I) Catalysts for Acetylene [2+2+2] Cyclotrimerization: Evidence of a Reverse Indenyl Effect. *Chem. Eur. J.* **2013**, *19*, 13337–13347. [CrossRef]
45. Dahy, A.A.; Suresh, C.H.; Koga, N. Theoretical Study of the Formation of a Benzene Cobalt Complex from Cobaltacyclopentadiene and Acetylene. *Bull. Chem. Soc. Jpn.* **2005**, *78*, 792–803. [CrossRef]
46. Dahy, A.A.; Koga, N. Trimerization of Alkynes in the Presence of a Hydrotris(pyrazolyl)borate Iridium Catalyst and the Effect of Substituent Groups on the Reaction Mechanism: A Computational Study. *Organometallics* **2015**, *34*, 4965–4974. [CrossRef]
47. Albright, T.A.; Hofmann, P.; Hoffmann, R.; Lillya, C.P.; Dobosh, P.A. Haptotropic rearrangements of polyene-MLn complexes. 2. Bicyclic polyene-MCp, M(CO)3 systems. *J. Am. Chem. Soc.* **1983**, *105*, 3396–3411. [CrossRef]
48. Schore, N.E. Transition metal-mediated cycloaddition reactions of alkynes in organic synthesis. *Chem. Rev.* **1988**, *88*, 1081–1119. [CrossRef]
49. Dalla Tiezza, M.; Bickelhaupt, F.M.; Orian, L. Group 9 Metallacyclopentadienes as Key-Intermediates in [2+2+2] Alkyne Cyclotrimerizations. Insight from Activation Strain Analyses. *Chem. Phys. Chem.* **2018**, *19*, 1766–1773. [CrossRef]
50. Bonifaci, C.; Ceccon, A.; Gambaro, A.; Ganis, P.; Mantovani, L.; Santi, S.; Venzo, A. Heterobimetallic heptamethylindenyl complexes of Cr0 and RhI: Trans-[Cr(CO)3-indenyl-RhL2] (L2 = COD, L = CO). *J. Organomet. Chem.* **1994**, *475*, 267–276. [CrossRef]
51. Groom, C.R.; Bruno, I.J.; Lightfoot, M.P.; Ward, S.C. The Cambridge Structural Database. *Acta Crystallogr. B* **2016**, *72*, 171–179. [CrossRef]

52. Rerek, M.E.; Basolo, F. Kinetics and mechanism of substitution reactions of eta 5-cyclopentadienyldicarbonylrhodium (I) derivatives. Rate enhancement of associative substitution in cyclopentadienylmetal compounds. *J. Am. Chem. Soc.* **1984**, *106*, 5908–5912. [CrossRef]
53. Fernández, I.; Bickelhaupt, F.M.; Cossío, F.P. Type-I Dyotropic Reactions: Understanding Trends in Barriers. *Chem. Eur. J.* **2012**, *18*, 12395–12403. [CrossRef]
54. Bonifaci, C.; Ceccon, A.; Santi, S.; Mealli, C.; Zoellner, R.W. Cofacial and antarafacial indenyl bimetallic isomers: A descriptive MO picture and implications for the indenyl effect on ligand substitution reactions. *Inorganica Chim. Acta* **1995**, *240*, 541–549. [CrossRef]
55. Kozuch, S. Steady State Kinetics of Any Catalytic Network: Graph Theory, the Energy Span Model, the Analogy between Catalysis and Electrical Circuits, and the Meaning of "Mechanism". *ACS Catal.* **2015**, *5*, 5242–5255. [CrossRef]
56. Te Velde, G.; Bickelhaupt, F.M.; Baerends, E.J.; Fonseca Guerra, C.; van Gisbergen, S.J.A.; Snijders, J.G.; Ziegler, T. Chemistry with ADF. *J. Comput. Chem.* **2001**, *22*, 931–967. [CrossRef]
57. Baerends, E.J.; Ziegler, T.; Atkins, A.J.; Autschbach, J.; Bashford, D.; Baseggio, O.; Bérces, A.; Bickelhaupt, F.M.; Bo, C.; Boerritger, P.M.; et al. ADF2016 SCM. In *Theoretical Chemistry*; Vrije Universiteit: Amsterdam, The Netherlands, 2016.
58. Becke, A.D. Density-functional exchange-energy approximation with correct asymptotic behavior. *Phys. Rev. A* **1988**, *38*, 3098–3100. [CrossRef]
59. Parr, R.G.; Yang, W. *Density-Functional Theory of Atoms and Molecules*; International series of monographs on chemistry; 1. issue; Oxford University Press: New York, NY, USA, 1994; ISBN 978-0-19-509276-9.
60. Lee, C.; Yang, W.; Parr, R.G. Development of the Colle-Salvetti correlation-energy formula into a functional of the electron density. *Phys. Rev. B* **1988**, *37*, 785–789. [CrossRef]
61. Johnson, B.G.; Gill, P.M.W.; Pople, J.A. The performance of a family of density functional methods. *J. Chem. Phys.* **1993**, *98*, 5612–5626. [CrossRef]
62. Van Lenthe, E.; Baerends, E.J.; Snijders, J.G. Relativistic total energy using regular approximations. *J. Chem. Phys.* **1994**, *101*, 9783–9792. [CrossRef]
63. Van Lenthe, E.; Baerends, E.J.; Snijders, J.G. Relativistic regular two-component Hamiltonians. *J. Chem. Phys.* **1993**, *99*, 4597–4610. [CrossRef]
64. Van Lenthe, E.; Ehlers, A.; Baerends, E.J. Geometry optimizations in the zero order regular approximation for relativistic effects. *J. Chem. Phys.* **1999**, *110*, 8943–8953. [CrossRef]
65. Van Lenthe, E.; Baerends, E.J. Optimized Slater-type basis sets for the elements 1-118. *J. Comput. Chem.* **2003**, *24*, 1142–1156. [CrossRef]
66. Bickelhaupt, F.M.; Baerends, E.J. Kohn-Sham Density Functional Theory: Predicting and Understanding Chemistry. In *Reviews in Computational Chemistry*; Lipkowitz, K.B., Boyd, D.B., Eds.; John Wiley & Sons, Inc.: Hoboken, NJ, USA, 2000; Volume 15, pp. 1–86. ISBN 978-0-470-12592-2.
67. Bickelhaupt, F.M.; Houk, K.N. Analyzing Reaction Rates with the Distortion/Interaction-Activation Strain Model. *Angew. Chem. Int. Ed.* **2017**, *56*, 10070–10086. [CrossRef]

© 2019 by the authors. Licensee MDPI, Basel, Switzerland. This article is an open access article distributed under the terms and conditions of the Creative Commons Attribution (CC BY) license (http://creativecommons.org/licenses/by/4.0/).

MDPI
St. Alban-Anlage 66
4052 Basel
Switzerland
Tel. +41 61 683 77 34
Fax +41 61 302 89 18
www.mdpi.com

Catalysts Editorial Office
E-mail: catalysts@mdpi.com
www.mdpi.com/journal/catalysts

www.ingramcontent.com/pod-product-compliance
Lightning Source LLC
LaVergne TN
LVHW070724100526
838202LV00013B/1165